REVEALING SECRETS

JOHN BLAXLAND is Professor of International Security and Intelligence Studies in the Strategic and Defence Studies Centre (SDSC), Coral Bell School of Asia Pacific Affairs at the Australian National University (ANU). He is a Senior Fellow of the Higher Education Academy and a Fellow of the Royal Society of New South Wales. He was also formerly a military intelligence officer, Head of SDSC and Director of the ANU Southeast Asia Institute. He is the author and editor of several publications on military history, intelligence and international security affairs.

CLARE BIRGIN's career in DFAT spanned 30 years, with a focus on national security and intelligence. She had postings in Warsaw, Moscow, Geneva, and Washington DC as the Liaison Officer of the Office of National Assessments, followed by postings as Ambassador in Hungary, Serbia, Kosovo, Romania, North Macedonia and Montenegro. Subsequently she was a Visiting Fellow at the ANU before joining John Blaxland's history writing team. She has been awarded the Polish Government's Knight's Cross Medal and the Benito Merito Medal by the former Polish Foreign Minister.

'George Orwell famously wrote during World War Two, "we sleep safe in our beds because rough men stand ready in the night to visit violence on those who would do us harm". Reading this superb history by John Blaxland and Clare Birgin on Australia's involvement with Sigint and cyber we can contemplate a new formula. We sleep safer because 24/7 intelligent, technologically competent patriotic men and women who work for our agencies, develop and work our electronic defence and offence capacities at worldclass standard. This in a world now in which we are constantly under attack. The work so secret it is proving impossible to produce an official history. This is the closest we can get and it is very good. If you are seriously interested in our defence and survival, or you would just like a good read, this belongs on your bookshelf.'

Kim Beazley, former Defence Minister

'A meticulous compilation of the largely unsung past achievements of our most consistently productive intelligence source. And a thoughtful analysis of how to approach the extraordinary challenges posed by the new cyber universe. Blaxland and Birgin make an important contribution to our understanding of issues needing much more open debate than our own and allied governments have traditionally allowed or encouraged.'

Gareth Evans, former Foreign Minister

'Australia has been part of sigint since the practice began, which has shaped its history in ways that Australians know little about. Their government likes to keep things that way. *Revealing Secrets* overcomes efforts to keep Australians ignorant about their sigint history, by discussing everything that can be said about it without access to secret records. Anyone interested in the past and future of Australia has much to learn from this book.'

John Ferris, author of *Behind the Enigma, The Authorised History of GCHQ, Britain's Secret Cyber Intelligence Agency*

'The most comprehensive and best informed account we have had of the history of signals intelligence in Australia. Essential reading for anyone seeking to understand not just our country's past, but Australia's strategic future as well.'

Allan Gyngell, author of *Fear of Abandonment: Australia in the World Since 1942*

'*Revealing Secrets* tells the remarkable, but little-known, story of how a small, back-room military office grew into a major Australian government agency. Deeply researched, authoritative and accessible, it is a valuable and timely contribution to understanding issues that have never been more important to national security.'

Emeritus Professor David Horner, author of *The Spy Catchers*

REVEALING SECRETS

AN UNOFFICIAL HISTORY OF AUSTRALIAN SIGNALS INTELLIGENCE & THE ADVENT OF CYBER

JOHN BLAXLAND & CLARE BIRGIN

UNSW PRESS

A UNSW Press book

Published by
NewSouth Publishing
University of New South Wales Press Ltd
University of New South Wales
Sydney NSW 2052
AUSTRALIA
https://unsw.press/

© John Blaxland and Clare Birgin 2023
First published 2023

10 9 8 7 6 5 4 3 2 1

This book is copyright. Apart from any fair dealing for the purpose of private study, research, criticism or review, as permitted under the *Copyright Act*, no part of this book may be reproduced by any process without written permission. Inquiries should be addressed to the publisher.

 A catalogue record for this book is available from the National Library of Australia

ISBN: 9781742237947 (paperback)
       9781742238746 (ebook)
       9781742239682 (ePDF)

Design Josephine Pajor-Markus
Cover design Luke Causby, Blue Cork
Cover images (top) Working in the 'Code Room', NAA7648055;
    (bottom) Ihor/Adobe Stock

All reasonable efforts were taken to obtain permission to use copyright material reproduced in this book, but in some cases copyright could not be traced. The authors welcome information in this regard.

CONTENTS

Introduction   1

Chapter 1   Early cryptology, secrets & intelligence   7
Chapter 2   From Federation & the Anglo-Boer War to the First World War   30
Chapter 3   First World War Allied Sigint   42
Chapter 4   Australian Sigint & the interwar years   73
Chapter 5   Sigint in the Second World War, 1939-41   102
Chapter 6   Sigint arrangements & the War in the Pacific   141
Chapter 7   Wartime Sigint successes, bureaucratic & other challenges   187
Chapter 8   Postwar Sigint to Vietnam   222
Chapter 9   Reform, computers & military Sigint since the 1970s   262
Chapter 10   Legislative reform & the coming of cyber   300

Conclusion & looking ahead   332

Appendix A   Central Bureau, Brisbane, work flowchart   345
Appendix B   Army & Air Force wartime special wireless units   350

Acknowledgments   353
Glossary   355
Acronyms   360
Bibliography   366
Notes   383
Index   432

INTRODUCTION

Why does Australia have a national Signals Intelligence (Sigint) agency? How important are the ties with the United States, the United Kingdom, Canada and New Zealand to this arrangement? What does it do and why is it controversial? And what has Sigint to do with cyber? This book addresses these questions, explaining the antecedents of Sigint and cyber for a contemporary audience.

In this book, we seek to present a broad panorama giving a sense of the scale and the scope of Sigint and its role in the advent of cyber. There are already comprehensive histories of the national Sigint agencies of the United Kingdom and the United States.[1] Parts of Australia's Sigint history have been covered by authors including Desmond Ball and David Horner. Others, including John Fahey, Craig Collie and David Dufty, have sought to address particular historical aspects in the first half of the twentieth century, notably during the Second World War.[2] This book, however, is the first to offer a consolidated history of Australian Sigint and cyber in Australia, covering the period all the way from Federation through to the present.

This account covers the antecedents and early days of Australian Sigint up to and including the Second World War. The structure of this book needs to be understood in terms of what happened to the Australian Signals Directorate official history project. Initially commissioned by the former Director-General, Mike Burgess, in 2019, this project was cut off in mid-stride by his successor.[3] The 'postwar' chapters covering the years since 1945, therefore, are derived from publicly available material, mindful that many of the records that would enable a fuller account remain inaccessible to the public. These later chapters are a preview, or 'taster', pointing to where further research is needed. We hope these final chapters will encourage greater openness in future.

It is fair to say that the secret of success in the espionage business lies in keeping one's successes secret. This definition of success has meant, however, that the uninitiated are unaware of Sigint's pivotal role in Australia's national intelligence. This book aims to set the record straight as far as possible, exposing much that was 'most secret' for decades after the event.

The establishment of the Defence Signals Bureau, or DSB, in 1947 is widely seen as the beginning of Australia's national Sigint capability. It was the culmination, however, of a series of earlier events. DSB was a cover name for the Melbourne Signal Intelligence Centre, to be used 'where it is necessary to conceal the nature of its functions'.[4] Today it is known as the Australian Signals Directorate or ASD. But DSB, or ASD as it became, was the central, but not sole component of the national Sigint enterprise which included Sigint assets in the three armed services. This consolidated historical account began as an official history of the ASD. In the process of writing, we became conscious that this was only a part of Australia's Sigint history, which does not make sense on its own.

So, what is Sigint? Sigint concerns the revealing of others' secrets and the protection of one's own. Sigint in this work refers to all aspects of information obtained by intercepting analogue or digital messages (without the sender's permission or knowledge) transmitted via radio waves using the electromagnetic spectrum or via line (including cable telegraphy and optical fibre). This process involves code-breaking, direction finding, analysis, translation, interpretation and then dissemination to customers other than the originally intended recipient of such messages. Sigint also involves the protection of the sources and methods used to acquire and exploit those intercepted messages.

Sigint-related activity has been considered part of a field known as electronic warfare (or EW). Sigint, in fact, is probably best understood as part of the suite of EW capabilities that armed forces have developed in the twentieth century, including electronic countermeasures (or ECM), electronic counter-countermeasures known as ECCM and electronic support measures (or ESM). Both ECM and ECCM have also been defined as components of electronic protection (EP). Much of what we understand to be Sigint – the collection and processing of foreign electronic emissions – falls under the category of ESM. Certain protective security aspects of

Sigint could be categorised as ECM. For the purposes of this work, however, we will principally use the term Sigint.

Many former practitioners kept their oath of secrecy until the end of their lives, never telling others 'we also served'. If Australia's national Sigint agency was, and still is, too little known and understood, the frontline work of Sigint forces in the three services is even less so.

Why is the story important?

The traitor unmasked at the end of John le Carré's novel *Tinker Tailor Soldier Spy* declares that 'secret services were the only real measure of a nation's political health, the only real expression of its subconscious'.[5] Alex Younger, former head of MI6, has said that 'you can tell a lot about the soul of a country from its intelligence services'.[6]

Sigint, the most secret branch of intelligence, has influenced Australia's history and people, largely without most of us knowing it or fully understanding it, but it has left an imprint on us all.

Traditionally, intelligence – and Sigint, in particular – has been shrouded in secrecy. In recent years, in the interests of transparency and accountability, leaders and heads of intelligence agencies in democratic countries, notably the United Kingdom and, to a certain extent, in Australia and elsewhere, have lifted the veil judiciously to reveal the purposes of some of these agencies and some of their work – in the case of the United Kingdom, Christopher Andrew on MI5, Keith Jeffrey on MI6 and John Ferris on GCHQ. In Australia, the work headed by David Horner in the three-volume history of ASIO.

This is to their credit, but to reveal only a part of an enterprise can be misleading. Former Prime Minister Malcolm Turnbull's revelation of the Australian Signals Directorate's offensive capabilities in 2016 revived fears in some quarters of mass surveillance. Revealing only part of what agencies do now can backfire, but at the same time, there are compelling reasons for secrecy, including the protection of sources and methods.

How is it possible, then, to maintain the necessary secrecy for Sigint while ensuring that the Australian people, who this branch of intelligence serves, have a sufficient understanding of its purpose and what it does? There is no perfect answer, but this book will argue that history which takes Sigint into account is an essential first step.

The beginning of Australian Sigint was not the establishment of the Defence Signals Bureau in 1947, nor the formation of Central Bureau, a combined Australian Army, Royal Australian Air Force (RAAF) and US Army organisation in 1942. It was not the effort in the 1920s championed by the Royal Australian Navy's Director of Signals and Communications, Commander F.C. Cresswell to establish an Australian naval Sigint capability, or the work of a brilliant young Australian codebreaker called Eric Nave, or even the Australian Commonwealth Naval Board's establishment of its own stations, using Marconi wireless sets in the early 1900s.

This book first introduces the reader to cryptology in the ancient world, in the millennium leading up to the Industrial Revolution, the effect of cryptological developments on the revolutions and wars of the eighteenth and nineteenth centuries, the discovery of electricity and the invention of the telegraph. It sets the scene for the revolution in technological and military affairs which laid the foundations for Australian Sigint today.

It seeks to shed light where history has been misunderstood, or only half understood, because the cryptographic dimension was unknown at the time or deliberately left out. The Babington Plot, for instance, which led to the execution of Mary Queen of Scots, involved ciphers, a precursor of Sigint. George Scovell, an anonymous codebreaker, contributed to the defeat of Napoleon at the Battle of Salamanca in 1812.

The book traverses the world wars, exploring how the role of cryptology is only now beginning to be recognised: had he not ignored Sigint, Lord Jellicoe might have defeated the German Fleet at the Battle of Jutland in 1916. If not for the decryption of the telegram in which Germany's Foreign Minister Zimmerman asked the Mexicans to keep American troops busy on their border, promising them New Mexico and Texas in return, the United States might not have entered the First World War.

Few Australians would know the story of the ANZAC Squadron's unskilled wireless operators who taught themselves to intercept Turkish wireless communications in Mesopotamia in the First World War. Fewer still would know that, according to Nigel de Grey, the British Sigint official-cum-historian, in the art of traffic analysis, the often crucial tactical level of Sigint that revealed, for example, the location of enemy forces, 'the Australian Army held the lead'.[7] In the Pacific War, early in his remarkable career, Eric Nave was praised by the British Admiralty for discovering the strategic basis for Japan's war-fighting plans.

The Coastwatchers was a 700-strong organisation established in the 1920s. They were not involved in Sigint per se, but their work provided a plausible alternative explanation, or 'cover,' for Sigint. This way, 'sanitised' material from Sigint reports could be used without fear of compromising the work of the codebreakers and, if necessary, explained to the uninitiated. Sanitisation involved removing indications of the Sigint source so as to avoid raising enemy suspicions that their codes had been compromised. Obscuring a report's tell-tale indicators of electromagnetic or electronic interception made it less sensitive and therefore able to be shared with those who needed it for operational contingency planning. Coastwatcher reports were the perfect plausible alternative explanation.

A notable member of Australia's Coastwatchers was Ruby Boye-Jones, who transmitted coded intelligence messages from an island in the South Pacific, braving death threats from the Japanese. Admiral Halsey would later declare that Coastwatcher intelligence saved Guadalcanal which, in turn, 'had saved the Pacific'.[8] This was high praise, but while the Battle of Midway is famous, the Australian Sigint contribution to the Allied victory is still virtually unknown. And, in the words of an officer of Fleet Radio Unit Melbourne, 'thanks to our information, submarines knew when and where to prey on Japanese ships'.[9]

FRUMEL, Central Bureau, D Special Section and Section 22, are all odd-sounding, deliberately nondescript names. But they, along with the service intercept stations and the Coastwatchers, played a role that was vital and generally unrecognised.

The author of Britain's *Official History of British Intelligence in the Second World War*, Professor Harry Hinsley, did not attempt to cover the war in the Far East 'when this was so much the concern of the United States'.[10] While US historians addressed aspects of it from their own perspective, there was no official history of Australian intelligence in the Second World War. As a result, Australia's story was not told.

For many years after the Second World War, the Australian public was none the wiser. This work outlines the antecedents and major developments up to and including the Second World War. For the reasons noted above, the final postwar chapters draw on publicly available information. These final chapters reveal that Australian forces were involved in Sigint in the Korean War, and particularly during Indonesia's *'Konfrontasi'*. In Vietnam in 1966, Australia's 547 Signal Troop located the North Vietnamese army unit ahead of the Battle of Long Tan, but the intelligence was not well understood. After the battle, Australian and Allied staff officers paid more attention to Sigint. In subsequent operations, notably in Iraq and Afghanistan, the support provided built on the expertise developed earlier and capitalised on emerging technology.

Gradually, the general public also began to pay more attention. In 1975, Australian Sigint came under public scrutiny over East Timor, with accusations of an Australian government cover-up of the murder of five Australian journalists and, again in 1999, when there were allegations of distortion and misuse of intelligence. A royal commission under Justice Hope and the report by Philip Flood in 2004, respectively, dismissed these particular allegations. But in Australia, as in other democratic societies, popular demand for transparency and accountability is on the rise.

Today, Australia's Sigint faces new challenges presented by cyberspace and greater public scrutiny than ever. The book concludes with some observations on contemporary challenges with managing the Sigint and cyber domains, and some reflections on what the future may hold. It notes that Sigint is also enabling more intense scrutiny of the private and public lives of citizens than ever before. This is causing public concern – hence the need for Australia's Sigint and cyber security functions to be better understood by the people they are intended to protect.

1
EARLY CRYPTOLOGY, SECRETS & INTELLIGENCE

Why, you might ask, would we begin our history of Australian signals intelligence (or Sigint) by reaching back to the origins of the history and development of all cryptology. A simple answer is that by doing so, we satisfy the first recommendation of the 2019 *Comprehensive Review of the Legal Framework of the National Intelligence Community*, also known as the Richardson Review, which calls on Australia's national intelligence agencies to ensure that their training addresses their history, background and principles.[1] This review is in the tradition of earlier royal commissions into intelligence and security which shaped the Australian intelligence community of today.

An overlooked art and science

Much of this history is poorly understood. Indeed, for various reasons, cryptology is an art and a science that historians have tended to overlook. Australian military history has been written largely without permission to draw upon stories relating to the role of cryptology in statecraft and warfare.

Second, in the decades after Federation, Australia relied upon Britain for diplomacy and technical military support – particularly during the Anglo–Boer War and the First World War. This allowed historians to disregard the role of Sigint in Australia's national security. In the case of Australia's official historian of the First World War, C.E.W. Bean, Sigint does not feature prominently because, barring the few exceptions highlighted in this work, Australian forces played a peripheral role in it; and yet Sigint had an enormous impact on Australia's national security which has not been properly understood or fully explained. Although by the Second World War, Australia was directly and extensively involved in Sigint and other intelligence activities, Australia's official historian, Gavin Long, omitted it, apparently due to national security constraints.

The brief survey of cryptography and Sigint that follows illustrates that not only is cryptology a normal part of statecraft, it is essential for those seeking an advantage in war and indeed in many aspects of international relations in circumstances short of war. While some elements of Sigint – including the interception of enemy communications and cryptanalysis – are as old as warfare, other features are markedly modern, transformed by electricity, the invention of wireless telegraphy and, eventually, digital technology. The Australian Sigint of today and its cyber offshoot is a product not just of its Second World War antecedents, but of technological breakthroughs and Sigint-related innovations. Australia was the beneficiary in large part thanks to its relationships with the United Kingdom and the United States.

Sigint is not simply a twentieth-century creation. Its history can be traced back a long way – before even the mid to late nineteenth-century inventions of the telegraph and the wireless radio. The notion of secret messages, cryptology and eavesdropping, now commonly considered to be part of Sigint, is as old as recorded history.

This chapter provides context for Australian Sigint in the twentieth century, demonstrating how again and again, because of the secrecy of cryptology, historical events so well known as to be part of our collective memory, have been only half understood, or entirely misread. It also demonstrates that cryptology and the interception of secret messages has been an age-old and normal, if little understood, part of statecraft. Nations have always, as covertly as possible, seized opportunities to discover what their competitors and adversaries were thinking and intending to do.

Ancient cryptology – in 'the West'

The collecting, organising, analysing and reporting of sensitive information meant to be hidden from adversaries and others in some form – what today would be called 'intelligence' – has been a part of war and society for millennia.[2] References to espionage can be found in some of the earliest records, both in civilisations identified broadly as being from 'the East' and 'the West'. As the British intelligence historian Christopher Andrew has observed, 'the Christian Old Testament (the Jewish Tanakh) … contains more references to spies than any history of Britain or of most other countries.'[3]

In ancient Greece, during the Peloponnesian War (431–04 BCE), in order to send encrypted messages to one another, the Spartans used a *skytale* – a staff used for writing code around which a strip of leather was wrapped slantwise and the message written lengthwise on it. When unrolled, the message was unintelligible, but rolled slantwise with a *skytale* of the same dimensions in possession of the intended recipient, it could be deciphered, with the key letters and words separated from the other meaningless letters assembled on either side.[4] Writer Craig Bauer describes this as a transposition cipher – the letters are not changed, only their order, in contrast to a substitution cipher. The *skytale* also appears to mark the origin of the commander's baton which, over time, became purely symbolic.[5] Dealing with transposition ciphers such as this remains a staple for cryptologists in Australia's national Sigint agencies to this day.

Around the fourth century BCE, Aeneas the Tactician, one of the earliest Greek writers on the character of war, wrote a collection of tricks. The original in Greek has been lost but a version was published in 1757.[6] The 31st chapter (*Ratio mittendi litteras occultas*) addresses various forms of secret messages, providing an important guide to concealing communications from one's adversaries which has set the standard for hundreds of years.

A few centuries later, Julius Caesar (100–44 BCE), employed a mono-alphabetic substitution cipher: that is, where a given letter is consistently replaced by the same cipher text letter. This method was used to disguise messages sent to and from the battlefield. Each letter was replaced by the third letter to follow it alphabetically. Upon reaching the end of the ciphertext alphabet, the first three unused letters, A, B and C, mark the end of the text. Caesar also used Greek letters as substitute for Latin text.[7] Another variant on the Caesar Cipher is called a reverse Caesar Cipher (for which the ciphertext alphabet is written backwards before being shifted).[8] Once again, this technique has stood the test of time, becoming a technique of interest to Australian cryptologists in the twentieth century.

Ancient cryptology – in 'the East'

Despite these examples of ancient cryptology, the first book on espionage, in fact, came not from Europe, but China. *The Art of War* (*Sunzi bingfa*), is traditionally ascribed to one of Confucius's contemporaries, a Chinese general known as Sun Tzu (c.544–c.496 BCE).[9] In a chapter entitled 'The Use of Spies', the Chinese sage asserts that all warfare is based on deception. Perhaps the most famous dictum attributed to Sun Tzu is: 'he who knows the enemy and himself will never in a hundred battles be at risk'.[10]

Today, Islamic scholar Yaqub ibn Ishaq al-Kindi (c.800–873 CE) is recognised as a leader in the invention of cryptanalysis – the art and science of decrypting enciphered messages without prior knowledge of the cipher, otherwise known as code-breaking. Al-Kindi took the next big step, working out that the statistical properties of language could be used in cryptanalysis. He discovered the frequency principle, whereby every alphabet has letters which are more often used than others. In English and French, for instance, E and T are the two most commonly used letters. It follows, then, that the two most frequently encountered enciphered letters are likely to represent these letters.[11] This frequency principle became the basis for European Renaissance cryptanalysis and cryptography. Its significance would continue throughout the twentieth century and remain a staple in teaching cryptology at Australia's national Sigint agency. The significance of early Arabic researchers to the history of cryptology is contained in the very word 'cipher' which has Arabic origins.

While China has tended not to feature in Western histories of cryptography, this is partly because there are few sources and partly because of the mistaken belief that there is nothing to write about.[12] Yet the Chinese invention of wooden block movable type printing (c.1041–48 CE) was one of the four great inventions of early China (along with paper, the compass and gunpowder), predating the development of cryptography in Europe.

In China, the development of this technology is a badge of pride. For instance, in the opening ceremony of the 2008 Olympics in Beijing, dancers performed in boxes of varying sizes with letters on top, representing the invention of movable type.[13] The applications of movable type printing have been immense, carrying with them implications for intelligence. Movable type printing facilitated the production of the world's earliest paper currency which was first issued in 1023.[14] The notes, called Jiaozi, carried a picture, a stamp and a cipher code.[15] With an expansion of writing came a greater need to conceal meaning in certain circumstances. During the Qing Dynasty, Tin Tei Wui was one of many secret societies involved in uprisings, including those in 1767 and 1769 in the south of Fujian Province. In order to survive, its members used a secret code for communications.[16] In the nineteenth and twentieth centuries in Jiangyong, in China's Hunan Province, women used a special script known as *Nushu* (which literally means 'female script') which no man could read or write. Created by women who had no right to education at that time, the writing system allowed these women to keep diaries, write poetry and communicate with each other.[17] During the Cultural Revolution (1966–76) *Nushu* was officially condemned as 'witches' writing'.[18] All of this demonstrates a key point: cryptological developments took place not just in Europe and the 'Middle East', but also in the 'Far East', Australia's near north.

European cryptology from the 1300s to 1500s

The papacy first used ciphers during the pontificate of John XXII (r. 1316–34). In 1467 the Renaissance polymath Leon Battista Alberti (1404–72) published *De Componendis Cifris* ('Concerning the Solution of Ciphers') – the first major European work on cryptanalysis. Like al-Kindi's work six centuries earlier, it showed how some letters are more frequently used than others, and this variation enabled cryptologists to unlock ciphers which relied on simple letter substitution. Alberti also developed a mechanism known as a cipher disk with two alphanumeric concentric circles that could be altered during the encryption process, changing the results in different parts of the same message. This practice came to be known as 'polyalphabetic substitution'.[19] Alberti's three remarkable firsts – the earliest Western exposition of cryptanalysis, the invention of polyalphabetic substitution, and the invention of enciphered code – make him the 'Father of Western cryptology'.[20]

In 1518 the first printed work on cryptology, *Polygraphiae* by a German abbott, Johannes Trithemus (1462–1516), was published. Book V contains Trithemus's contributions to 'polyalphabeticity' or polyalphabetic substitution. Trithemus details the way in which a sequence of letters is shifted to different positions in relation to the plaintext alphabet, with each letter in turn enciphered by separate progressive alphabet keys that are exhausted before being repeated.[21] Knowledge of the length of the key makes breaking this system relatively easy.[22] Modern cipher machines often embody this type of key progression.[23] This method is also known as the Vigenère Cipher, named after Blaise de Vigenère, who published the *Traicté des Chiffres* (or 'Treatise on Ciphers') in 1586, although credit properly goes to the earlier work of Alberti, Trithemus and Giovanni Battista Porta (1535–1615).[24] Nowadays, the Vigenère is considered a class of polyalphabetic substitutions, of which one is a running key using mixed alphabets. It uses a table of shifted direct standard alphabets and encrypts letter by letter using a short, repeating keyword.[25]

The second book on cryptology, published in 1526, was *Opus Novum* (or 'New Work') by Jacopo Silvestri. It taught the reader to encipher and decipher messages using a cipher wheel.[26] Centuries later, cipher wheels would be made famous by the German Enigma machine and such techniques would feature in machine-generated cipher equipment which would be the focus of Australia's national Sigint authorities in the second half of the twentieth century.

In France, Cardinal Richelieu (1585–1642) established the *cabinet noir* (or 'black chamber'), the first cryptanalytical agency of its kind. This term was used to describe agencies involved in encoding and decoding messages, which came to be known as cryptography.[27] Richelieu's *cabinet noir* has had a lasting influence, being the precursor to national cryptologic (Sigint) agencies, including the Australian Signals Directorate (ASD). Even the black, glass-encased main building of the modern-day National

Security Agency (NSA) in Maryland in the United States is reputed to draw inspiration from the *cabinet noir*.

During the Renaissance struggles between the Catholics and the Protestants in France, Richelieu recruited Antoine Rossignol, who deciphered a message from starving Huguenots trapped in the seaside fortress town of Réalmont appealing for help, which persuaded Richelieu in April 1628 to maintain the siege. A month later, the Huguenots capitulated. Over time, the role of the decipherment of such messages in success in battle would grow exponentially, particularly once driven by the technology of the Industrial Revolution.

Early English cryptology

In the 1580s at a time of uncertainty over royal succession, Britain established a secret intelligence service directed by Elizabeth I's spymaster, Sir Francis Walsingham, to collect political and military intelligence in Europe. Walsingham established a cipher department in his London home.[28] One of the best stories in cryptanalysis concerns the role of Walsingham and his crew in foiling the 'Babington Plot' – the work of the Catholic chief conspirator and Mary Stuart's devoted supporter, Anthony Babington.[29] This plot aimed to assassinate Elizabeth I and replace her on the English throne with Mary, Queen of Scots. The plot was uncovered thanks to the combined work of agents, double agents (Mary's courier), and code-breaking of a mono-alphabetic cipher by Walsingham's cryptanalyst, Thomas Phelippes. One deciphered message sought Mary's approval for Elizabeth's assassination by 'six noble gentlemen ... who will undertake that tragical execution'. On its own, this message was not conclusive, but Mary's ambiguously worded reply (intercepted and doctored by Walsingham to make her response appear more incriminating) was presented as evidence to warrant the executions of Mary and Babington.[30] Had Babington's plot succeeded, England may well have reverted to Catholicism. This serves as one of the most important early examples of the consequences of timely Sigint! It also demonstrated how cryptology can complement human intelligence (Humint) in statecraft.

Walsingham's network of spies would subsequently play a major role in defeating the Spanish Armada in 1588, and over the next three centuries, such a service would be reactivated at various times. In the nineteenth century, it was known as the Secret Intelligence Service, responsible to parliament through the Foreign Office.[31] This precedent for parliamentary control would be followed by the counterpart organisations established in twentieth-century Australia.

During the English Civil War (1642–51), a domestic chaplain, Dr John Wallis, deciphered some captured correspondence from King Charles I. Wallis continued deciphering documents for the Parliamentary side and after the Restoration of the monarchy, while doubling as a royal chaplain. Working closely with the decipherers of the Secret Office, inside the Royal Mail Post Office, he continued his work even after the Revolution of 1688, which saw King James II replaced by his daughter Mary II and his Dutch nephew (and Mary's husband), William III of Orange, in a bloodless coup which also saw primacy of Parliament established over the Crown. On Wallis's death in 1703, his grandson, William Blencowe, became the first official decipherer – an office that would be filled by a succession of decipherers for the next century and a half.[32] The art and science of cryptology that had developed in Britain would, in effect, form the genesis of the capability in Australia over two centuries later.

European cryptology in the 1700s

During the 1700s, black chambers were common, with Vienna's *Geheime Kabinet-Kanzlei* (or secret cabinet chancellery) considered the best in Europe. It ran with 'almost unbelievable efficiency': letters opened, seals melted and opened, letters transcribed for later translation and returned, resealed with forged seals and ready for release back to the postal service within a couple of hours.[33]

In England, the Hanoverian kings, following the accession of King George I in 1714, saw a succession of royal decipherers appointed to break diplomatic ciphers, and to uncover and report on plots.[34] Diplomatic intercepts were read avidly by prime ministers and kings alike, giving the

British an advantage in peace negotiations with France and Spain that culminated in the Treaty of Paris in 1762.[35]

Ciphers grew in size and complexity, becoming an accepted and normal part of statecraft. In the late sixteenth century, Spain introduced a 500-key-length character cipher. During the English Civil War, King Charles I used one with 800 characters. And in France, Louis XIV had a *Grand Chiffre* (or Great Cipher) of 600.[36] All were designed to defeat the approach set out by David Arnold Conradus in 1739 in a book entitled *Cryptography Unmasked* or *The Art of Decyphering*, which drew conclusions from the frequency of different code numbers. His 'General Theory' listed three propositions and three rules as follows:

> Proposition 1: the art of decyphering is the explanation of secret characters by certain rules.
>
> Proposition 2: Every language has, besides the form of characters, something peculiar in the place, order, continuation, frequency and number of the letters.
>
> Proposition 3: In a writing of any length, the same letters recur several times.
>
> Rule 1: In decyphering regard is to be had to the place, order, combination, frequency and number of letters.
>
> Rule 2: In decyphering nothing is to be left to conjecture, where the art shews the way of proceeding with certainty.
>
> Rule 3: Writings of any length are most easy to decypher from the frequent recurrence and combination of the same letters.[37]

The propositions that Conradus laid out here would go on to guide cryptologists through the centuries.

What Conradus's work did not directly expose, but hinted at, was that the only way of ensuring that a cipher was entirely unbreakable was for the message to be sent using a one-time pad, with a random key as long as the message. The concept, simple in hindsight, eluded cryptologists for centuries and was only developed in the period 1917–18 by Americans Gilbert Vernam (1890–1960) and Major Joseph O. Mauborgne (1881–1971).[38]

In 1693, France developed a flag vocabulary for transmitting messages at sea. Over 80 years later, Britain's Admiral Lord Richard Howe produced a signal book for his West Indies fleet and in 1790 a *Signal Book for Ships of War*, which allowed ten flags to signify 9999 possible meanings. Lord Horatio Nelson would use this system to his advantage against the French fleet, commencing with the Battle of the Nile in 1799.[39] Nonetheless, he had no access to the official British encryption system and complained frequently about the risks of passing un-enciphered messages. The widespread use of flag hoists and the growing use of shore-based semaphore telegraphy made government and military communications public, adding to the demand for the encoding of such messages.[40] At the turn of the twentieth century, this concept of communicating enciphered messages over distances at sea would be dramatically improved thanks to new technologies.

Cryptology and the American Revolutionary War

Given the impact of American cryptology on Australia's Sigint enterprise in the twentieth century, understanding Australia's circumstances requires an understanding of the American experience. That story begins in the late eighteenth century, with the American Revolutionary War and particularly the American Revolutionary War leader, George Washington (1732–99) who outclassed his British opponents by the deft use of Humint and Sigint.[41]

British commanders and the American rebels both used a variety of methods for clandestine communications, including invisible ink, hidden messages and secret writing in the form of ciphers and book codes, sometimes called dictionary codes.[42] In 1775, Charles William Frederic Dumas designed and dispatched the first secret diplomatic cipher to

America's envoy in France, Benjamin Franklin, with the intention of masking Continental Congress correspondence from foreign agents in Europe.[43] Historian Ralph Weber observed that

> At the time of the American Revolution, the American founding fathers did not believe code and ciphers 'were employed for purposes of evil and cruelty'. Rather, they viewed secret writing as an essential instrument for protecting critical information in wartime, as well as in peacetime.[44]

In the end, American independence from Britain was achieved with the help of codes, ciphers, invisible ink, visual communications and hidden messages. The combination of techniques protected information passed between the United States and its representatives in Europe, while also helping George Washington to plan his strategy.[45] This experience would set an important precedent for American cryptology in later conflicts, although Washington's immediate successors paid little attention to maintaining the capability. Indeed, their cryptological skills atrophied and had to be revived during the American Civil War.[46]

Cryptology and the Napoleonic wars and beyond

French emperor, Napoleon Bonaparte (1769–1821), was dismissive of his adversaries' intelligence efforts. Intelligence only impressed him if it confirmed his own views. More interested in domestic than foreign intelligence, Napoleon used his *cabinet noir* to monitor intercepted correspondence from his government ministers.[47] At first, the popular army that had been raised by the *levée en masse* and Napoleon's extraordinary gifts as a military strategist were enough – particularly before his adversaries adopted similar organisation and tactics. In October 1806, Napoleon won the Battle of Jena against the King of Prussia, despite the Prussians having advance warning of the French army's intentions.[48] Napoleon's success is attributed to the inspiration he derived largely from map study.[49] But his enemies responded by raising their own mass armies, and from 1812 and the invasion of Russia, French victories would prove elusive.

While Napoleon himself may have not fully appreciated the significance of getting cryptology right, insights and practices from the Napoleonic era influenced twentieth-century cryptologists, including those in Australia. Napoleon's army tried using some simple codes, known as *petits chiffres*, which transposed letters into numbers from one to 50. While this method was fast, it was also easily broken, so it could only protect a message for a few hours. It had to be useable by an aide-de-camp racing on horseback to deliver the message. A more complex cipher table, like the one used in the foreign ministry, was the size of a large map and therefore unsuitable for use in the field.[50]

Napoleon's martial prowess could only take him so far against opponents who knew the value of intelligence. Arthur Wellesley (1769–1852), who was appointed Duke of Wellington following his success against Napoleon's forces in the Iberian Peninsula campaign, had learned the importance of military intelligence on campaigns in India.[51] In the Peninsular War, he relied on intelligence and cryptanalysis far more than Napoleon and his commanders.[52]

In 1809, under the Duke of Wellington, a corps of guides was formed. Drawing on Wellington's experience with an Indian corps of guides, the new corps gathered intelligence for British columns moving into unfamiliar territory. This concept of a corps of guides would shape the formation of an Australian intelligence corps – the oldest in the British Commonwealth.

Meanwhile, Wellington employed Sir George Murray as his principal intelligence staff officer and George Scovell, a code-breaking genius, as commander of the Corps of Guides.[53] In April 1811, Scovell broke into the 'Army of Portugal' Cipher. A more complex cipher introduced early in 1812, the Great Paris Cipher, involved 1400 numbers that could be applied to a wide range of combinations of words. The frequency principle enabled Scovell to make some initial breakthroughs, with certain numbers appearing more frequently than others, such as the French word '*et*' (and). Thereafter, the frequency principle would be key for cryptologists to understand and apply, including Australian cryptologists who used it from the Second World War onwards.

Captured French despatches which combined enciphered and unenciphered phrases assisted Scovell in his work. By the middle of 1812, he had largely solved the cipher. Thereafter, Wellington's campaign strategy was informed by French operational plans obtained from intercepted despatches.[54]

Scovell's contribution was decisive in the Battle of Salamanca on 12 July 1812, giving Wellington a distinct advantage as the campaign reached its climax and sealing the fate of Napoleon's forces in the Iberian Peninsula. Scovell's deciphering revealed that the army of French Commander Marmont would not be reinforced from the north, and also indicated when Napoleon would attack. Scovell was mentioned in Wellington's Salamanca dispatch and promoted to lieutenant colonel; achieving his advancement 'not by drawing blood but by the application of science and intellect'.[55]

The French defeat in the Iberian Peninsula can be attributed in large part to Napoleon's cavalier disregard for Spanish and Portuguese resentment of French rule. But cryptanalysis was instrumental in shortening that rule. Thereafter, the role of cryptology in British victories would be a legacy of the campaign. And, eventually, Australia's Sigint enterprise would be one of the beneficiaries of this legacy.

While the technology of the telegram was still decades away, the British Navy's Horatio Nelson developed a system of line-of-sight semaphore stations, building 15 by 1796. By 1806, the Admiralty in London could pass orders to the ships in Plymouth on Britain's south coast and receive a transmission acknowledgement, covering a distance of 200 miles (320 km) in three minutes. Although this system was only reliable during daylight hours in the summer months, as fog and long winter nights limited its effectiveness, it did foreshadow how technology would transform the collecting of intelligence.[56]

Like the King of Prussia, and Wellington in the Iberian Peninsula, the Tsar of Russia maintained a *cabinet noir* as one of his weapons against Napoleon.[57] Napoleon underestimated Russia's ability to break French ciphers, and his intelligence disadvantage was compounded by his generals' reluctance to give him bad news. With the benefit of intelligence, Russia adopted a strategy of attrition and harassment, while avoiding the decisive battle Napoleon sought. This led to Napoleon losing half his army before reaching a wintry and deserted Moscow in 1812. He escaped and returned to France having suffered massive losses, with only 6000 remaining of his *Grand Armée*.[58]

Napoleon's defeat at Waterloo on 18 June 1815 followed the interception of encrypted French battle plans by the head of Wellington's Intelligence Department, Colquhoun Grant, three days earlier. But the importance of this intelligence insight can be overstated, as Wellington, sensing what might happen, had anticipated Napoleon's move. Meanwhile, Napoleon delayed his final assault against Wellington's forces, having dismissed an intelligence report predicting the arrival of Prussian forces later in the day. Napoleon and his forces were duly surprised by the arrival of Prussian forces commanded by Gebhard Leberecht von Blücher.[59] The rest, as they say, is history.

Carl von Clausewitz, a Prussian general and renowned philosopher of war, lived through this period. He also overlooked the significance of cryptology and disparaged military intelligence, perhaps taking his lead from Napoleon who, fortunately for Britain, consistently underestimated the ability of his enemies to collect and use intelligence against him.[60]

At the 1815 Congress of Vienna, which demarcated territories and authorities of the European powers, more spies were present than at any previous congress. Competing for an intelligence advantage in the negotiations, they used eavesdropping and the interception and translation and or decryption of messages to an unprecedented extent and thereby 'set a new standard for the level of intelligence collection' in Europe. The Austrian foreign minister, Prince Metternich, presided over relentless interceptions of correspondence, stealing of documents, as well as the collection of agents' reports and intelligence obtained directly from officials and policymakers.[61] A high bar was set for diplomatic intelligence.

So far, the story has focused on developments before the effects of the Industrial Revolution were fully recognised. Many of the concepts and practices would be enhanced and accelerated by the technological breakthroughs that followed, including the discovery and harnessing of electricity.

Electricity

Technology played a major role in facilitating the advances which culminated in Sigint. Six decades before European dignitaries assembled at the Congress of Vienna, the American inventor and statesman, Benjamin Franklin, conducted his famous kite experiment. The experiment, undertaken in 1752, was carried out in order 'to determine the Question, whether the Clouds that contain Lightning are electrified or not', as Franklin himself put it. The experiment succeeded, and thereby was 'the *Sameness* of the electric matter with that of lightning compleatly [*sic*] demonstrated'.[62] Franklin's insights would have a profound effect on many fields, including Sigint.

Half a century later, in 1800, the first electric battery was invented by Alessandro Volta, but it was not until after the Napoleonic Wars, in 1821, that the first electric motor was invented by Michael Faraday.[63] Thereafter, the telegraph was developed in the 1830s and 1840s by the American Samuel Morse and other scientists, enabling the rapid transmission of information over vast distances.[64] These breakthroughs would transform society, business, industry and military organisations, as well as the collection, analysis and reporting of intelligence. They laid the foundations for modern technologically advanced national Sigint agencies.

Meanwhile, in 1844, a security breach revealed to the general public the practice of deciphering, prompting a storm of protest in the British Parliament. Britain's unchallenged naval supremacy, growing material prosperity and Victorian moral values encouraged the belief that a secret deciphering bureau was no longer necessary. As a result, the Deciphering Branch, a precursor to GCHQ, was closed that year. Thereafter, interception and deciphering appeared to die out in Britain, although the secret interception of mail in the postal service continued intermittently under the authority of the Secretary of State.[65] Ironically, this decision was made on the eve of a technological breakthrough that would greatly facilitate Sigint: the telegraph and Morse code.[66]

Telegraphy and the Crimean War

Technically, Samuel Morse's code was a cipher, which allowed messages to reproduce language by allotting each letter of the alphabet a separate identity in the form of a combination of short and long electrical pulses, dots and dashes, transmitted over an electrical cable. The simplest example of a combination of Morse code letters represented by 'dits' (•) and dah's (–) came to be the signal of distress: • • • – – – • • • (or 'SOS'). The use of a metallic cable for long distance communication was pioneered in 1838. Before long, the properties of latex rubber were identified as having an insulating effect, enabling latex-enclosed wires to pass through water without loss of electric current.[67] The first successful telegraphic Morse code message was passed between Washington and Baltimore in 1844.[68] These discoveries had a profound effect on society, the conduct of war and Sigint.

The Crimean War (1853–56), the first major conflict between European great powers since the Battle of Waterloo, was also the first war in which the telegraph played a major role, greatly facilitating the transmission of news. *The Times* journalists' reports of battles arrived in London a few hours after the events on the ground.[69] But messages from London still took as long as 24 hours to arrive at expeditionary headquarters because of the need to relay in several stages; and commanders in the field could only be contacted if they were near the cable link.[70] While Britain's commander in Crimea blamed *The Times* war correspondent for making it unnecessary for Russia to employ spies, in reality Britain's clumsy and drawn-out campaign had more to do with inadequate intelligence work than Russian reading of newspapers.[71]

In the following decades, the effect of long distance communications using the electromagnetic spectrum would be increasingly felt, affecting speed, coverage, cost and reliability. The speed of electricity promised real-time information. Increased coverage meant the telegraph acted as a node in an expanding network. The cost, initially staggering, would drop over time as investment in communications infrastructure and more advanced technology grew, allowing for more competition. Reliability, initially patchy, increased, as did the ability to intercept messages in transit: to eavesdrop on conversations – and to undertake actions based on the content of the intercepted message.[72]

In 1854, as the Crimean War was being waged, Charles Babbage, a Cambridge mathematics professor, discovered a general solution to Vigenère ciphers. The principles he outlined include those on which modern computers are designed.[73] In 1863, a Prussian infantry officer, Friedrich Kasiski, discovered how to break into Vigenère ciphers independently of Babbage. Their method was based on the fact that with a sufficiently long plaintext message, and with a high degree of probability, a Vigenère cipher text will repeat a part of the key with an identical part of the plaintext and this repetition would be a multiple of the key length. Once the cryptanalyst guessed the key length, they could then separate the cipher text into a set of mono-alphabetic substitution cipher texts and solve them.[74] This meant that the standard Vigenère cipher was safe only for very short messages with longer keys, the value of which could be measured in not more than some hours.[75] As technology enabled more messages to be transmitted, the significance of this discovery would become clearer. For twentieth-century Australian Sigint work, this breakthrough would be of great importance, providing the conceptual framework for decryption of enemy codes.

Shortly after the Crimean War, the significance of the telegraph for timely intelligence reporting became evident when it enabled a rapid response in India by British forces sent to quell the Sepoy Mutiny of 1857.[76] Expensive mistakes were made in the 1850s, but by the 1860s long distance telegraphy was viable and affordable.[77]

From the mid-1850s onwards, Britain's War Office maintained a small team for intelligence related tasks. While their staffing and duties varied, there was a consistency of function for much of the next half century.[78] Australia's colonies were not involved in such matters until well after 1901, when Federation compelled closer examination of the requirements for the defence of Australia.

In the mid-nineteenth century, after France closed its *cabinet noir* in 1848, Russia and Austria were the only great powers to maintain codebreaking traditions.[79] The telegraph, however, with its messaging of sensitive, often encoded information over cables, led to the re-emergence of black chambers in the late nineteenth and early twentieth centuries. Before the First World War, the most successful of these were those in France and Russia.[80] As these cryptographic centres were revived, there was much interest in a certain conflict across the Atlantic.

Technological innovations and the US Civil War

In North America, the telegraph played a major role in the US Civil War from 1861 to 1865.[81] In the Confederacy, simple ciphers, codenames, and book codes, known as 'dictionary ciphers', prevailed until a polyalphabetic cipher became the accepted standard. In the North, a route or word transposition system was used to protect military telegrams transmitted by US Military Telegraph controlled by the US Secretary of War.[82] Historian David Kahn has observed that

> The men in gray, who sometimes could not read their own messages, could never solve the Union's ... Even the capture of two of the ciphers themselves [in mid-1864] failed to help. The Yankees simply got out a new list of routes and jargon words, and the result was always more than the rebels could handle.[83]

As the Civil War unfolded, President Lincoln became an avid consumer of intelligence reports; he monitored developments closely, undertaking his own analysis of the intelligence.[84] For the first few years of the war, however, field commanders derived little benefit from the telegraph. Not surprisingly the letter remained the main channel for delivering messages, often with several days' delay. In due course, by 1864, when Ulysses S. Grant took command of the Union armies, the telegraph was the principal means of managing his army at the higher levels, although it remained unfeasible for management of the fast-flowing tactical aspects of battles.[85]

Civil War technological innovations included balloons for aerial observation, the precursor of imagery intelligence (or Imint).[86] By early 1863, the Bureau of Military Information was established. Under General George G. Meade (after whom Fort Meade, the home of America's NSA, is named) the Bureau helped defeat Meade's rival, General Robert E. Lee, at the Battle of Gettysburg (1–3 July 1863). This forerunner of the later Military Intelligence Division of the US Army was disbanded after the war.[87] The Union Army also quickly contracted, with the wartime US Military Telegraph service absorbed into the US Army Signal Corps.[88]

In 1869, aware of the lessons of the US experience, the British Army established a signal division at the School of Military Engineering to teach electric as well as visual signalling. By the time of the Franco-Prussian War of 1870–71, the impact of telegraphy on military intelligence systems was becoming obvious.[89] Japan and other powers were also paying close attention, seeking to modernise and capitalise on Western technological innovations such as the use of telegraphy, machinery and also in the field of cryptology. The proliferation of such ideas and techniques would have a dramatic effect, not just on the battlefields of Europe, but also on the power dynamics in Australia's neighbourhood.

Telegraphy and the reach of Britain's empire

Meanwhile, as Britain's empire expanded, telegraphic cable stations were established in India in 1870 (via Lisbon, Gibraltar, Malta and Alexandria), Canada, South America, and down the African coast. They would also stretch across to Australia and New Zealand, connecting Britain's colonies and principal trading partners to the empire's commercial hub in London.[90]

To Australia's north, the arrival of telegraphy in China in 1871 imposed a global alphabetical structure on a non-alphabetic Chinese writing system. The Chinese telegraph code of 1871, invented by two foreigners, consisted of a group of 6800 commonly-used characters organised according to the Kangxi radical stroke system and then assigned a series of four-digit numerical codes.[91] At first, this put China at a disadvantage, but in the longer term it made Chinese telegraphers code conscious in a way that telegraphers working with alphabets were not. On the one hand, Chinese was a character-based script to be transmitted over telegraph cables. On the other, it was a code-sequence of digits that required a telegraph code book to decipher. After hearing the Morse code pulse sequence, Chinese telegraphers (unlike telegraphers in countries with alphabets) could not skip the step of numerical encipherment and write out the message in plaintext. They would first have to write out a code, the symbols to which the Morse code pattern corresponded. Only then could they translate this code into 'plaintext' by looking up its corresponding Chinese character in the 1871 telegraph code book.[92]

Standard Telegraphic Code (STC) is still in use today, and some linguists were reportedly so familiar with it that they could read it directly without first rendering it into characters. Although it was not intended for secrecy, it is still a code in every sense. The great British cryptanalyst John Tiltman, who will reappear in this story, not being familiar with STC, came across some traffic in Hong Kong and was half way to recovering it when a colleague pointed out that you could buy the complete code book in any post office![93]

The phenomenon of being forced to 'stay in code' affected the way Chinese telegraphers worked. Curiously, rather than opting for the simplest version of what they had to do, they added additional practices of their own design between themselves and the Chinese code to make their relationship with it more personal, workable and easy to remember. In effect, the Chinese telegraphers began the process of surrounding a system that had once surrounded them.[94] Reflecting on these circumstances a century and a half later, it appears China's difficult history with the telegraph and then the typewriter prepared it to take fuller advantage of computer software. What seemed to have been a great disadvantage for a non-alphabetic writing system would eventually turn out to be a greater technological advantage.

The year after the telegraph reached China, it arrived in Australia. On 22 August 1872, the overland telegraph from Adelaide to Darwin was complete. Within months, the connection was made across to Java in the Dutch East Indies (now Indonesia), which by then was already connected by telegraph cable back to Europe and the United Kingdom. That connection reduced the time for two-way communications from the United Kingdom to the Australian colonies from months to hours. This would revolutionise the linkages between the Australian colonies and the rest of the world.[95] In addition, Britain's central place in the emerging transatlantic telegraphic network gave it privileged access to international telegraphic messaging and led to a growing interest in cryptography and to the development of experts capable of deciphering intercepted but encoded telegrams.[96] Britain's central position in the global telegraphic network would later prove remarkably useful, particularly during the world wars of the twentieth century.

The wireless radio and machines

The invention of the wireless radio would further transform intelligence. The concept of wireless communication was first proposed in the 1860s by British physicist James Clark Maxwell, who postulated that electromagnetic waves could be propagated in space and would travel at the speed of light. In 1888, the German scientist Heinrich Hertz demonstrated electromagnetic wave propagation and reception, but only over a very short distance.[97] For greater distances, a mechanism to add power and range had to be invented. Then, in 1895, an Italian scientist, Guglielmo Marconi, managed to send and receive his first radio signal in Italy over a wireless radio device. Marconi recognised the potential for the use of such technology at sea, where a naval arms race between the great powers was underway.[98] Marconi foresaw how wireless communication would transform warfare at sea. As we know now, it also transformed the way one could eavesdrop on such communications.

During the late nineteenth and early twentieth century, British governments collected intelligence routinely. Australia, being dependent on Britain for defence and foreign policy concerns throughout this period, relied on the services Britain provided and judgements it reached from this work. Agencies were set up to deal with specific problems and disbanded once they were solved.[99] Victorian officials during the 1870s and 1880s, for instance, practised virtually all the elements of modern Sigint then possible, short of the regular acquisition and solution of encoded foreign messages.[100]

The interception of letters was a 'people' activity, in the same way that couriering or posting was. The cryptanalysis of hand-based systems was traditionally a hand-based activity as well, requiring meticulous record keeping and rigorous analysis. Radio transceivers were wireless devices for transmitting and receiving messages. Improvements in reliability and range enabled radio use for military, diplomatic and civilian communications, which could equally be applied to the interception of those same messages.

As ciphers became harder to break, and were intercepted in greater volume, the use of machines to relieve this burden, and in a continuing thread, would make that which was impossible through hand exploitation, possible. While special purpose, hand-driven 'machines' had been trialled to duplicate the cryptosystem under attack, it was not until the early twentieth century that more general purpose, large-scale electro-mechanical machines were applied to cryptanalysis. In particular, accounting machines such as tabulators, sorters and collators were identified as having intelligence application.

Computers would not emerge until well into the twentieth century, but the first tabulator was invented in the late nineteenth century and this would come to play a significant role. This tabulating machine was designed by Herman Hollerith (1860–1929) to compile the American census of 1890. His invention needed specially designed, punchable cards. Essential data collected for each person could be coded and then entered by punching holes into given spots on each card, storing individual data on separate cards.[101]

Punched cards could not conduct electricity, but the holes would enable electric contacts to be made. Cards, therefore, could be 'read' at high speed by wire brushing against them, allowing electrical current to pass through the holes, and relay switches would produce the desired effects. Once the cards were punched in this way, determining, say, the number of six-year-old girls resident in a certain location would be achieved by running all the cards through a machine wired to count those with the appropriate combination of holes punched. The tabulators were fitted with plugboards which could be wired to achieve the desired count. Eventually it became obvious that one tabulator would be more efficiently used if a number of plugboards were available as interchangeable accessories. Any standard use would require an appropriate pre-wired plugboard for insertion into the tabulator. The pre-wired plugboard would not emerge until the mid-twentieth century, but these plugboards for use with Hollerith's contraption were the predecessors of modern computing software. Two and a half decades after the 1890 census, the British Royal Navy Sigint cryptanalysts of the First World War (Room 40 of the Admiralty) were the first to understand and apply Hollerith's invention for Sigint purposes. It would be late in the interwar years, however, before plugboards were in circulation.[102]

With a growing need to muster different sources and types of information, a British 'Topographical and Statistical Department' was established in 1855 under Major Thomas Jervis. The department became the Intelligence Branch in 1873. Enlarged in 1888, it was headed by the Director of Military Intelligence (DMI). Until the Anglo–Boer War or South African War (1899–1902), however, the DMI's domain was regarded by the rest of the War Office as little more than a useful reference library.[103] Up to 1898, Britain was a cryptological novice, having only just begun to consider the use of cryptanalysis, and had low standards of cipher security.[104]

As these events unfolded, British colonies and dominions, including Australia, were following Britain's lead. These developments were highly significant for Australia, where a national army and navy, requiring an intelligence capability, were being established after Federation.

British troops garrisoning Australian colonies had left in 1870. Reflecting trends in the British Army, a visiting British officer had suggested in 1878 that scouts or guides prepare plans and maps for the possible defence of Brisbane, but there was little enthusiasm to pay for this. Germany's annexation of New Guinea in 1884 was an added incentive for the Australian colonies to form a federation and take a more strategic approach to defence, including intelligence. That momentum would grow following the onset of the Anglo–Boer War.

2
FROM FEDERATION & THE ANGLO-BOER WAR TO THE FIRST WORLD WAR

Australia's Federation on 1 January 1901 was a time when imperial defence was uppermost in policymakers' minds. Indeed, concerns over defence lent momentum to the Federation project. Linked to the defence imperative was the demand for more information about the potential threats to Australia. Imperial defence was the cornerstone of Australian defence, but this did not stop the Australian government from sending out its own scouts from time to time to learn more about possible threats.

The development of national intelligence arrangements would at first be limited by the prevailing view that intelligence was best conducted from London, with Australia providing support when necessary. Even before Federation, there had already been requests for such support, and these had added impetus to the Federation project. Australia's experience at war as part of the empire would be a formative one.

The Anglo-Boer War

The outbreak of the Anglo–Boer War, in October 1899, between Britain and the Boer republics of Transvaal and the Orange Free State, inspired a groundswell of support for the empire in the Australian colonies, with volunteer forces raised to join the fight in southern Africa. Initially, Australians were despatched to support the British-led campaign. The colonies were only required to provide combat forces, while logistics and intelligence support were left to Britain.[1]

The Anglo–Boer War raised awareness in London of the value of a military intelligence organisation. Although prewar intelligence assessments warned of Boer capabilities and intentions, early enemy successes revealed British intelligence shortcomings. There was a compelling argument for intelligence reform and expansion.[2] The South African republics had a sophisticated telegraph and heliograph network and a Vigenère substitution cipher system as complex as those used by British units.[3] Having volunteered only infantry and mounted troops to the conflict, Australia had little exposure to this more sophisticated field of endeavour.

Early communications intercepts and telegraphy

During the war, the British Army experimented with wireless radio communications, but since the Boers had little or no wireless equipment, there were few opportunities to intercept and exploit their transmissions. Instead, the focus was on interception of the heliograph and the telegraph, and the improvement of ciphers for use in the field.[4]

British commanders sent important operational messages by telegraph in clear text, which the Boers could tap. This intelligence was exploited by Boer forces, particularly early in the war. British forces subsequently tightened their own security by encoding telegrams. But even enciphered British traffic was vulnerable, due to the primitive field ciphers they used and Boer exploitation of captured written keywords.[5]

At the War Office, 'Section H' was responsible for cable censorship and the development of ciphers for field use. The covert intercept and deciphering of international telegraphic messages by British officials continued until the end of the war. The study of ciphers, and the composition and issue of new ciphers remained the responsibility of a special duties section of the Military Operations Directorate.[6]

Britain's commander-in-chief, Lord Kitchener, was kept abreast of Boer leaders' intercepted telegraphic communications.[7] In the field, however, Britain probably lost more than it gained through Sigint. Britain's early success in breaking up Boer communications networks actually stymied its interception of Boer communications traffic. Once the campaign degenerated into guerrilla warfare, there was little Sigint to collect, simply because the Boer forces could not use heliograph or telegraph circuits.[8]

Tactical field communications intelligence might not have played a critical role, but at higher levels, it proved useful. In May 1901, decoding of an encrypted telegram from a Boer general to Boer President Kruger, then visiting Holland, provided Lord Kitchener with valuable information in the peace negotiations.[9]

The British Army deployed a field intelligence department, working closely with the corps of scouts and guides which included a number of Australians. The Anglo–Boer War experience with field intelligence, including detainee interrogation and reconnaissance patrols, paved the way for an intelligence corps for the Australian armed forces, although not one focused on Sigint.[10] That would happen later. Australia's approach at the time reflected that of Britain's War Office and the general staff, which maintained an intermittent interest in code-breaking through the first decade of the twentieth century. While the British Army broke Boer codes during the Anglo–Boer War, until the war approached, it had maintained only a basic capacity.[11] Indeed, much of the intelligence apparatus of the British Army, along with its accumulated experience, was dismantled after the signing of the Treaty of Vereeniging in May 1902.[12]

Overall, while cryptology was growing in importance among the great powers of Europe, the experience of the Anglo–Boer War left military practitioners with little appreciation of the future importance of Sigint. The British Army's *Field Service Regulations* of 1909, for instance, failed to mention radio intercept as a means of intelligence collection. David Henderson's *The Art of Reconnaissance*, published in 1914, similarly overlooked it. Nonetheless, by 1912 the British Army had acquired wagon-mounted wireless transmitters for use with cavalry units. And by the end of 1914, these would be used exclusively for the interception of enemy transmissions; a task performed 'with appreciable success'.[13] Australia would find itself dependent on Britain's support in this field.

Introduction of the wireless radio

Recognising the significance of the invention of the wireless radio, the British Admiralty decided in 1900 to adopt wireless as the principal means of communication and purchased 50 Marconi sets. These high-powered wireless radios transmitted messages over unprecedentedly long distances. Drawing on the electricity from steam engines, along with large antenna rigs, the radios took advantage of 'ground wave' propagation, which worked well over the ocean surface, following the curvature of the earth, enabling transatlantic communications and the expansion of commerce. The network came to be seen by some as more important for Britain's imperial position than even the Royal Navy.[14]

The first significant interception of wireless radio communications in battle occurred during the Russo–Japanese War of 1904–1905. The interception and exploitation of Russian wireless radio messages by the Imperial Japanese Navy (IJN) enabled them to defeat the Russian fleet.[15]

Eager to keep up with Britain's technological advances, by 1914 the warships of the United States, France, Italy and Russia were also equipped with wireless radios, as were ships acquired for the Royal Australian Navy (RAN) in 1913.

Marconi's wireless radio operated using spark machines, but in 1909, Cyril Elwell, an Australian-born American inventor, discovered that an electric arc, as used in lights for public places, could generate 'continuous-wave' emanations. Another method of producing continuous waves was the vacuum tube. But war interrupted further research and continuous-wave transmitters did not replace spark machines until after the First World War.[16]

Meanwhile, Germany was also investing in new wireless technology. The leading German company involved, Telefunken, increased transmission ranges and pioneered research and development of continuous-wave transmission, enabling numerous separate channels to operate simultaneously, and thus increasing the volume of traffic that could be broadcast concurrently.[17]

Wireless interception – early days

Not far behind the wireless radio, the scope for Sigint increased. Wireless radio communications gave impetus to the integration of three separate fields – interception, direction finding and code-breaking. As wireless radio communications traffic could be intercepted, codebreakers were supplied with a steady flow of messages to decrypt. With greater volume came a greater chance to solve the codes.[18] For direction finding, triangulation was key. This involved taking a bearing towards a target from two or more locations and measuring the relative signal strength to identify the location from which the signal emanated. This would prove critical to locating enemy positions and to capitalising on the intercepted signal, even if the signal content could not yet be decrypted.

France was quick to appreciate Marconi's invention. Not surprisingly, Sigint is central to an understanding of French policy towards Germany before the First World War, notably during the First Moroccan Crisis of 1905 and the Agadir Crisis of 1911.

Russian policy towards Germany was similarly influenced by decrypted diplomatic telegrams. In 1901, for example, Count V.N. Lamsdorff, the Russian foreign minister, concluded after studying German intercepts, that despite 'friendly verbal outpourings', Germany sought 'to hamper the realization of Russia's tasks within the Ottoman Empire'. The *cabinet noir* at the Russian foreign ministry, directed from 1901 to 1910 by Aleksandr Savinsky, may have been even more influential than its French counterpart with which it cooperated from time to time.[19]

Russian diplomatic cryptology expertise during the first two decades of the twentieth century was revealed when the Bolsheviks published tsarist diplomatic documents in the interwar period. These documents showed that Russia's *cabinet noir* successfully decrypted British diplomatic traffic up to and after the outbreak of war. In fact, Russian cryptanalysts successfully broke the codes and ciphers of all the great powers except Germany, which improved its codes and ciphers after discovering, during the Agadir Crisis, that France could read their telegrams. Thereafter, as the war progressed, the Russians found decrypting British and French traffic embarrassingly easy. In 1916, a Russian official even warned the intelligence officer (and future foreign secretary) Sir Samuel Hoare to change British ciphers as the existing ones were as easy to read as a newspaper.[20] Ironically, Russia's own field military commanders did not always appreciate the importance of codes and ciphers – with disastrous results, not just during the Russo-Japanese War, but in the early battles of the First World War.

Early Australian espionage

When the Commonwealth of Australia came into being as a federation on 1 January 1901, Australian colonial troops were operating in the South African War. But it was not until a year later, in January 1902, that Major General Sir Edward Hutton took up his appointment as General Officer Commanding (GOC) of the Commonwealth Military Forces.[21]

While Australia's security in the region was the focus of the military, the government was also keen to protect Australia's regional commercial interests. Wilson Le Couteur had acquired extensive experience of the New Hebrides (now Vanuatu) while working as an agent there for the Australasian United Steam Navigation Company and Burns Philp Pty Ltd.[22] Le Couteur offered his services to Australia's Prime Minister Edmund Barton to obtain 'reliable information' on the state of affairs in the island group.[23] The decision by France in 1900 to promote the interests of its nationals in the New Hebrides (beginning a process which saw the European population becoming predominantly French)[24] had revived Australian government fears about occupation by a foreign power in its near neighbourhood. Australian businesses, notably Burns Philp, were also concerned that if French interests prevailed, they would be unable to compete.[25] The Barton government therefore stood to gain from acquiring a better understanding of the situation on the ground, and in promoting British settlement, in order to protect Australia's interests.[26]

After some months' careful consideration, Le Couteur's offer of service was accepted and he became Australia's first spy. He was given tight riding instructions by Atlee Hunt, Secretary of the Department of External Affairs, who requested, inter alia, that he 'take every possible care that the object of your visit is not made public in any way, either in Sydney or in the islands', and that he ascertain the number of French and British residents and the respective extent of their trade interests and claims to land.[27]

In late May 1902 Major William Throsby Bridges became Australia's first military spy when he was engaged to gather information of military significance about New Caledonia.[28] Bridges later became Chief of Intelligence (1905–09), responsible for military training and military intelligence,[29] among other responsibilities, including membership of the Council of Defence.[30] Then, in 1909, Bridges became Chief of the General Staff (CGS).[31]

The Army's Intelligence Corps

In 1906, Anglo–Boer War experience and concerns about lack of knowledge of Australia's neighbourhood resulted in the establishment of an Australian corps of guides to gather topographical information for mapping – an echo of Scovell's corps of guides established during the Napoleonic Wars a century earlier.[32] Also in 1906, with the lessons of the Anglo–Boer War in mind, and following the Prussian military a century earlier, the British General Staff was established on a permanent basis.[33] Canada had founded a corps of guides in 1903, so there was a clear precedent for such a formation among the self-governing dominions of the British empire. Major Bridges, who was schooled at the Royal Military College of Canada,[34] was instrumental in establishing a similar body in Australia. On 6 December 1907, the Australian Intelligence Corps came into being after Defence Minister Thomas Ewing and the governor-general approved its creation.

The governor-general approved the transfer and promotion of the Honourable James Whiteside McCay 'to be Colonel and to command'.[35] McCay had an extraordinary career. The Irish-born migrant son of a Presbyterian minister, married to the daughter of a Catholic police magistrate, McCay was a Melbourne University graduate and polyglot (speaking French, Italian and Spanish). Witty and argumentative, he entered politics supporting Edmund Barton as first prime minster and was made defence minister (from August 1904 to July 1905). In 1886 McCay had been commissioned into the 4th Battalion, Victorian Rifles.[36] He proved to be a highly capable commanding officer, but with a talent for exasperating others, including his superiors. Even Bridges, who recommended his appointment, complained later that McCay 'gave too much trouble'.[37] A letter from Field Marshal Lord William Birdwood recalls that McCay 'apparently wished to regard all orders from the point of view of the lawyer and to argue about them', noting with relief that he seemed to have got over this.[38] McCay's 1911 Standing Orders for the Intelligence Corps include the reminder that 'information gained in peace is of little value unless organised, collated and recorded in a form instantly available for use in war'.[39] In 1908, the duties of the Intelligence Corps were significant, but they were directly related to the intelligence function. They included the training of officers and other ranks in intelligence work; collecting information about the topography of the Commonwealth and its dependencies and about foreign countries, particularly in the Pacific; preparing maps and plans; and compiling and recording all information ready for immediate use.[40]

By 1912, however, McCay and his corps were struggling with Military Board members who refused to recognise the value and special nature of intelligence work, insisting that intelligence staff be treated like all others. McCay was now, in addition to being commander of the Intelligence Corps, the head of the Intelligence Section of the Operations Directorate of the Commonwealth Section of the Imperial General Staff. The responsibilities of the Intelligence Directorate of the Australian Military Forces expanded to include non-intelligence functions. As well as collection of information on the Commonwealth and its dependencies and on foreign countries, particularly those in the Pacific, preparation of summaries, histories, military handbooks and route books, supervision of training in intelligence duties, and war ciphers, intelligence staff were now responsible for newspapers, statistics, accounts, censorship, telegraphs, cables, production and issue of maps, plans, photographs and so on, the General Staff Library at Headquarters, supervision of district reference libraries and aviation.[41] Cryptography (war ciphers) was just one item on a long and varied list.

Nevertheless, the Intelligence Corps led by McCay proved to be one of a number of examples in Australian intelligence history of a gifted individual making a remarkable impact, notwithstanding inadequate resources, bureaucratic infighting and personality clashes. The report on an intelligence staff tour held in New South Wales, under the direction of McCay as Director of Intelligence and John Monash (then a lieutenant colonel) as Assistant Director, shows that these leaders devoted considerable thought and planning to Australia's intelligence function. The tour simulated the invasion of Australia by a foreign power and what the Intelligence Corps would be required to do in response.[42]

But only two years later, a military order proclaimed that the 'existing forces of the Australian Intelligence Corps as a separate body' would cease from 30 September 1914. After that date, its duties were to be 'performed by officers of the Citizen Forces selected for duty with the Intelligence Sections of the General Staff in each Military District'.[43] The two most involved in establishing the erstwhile corps, McCay and Monash, would subsequently command divisions on the Western Front, each thereby earning a knighthood. Parliamentary tributes on the death of McCay in 1930 included one from Senator Sir George Pearce, Minister for Defence, who noted that 'only those associated with the building up of our defence forces can realize the part played by the Intelligence Corps in preparing Australia to take her part in the Great War'.[44]

Naval Intelligence Department and wireless stations

Demand for radio-telegraphy installations led to the convening of a conference in Melbourne in December 1909, which included representatives from New Zealand and Fiji. The conference recommended building high-powered stations on the east and west coasts of Australia, in New Zealand, Fiji, Solomon Islands and New Hebrides. The German company Telefunken had tendered to erect stations at Sydney and Fremantle. The tender was initially accepted, but then (in a decision that would find an echo a century later over Huawei technology) was blocked by the Department of Defence. This delayed the construction of the network until an alternative supplier could be found. The Post Office, disregarding the risk of war, maintained its right to sole control. The Australian Commonwealth Naval Board

(ACNB), however, began to establish its own stations, using wireless sets from the Marconi Company; the same company which would equip the Australian warships.[45] The Navy thus led the way for Australia in laying the foundations for a national Sigint enterprise.

The RAN's Naval Intelligence Department (NID) focused on China, Japan, the Philippines, the Dutch East Indies and Dutch New Guinea.[46] Ian Pfennigwerth, historian and biographer of leading Australian codebreaker Eric Nave, points out that intelligence requirements revolved around two questions: what do I know about the enemy or potential enemy? and what does the enemy know about me? Sought-after categories of intelligence were enemy weapons and fuel supplies and any prospects of hindrance to operations; the location, composition, aim and objectives of enemy forces; the capability of their weapons and systems; the nature of their tactics; background enemy situation appreciations, enemy knowledge of friendly force dispositions and capabilities; weather and hydrographic intelligence; and enemy electronic and other signatures.[47]

In December 1912, a former Royal Navy officer, Walter H.S.C. Thring, was appointed Lieutenant Commander in the RAN, responsible for naval defence and intelligence matters under Rear Admiral William R. Creswell, the first Naval Member of the ACNB. In 1913, the Admiralty also posted a Royal Marine Light Infantry officer, Major Percy Molloy, to assist with the setting up of the NID in Melbourne;[48] a body in which Sigint would play a prominent role.

While naval intelligence concerns were broad ranging, the technological constraints of old meant that for early twentieth century naval intelligence purposes,

> the circle of sea over which control needed to be exerted was only the radius of the best lookout's vision from the tallest structure in the ship. This could be extended by the employment of scouting forces, the use of balloons and, later, by ship-launched aircraft. Knowing the enemy was within this wider circle was, naturally, most useful. But any contest of wills over control of the sea could only begin at the maximum effective range of the biggest guns in either force, about 13 nautical miles (nm) for battleship encounters. Aerial attack was

in its infancy and proved largely ineffective against ships. In naval operations of the day the emphasis was on tactics, which made local intelligence a key factor in operational success.[49]

For the RAN, collection of information was always important, in peacetime as well as at war. Replenishment and victualling services at ports, and political developments in the region were important information to be collected and shared with other stations of Britain's Admiralty, particularly with the China, East Indies and Cape stations.[50] Messages were passed by wireless radio telegraphy (known in shorthand as W/T). As part of their remit, warship wireless operators listened to messages carried by wireless telegraphy. Thus, on the eve of war, the RAN W/T Service included 30 operational stations intercepting and reporting on enemy and other wireless messages.[51] They were poised to play a prominent role relating to Sigint at the outset of war. This would involve seizure of German merchant ships and their secret code books for exploitation by the Admiralty.

Prewar developments in the United Kingdom

While wireless radio communications proliferated, Britain, with concerns about the security of its colonies and naval bases in mind, invested heavily in commercial cable networks. In the first decade of the twentieth century, Britain had the best and most secure cable network. With many neighbouring nations reliant on cables which transited the United Kingdom, it was ideally placed to intercept and censor them in an emergency. When Britain laid a cable across the Pacific, almost the entire line, save for a segment through the US state of Maine, transited British empire territories from Canada to Fiji, New Zealand and Australia.[52]

In 1911, the War Office published a *Manual of Cryptography*, which omitted reference to the transformative effect of wireless radio communications on Sigint. On the eve of war, that function was still performed by a lone 'cipher expert'. In addition, the law still precluded Britain eavesdropping on diplomatic messages in peacetime. No one yet had responsibility for meeting the military cryptanalysis needs of deployed land or naval forces.[53]

In the lead-up to war, however, the British Army and the Royal Navy had managed to practice Sigint on field manoeuvres and in exercises at sea, although mostly through the interception of plain language traffic.[54]

For the British Army, field military exercises in 1912 revealed that the 'enemy' could overhear nearly all of their wireless radio messages. The British Army's headquarters in London was known as the War Office and the Special Duties Section in the War Office (known as MO5) studied foreign codes and ciphers, but did not engage in organised code-breaking of the encrypted communications of foreign powers before August 1914.[55] In the ten years before the First World War, Britain had nothing comparable to France's *cabinet noir,* the *Bureau de Chiffres* (Military Cryptography Commission or Cipher Office).[56] Australia's forces were even further behind.

Nevertheless, by 1912 naval commanders and signals and intelligence staff officers recognised the value of intercepting wireless radio messages. They also appreciated the potential of direction finding and communications traffic analysis – all skills that would become fundamentally important to a national Sigint agency.[57] In 1912 the British Admiralty's NID began recording intercepted German naval wireless radio communications traffic, and, by early 1914, was attempting to decrypt their encoded messages.

Meanwhile, the cryptographic systems of Britain's Foreign Office were vulnerable to any attacker. The Royal Navy's systems were worse than those of the Kaiser's Navy. The British Army's ciphers were mediocre, although considered 'no worse than their peers'.[58]

3
FIRST WORLD WAR ALLIED SIGINT

When war came in August 1914, Australia, like Canada and New Zealand, was still a subordinate part of Britain's empire. This meant that when Britain was at war, Australia and its fellow dominions were at war; Australian foreign policy remained within the remit of London. Australia also relied on Britain to manage the machinery of intelligence.

While much of Australian defence policy was formulated by the Australian government, Britain's dominant influence remained. Britain enjoyed a decisive advantage in the nineteenth century via its collection of information through the institutions of the empire, including the Foreign Office, the diplomatic service, the Colonial Office, the Post Office, the Treasury, the East India Company and the Bank of England. Political, economic and military information from outside the empire was the domain of the Foreign Office and the empire's diplomatic service.[1] Access to this information was part of what made Great Britain a great power. Australia was a beneficiary of these arrangements, which would be transformed by events between 1914 and 1918. It was an experience that created common views, trust and respect – all of which would later prove essential to the intelligence arrangements of the Second World War.

Initial arrangements

In 1914, Australia's government bureaucracy was unrecognisably small compared to its greater size following the two world wars. At that time, Australia had no diplomatic presence abroad, and its trade was overwhelmingly with Britain and other parts of the British empire. The limitations of its national intelligence functions need to be seen in this light. The Australian Imperial Force (AIF), like its Canadian counterpart, the Canadian Expeditionary Force (CEF), would fight as a subordinate formation within the command structure of the British Expeditionary Force (BEF).[2] This meant that Australia did not acquire many of the specialist capabilities normally retained above corps level (that is, a level of command incorporating two or more divisions, each with two or more brigades).

The RAN was already an integral part of the fleet at the disposal of the British Admiralty and remained so for the duration of the war. This meant that intelligence capabilities and functions of the BEF and the Royal Navy (RN), while managed centrally, were conducted on behalf of, and with forces from, the outposts of empire.

Technically in 1914, despite improved intelligence as well as command and control facilities, having to operate across the globe rather than just in the Mediterranean Sea meant that things were no easier for the admirals of 1914 than they had been in Nelson's day, more than a century earlier. This suggested that 'the meeting – and missing – of fleets that took place was almost as haphazard as theirs'.[3] During the First World War, however, with the help of technological innovations, they would get much better at it.

The approach to collecting, analysing and disseminating sensitive information would evolve with new technologies and techniques, incorporating specialisations in three areas: photographic or imagery intelligence, principally aerial photography and mapping; human intelligence, covering prisoner interrogation, document exploitation and human espionage; and Sigint, also known as cryptology, covering cryptography and various facets of wireless (y–intercept) and telegraphic and telephonic line traffic analysis.

During the First World War, Sigint involved monitoring electro-magnetic communications – the wireless radio, telegraphy and various manifestations at sea, on land and in the air, each being susceptible to some degree of interception. The years from 1914 to 1918 would witness extraordinary technological developments, not least in communications technology and capabilities relating to Sigint.

According to John Ferris, the establishment of code-breaking bureaus in 1914 was the logical culmination of a longstanding trend. Ferris writes that

Few practices save cannibalism were beyond the pale for British statesmen, subject to the principle that they not be caught publicly in the act. They had differing appetites for the fruits of espionage. Some found the taste repugnant; others deemed it a delicacy beyond compare; most sampled the dish in a pragmatic spirit, according to hunger or need.[4]

Yet it was only on war's approach that the Admiralty established stations in Britain, Italy and Malta to intercept wireless telegraphy. Thus, by July 1914, British statesmen had some experience with their own intelligence departments.[5]

On the outbreak of war, the Admiralty and War Office each established small cryptanalytic sections. The Admiralty's was given a deliberately nondescript designation: 'Room 40 Old Building', although it was later named Naval Intelligence Directorate Section 25 or NID-25. Some called it the Royal Navy's black chamber.[6] For land forces, the War Office established MO5b, which it later renamed MI1b. Both expanded with the recruitment of specialist academics and would grow to be 100- and 80-strong, respectively.[7]

Even compared with French and Russian prowess in Sigint, by July 1914, British expertise was impressive. Indeed, during the First World War, Britain was arguably the world's leader in Sigint. Developments in and around Australia would be among the early successes, often due to the skilful use of Sigint.

Australia did its share; for example, the destroying of German raider, SMS *Emden,* discussed below. But Australian naval Sigint at this stage was largely overlooked. This was partly because no formal Sigint organisations had yet been established by Australia's forces. It was also because, on Australian ships, the work of a signaller, monitoring own-force communications networks and, when required, monitoring enemy signals traffic, only became recognised separate functions after the First World War.

Among those who played a contributing role in Australian naval Sigint was signalman John Varcoe, who served on the destroyer HMAS *Parramatta* in the Mediterranean from 1917 to 1919. Immortalised as the sailor in the statue at the Cenotaph Memorial in Sydney's Martin Place,[8] Varcoe was one of 1500 Australians and New Zealanders who served in a little-known battlefront alongside the Serbs in the First World War. Australians and New Zealanders accompanied the Serbian Army on a fighting retreat to the Adriatic coast in 1915. When the fighting shifted to the Salonika Front, many served with the British Army, the Royal Flying Corps, two AIF units and six Australian navy destroyers.[9] On 15 November 1917, when the Italian passenger liner *Orione,* on its way to Brindisi, was torpedoed, *Parramatta* came to the rescue. For his gallant efforts, Varcoe was awarded the Distinguished Service Medal.[10]

Maritime successes

One of the first hostile acts undertaken by Britain in August 1914 was to sever the five cables that linked Germany to the outside world. The British cable ship *Telconia* hauled up and cut the cables, forcing Germany to rely on British telegraph lines or wireless radio. This gave the Allies a remarkable advantage from the outset, one that would be exploited with increased success in the years ahead. Germany would reciprocate, cutting cables to Russia to the north and, once the Ottoman empire joined, via the Black Sea route. Britain, however, was never as affected by this disruption as Germany and its Central Power's allies.[11]

The maritime blockade rested on British sea power and Anglo–French control over transatlantic cables. When German cables were cut and the United States insisted transatlantic wireless messages be sent in the clear, the Anglo–French Entente was able to exploit the opportunity. Looking back, it is hard to measure the effect of the blockade, but communications intelligence was a significant factor, helping to minimise the damage to Britain while maximising that to Germany.[12]

On 2 August, three days before Britain's formal declaration of war, but after Germany's declaration of war with Russia, the Admiralty in London sent orders for search and seizure operations that had been planned in the event of war. The RAN Intelligence Department identified seven German vessels in Australian ports or waters, and boarding parties prepared to seize the ships and their code books. As war was declared, men were ready to pounce.[13] Shortly after Britain's declaration of war, the German-flagged merchant ship SS *Hobart* contacted Esperance, Western Australia, 'apparently ignorant of outbreak of war'.[14] As the vessel steamed towards Melbourne, all southern coastal radio stations jammed its signals, concealing the declaration of war from the *Hobart*, even though it meant that radio transmissions from all other German ships in the area were jammed, too. As a result, the hapless *Hobart* was boarded by Australian authorities as it arrived in Melbourne. The ship's master at first denied the existence of a code, but in the early hours of the morning the code book was found in a small special safe.[15] The code book, the *Handelsschiffsverkehrsbuch* (HVB), was obtained intact. When the Admiralty learned that German merchant marine messages were being read in Australia, they requested copies of the code book. In the meantime, an academic from the naval college in Geelong, with a mostly female team, set about the work of decryption and translation. For a while, messages intercepted elsewhere were forwarded for decryption in Melbourne.[16] An account of this extraordinary set of circumstances is found in James Phelps's *Australian Code Breakers: Our Top-secret War with the Kaiser's Reich*.[17]

In the end, the plan worked so well that four sets of the German merchant fleet's HVB code book were captured and used to intercept German merchant marine communications. The Royal Navy relied on copies being passed back to the Admiralty in London from Australia. Copies of the code book were also sent to the East Indies, Cape of Good Hope, Mediterranean and China stations, as well as to New Zealand. The Imperial German Navy, displaying remarkable complacency, would continue to use these compromised codes until early 1916, making it relatively easy to identify colliers tasked to rendezvous for the refuelling of coal-dependent warships. Germany failed to alter its algorithms, thus allowing the replacement codes to be readily broken. This would not be the last time that naval and military commanders would underestimate the code-breaking capabilities of their enemies.[18]

The code books sent to the British Admiralty in London were received by Room 40 OB in October 1914. In the same month, by the good offices of the Russian Imperial Fleet, the *Signalbuch der Kaiserlinchen Marine* (SKM) salvaged from a light cruiser, SMS *Magdeburg* in the Baltic, was delivered personally to the First Lord of the Admiralty, Winston Churchill. Then, in December 1914, the Room 40 cryptanalysts received the *Verkerhrsbuch* (VB), a lucky discovery by a British trawler. These three code books gave Britain a major advantage over the enemy. Further exploitation of the code books was aided by rigid and predictable German naval command and control procedures which included daily reporting of positions by all units. While it took some time for the intelligence provided by Room 40 staff to be trusted, it would eventually relieve the British Grand Fleet in the task of trawling the North Sea for the German Navy.[19]

The experience of gaining access to German codes was a pointer to the growing role of Sigint in naval warfare, not just to the great powers in Europe but in Australia's neighbourhood, too. Capitalising on Sigint for the direct defence of Australia would be important in the Second World War and at Midway in 1942, but events in the early stages of the First World War already foreshadowed its future significance for Australia.

In early August 1914, the German East Asiatic Cruiser Squadron was detected by Sigint off New Britain (now part of Papua New Guinea). The German warship SMS *Scharnhorst* signalled German stations on the Pacific Islands of Yap and Nauru. The communications traffic with three other German warships enabled direction finding (DF) from Australian stations to triangulate the bearings and plot the rough positions of the flotilla.[20] Focused on escort for the initial batches of New Zealand and Australian troop transports, the Australian station was instructed not to pursue until the German ships had headed east towards Chile. Subsequently, an Australian task force with leading warships, HMAS *Australia*, *Sydney*, and *Yarra*, and three destroyer escorts, formed the Australian Naval and Military Expeditionary Force (ANMEF). This force was tasked with landing troops at Rabaul, to destroy the local radio transmitter and capture German New Guinea.[21] The operation was a complete success, although lack of confidence in the Sigint and the difficulty of locating a fleeing enemy across the vast expanse of the Pacific meant the German flotilla was not pursued, even though – had there been an encounter – the Germans would have been outgunned by the Australians.[22] The *Scharnhorst* and her escorts steamed eastwards to avoid Australian warships, heading for Germany through the Atlantic via South America. On 1 November, off the coast of Chile, the Germans successfully fought off British warships at the Battle of Coronel, sinking three British cruisers. Sigint provided leads, but limited communications, transmission delays and the lack of support to outgun the German flotilla left the Royal Navy with the biggest defeat at sea in over a century. After a brief stop in the neutral Chilean port city of Valparaiso, the Germans rounded Cape Horn into the Atlantic where fate (and good Sigint) would catch up with them at the Battle of the Falklands.[23]

HMAS *Sydney*, the *Emden* and the underestimation of Sigint

Sigint would also play a crucial role in an early victory by HMAS *Sydney*: the defeat of the German commerce raider SMS *Emden*.[24] From August through to early November 1914, the *Emden* had a remarkable run of successes, intercepting and sinking Allied commercial and military shipping in Southeast Asian waters, around the Dutch East Indies (now Indonesia), Malacca Strait and the Bay of Bengal. The *Emden* sank vessels with cargo of no use and co-opted those carrying coal, but treated prisoners with dignity and released captives where possible. The reputation of the ship's captain Karl von Müller rose, as did the frustration of the Royal Navy's First Sea Lord, Winston Churchill.[25]

Von Müller's luck ran out when he set out to rendezvous with one of the captive coaling ships near the Cocos (Keeling) Islands. There, before the *Emden* jammed its signal, a British wireless station recognised a transmission as irregular and managed to transmit several messages including 'strange ship in entrance' and then 'SOS, *Emden* here'. HMAS *Sydney* was in the escort flotilla for the first convoy of troopships transiting the Indian Ocean, carrying soldiers for the Australian Imperial Force, and was well placed to intercept the *Emden*. The convoy commander ordered HMAS *Sydney* (via visual signal, so as to avoid breaking wireless silence) to leave the convoy and pursue the *Emden*. *Sydney* accordingly departed towards Cocos Island at 26 knots. Like *Emden*, *Sydney* was a light cruiser, but with greater speed and firepower. Von Müller recognised his error, weighed anchor and, in haste, left his landing party behind, including ten of his principal gunlayers. Battle ensued mid-morning on 9 November 1914, with an exchange lasting over one and a half hours. Short on critical manpower and facing *Sydney*'s eight 6-inch guns against its own ten 4.1-inch guns, the German ship suffered catastrophic blows. Von Müller drove the wreck of the *Emden* onto the reef on North Keeling Island. The *Emden*'s crew was detained by *Sydney*'s crew the next day.[26] Jack Ryan, who features later in this story, was a wireless operator and youngest member of the ship's company on HMAS *Sydney* when the *Emden* was destroyed.[27]

Reflecting on this period of naval warfare involving the battles at Coronel, the Falklands and Cocos Islands, military historian John Keegan observed that

> cruiser warfare failed because the Germans could not conceal the movement of their ships. A steady stream of clues as to their whereabouts were picked up, often with great rapidity, sometimes in real time, and circulated with efficiency by the British between the Admiralty, local commands and pursuing naval units on the worldwide wireless and cable network.[28]

When naval commander Graf von Spee's East Asia Squadron finally left the Pacific and the other German warships had been interned, captured or destroyed, Australia's official navy historian, Arthur W. Jose, recorded that

> Australian wireless stations became engaged in intercepting German messages passing through neutral stations. They thence obtained a good deal of useful information concerning the enemy merchant vessels that were sheltering in ports in the Dutch East Indies. At one time it was noted that several of them were coaling as if to emerge from their shelter, and the necessary precautions were taken to let them know that they were being watched; while no one wanted to capture them, no one desired to have a possible raider let loose in the Archipelago.[29]

Jose made a telling observation on the deliberate understatement of the importance of Sigint during these encounters: 'It is impossible to give many details of the results of this branch of the radio-telegraphists' work; its great value to the naval intelligence service cannot be overestimated, but must here be taken for granted'.[30] Jose's restraint matched the approach of other historians who, through ignorance or at the behest of security officials, overlooked in their narratives the central role played by Sigint in countless operational successes. This secrecy was a hallmark of success, protecting sources and methods. But it would carry with it serious implications, leading to an underestimation by the uninitiated of the role of Sigint and often to a misunderstanding of events.

Meanwhile, several opportunities to exploit this early capability were missed. In December 1914, for instance, the presence and movement of German ships raiding towns along the English east coast was misjudged and its reporting delayed, allowing the German fleet to escape. Again in January 1915, at Dogger Bank in the North Sea, the Admiralty failed to use Room 40 intelligence reports on the location of German submarines. The oversight enabled German battle cruisers to escape an otherwise fatal encounter with superior British ships.[31] Problems with interpreting and communicating Sigint in a timely, effective manner persisted into 1916.

The Battle of Jutland

The next major naval exchange was the Battle of Jutland in May 1916. The Imperial German Navy had sought to catch up with Britain's naval capacity in the lead-up to the war, and the confrontation in the North Sea had been anticipated as a climactic battle. In the end, it was not as decisive as either side had hoped, but Germany's close shave left its war leaders apprehensive about a repeat encounter. This meant that the fleets would spend much of the rest of the war making little difference directly to its outcome.

Sigint was prominent at Jutland, but was not used as effectively as it could have been. The Admiralty's Room 40 (that is, NID-25) delivered intelligence on German submarines hunting ships in convoy transiting the Atlantic. Room 40 allowed direction-finding fixes, even when ciphers were changed and decryptions delayed. By war's end, Room 40 had become an industrial-scale naval Sigint centre. The impact of this experience with Sigint on the postwar Royal Navy and, in turn, Dominion navies, including the RAN, would be profound.

At the time of the Battle of Jutland, however, Room 40 had not yet reached its full strength. The RAN had only a minor role in the battle, but it was one that would influence the development of Australian naval Sigint.

On 30 May, Room 40 reported, based on decrypted signals and traffic analysis, that the German fleet, under Admiral von Scheer, would put to sea the next day.[32] Room 40, however, was restricted to issuing decrypts without comment or assessment from the full range of signal activity.[33] It could not communicate directly with the fleet, relying on the Director of Naval Operations (DNO), Rear Admiral Thomas Jackson, to act as intermediary. The following day, after the British fleet had departed Scapa Flow and Rosyth, the DNO ascertained that the principal call sign for the German fleet indicated it was still in port at Wilhelmshaven. Had they been consulted, Room 40 could have told the DNO the German fleet's practice was to leave Scheer's call sign behind and that he used a separate call sign when deployed at sea. Not for the last time would Sigint be discounted by those with insufficient understanding of the capability.

Oblivious to this crucial intelligence, the DNO advised Admiral John Jellicoe that the German fleet was still in port.[34] Complicating matters further, an intercept plotted a German warship incorrectly as being in the same location as Jellicoe. The error was the fault of the German officer transmitting the intercepted message, not Room 40.[35] But on the strength of the advice that von Scheer and the German fleet were still in port, Jellicoe delayed his departure, leaving Vice Admiral David Beatty's ships from Rosyth some 70 nautical miles out in front. There, to his surprise, Beatty then met the entire German fleet, suffering significant damage before Jellicoe's ships arrived from Scapa Flow. Jellicoe missed a real opportunity to defeat the German fleet.[36]

Ferris argues that eventually, Room 40 wrecked Germany's only chance to win the naval war, particularly through its 'whittling' strategy, provoking German warships into actions against larger but hidden forces or over submarine traps. Increasingly effective use of Sigint prevented

Germany from deceiving Britain; in addition, early on, Britain's detection of German code-breaking deprived the Germans of the chance to match Britain's capability. This, Ferris asserts, was 'a classic case in which intelligence multiplied the power of the stronger side'.[37]

Sigint and Australian land forces

Unlike the RAN, which was an integral part of the Royal Navy and automatically benefitted from its intelligence, including Sigint, during the war Australian land forces had to be raised and organised to facilitate the closest possible cooperation with Britain.

Accordingly, in the Australian military forces – of which the all-volunteer AIF constituted the largest part – the general staff structure followed the British model. The head of the general staff intelligence organisation throughout the war was the former intelligence corps officer Major Edmund Piesse. His office, known as MO3, became a directorate in March 1918, and Piesse himself Australia's first Director of Military Intelligence (DMI); but his duties were censorship and counterespionage, rather than Sigint.[38]

In the field, the AIF division headquarters structure again largely followed the British model, with the 'G' (operations) Branch dividing responsibilities between the General Staff Officers (GSOs) I, II and III. GSO IIIs attended to intelligence work. When the former Director of Intelligence, now Major General Sir William Throsby Bridges, raised the 1st AIF Division, he appointed Thomas Blamey as GSO III (Intelligence) on the staff of his headquarters – an appointment he held at Gallipoli.[39] Blamey's duties included handling material mostly from Humint sources, with no Sigint assets at the disposal of the division.

During the war on land, military intelligence was collected by combat units through reconnaissance, observation, the capture and interrogation of prisoners and exploitation of documents. Australian forces took part in these activities. The second main source was agent networks behind enemy lines, principally reporting on troop movements. The third means was technical information collection from three sources: aircraft photography, revealing the structure of enemy defences and force dispositions; 'flash spotters' and 'sound rangers' to locate enemy artillery pieces in order to guide counter-battery artillery fire; and eavesdropping on enemy communications – or Sigint.[40]

Sigint's increasing importance to land-based operations

While British naval Sigint was significant, its land counterparts took some time to catch up. The British Army started the war with ten radio sets, only one of which was in France in August 1914. Germany, on the other hand, had dozens of transportable radio stations as backup to their reliance on telegraph lines.[41]

The challenges of sustaining military operations on land drove technological advances. These transformed land operations and produced the intelligence needed for successful campaigns. In 1914, Britain's military intelligence staff consisted of around 20 staff-college-trained officers and supporting personnel. During the war, the number of intelligence personnel increased exponentially as the War Office expanded into counterespionage and censorship, both of which required large staffs.[42]

As for Sigint, however, there was no true centre in the War Office during the war, although a small section was established to address German ciphers: MO5b, under Brigadier-General Anderson, who drew on his experience with Sigint during the Anglo–Boer War.

In fact, cooperation between MO5b and Room 40 broke down, reportedly after Churchill showed Kitchener, his opposite number at the War Office, the contents of a military intercept before his own section had managed to pass the information to him. In 1916, the new post of DMI was created and MO5b was renamed MI1b. By war's end, it included 45 men and 40 women. Only in the spring of 1917, after MI1b succeeded in breaking the ciphers of the German Army's political section, was limited cooperation between the two agencies resumed.[43]

Initially, the BEF lagged behind the French Army in Sigint and imagery intelligence (Imint). The BEF had the capability to intercept

German wireless communications from the outset, in August 1914.[44] As the war progressed, British Army Sigint capability developed, and it was shared with the imperial forces deployed, including the Australian Corps, at times involving Australian personnel directly. Sigint entailed intercepting messages; analysing the communications traffic patterns, procedures, enemy force locations (known as traffic analysis); deciphering enemy codes and ciphers; maintaining signals security for British and Dominion forces (to reduce the risk of enemy eavesdropping); and signals deception (to mislead the enemy about own-force intentions).[45] The wagon-mounted wireless teams that deployed in August 1914 with the cavalry units proved particularly useful. The Marconi Wireless Telegraph Company assisted, experimenting in locating enemy positions with wireless radio DF.[46]

Eavesdropping in the trenches

Sigint eavesdropping in the trenches was developed by the French and copied by the Germans after they captured a French instrument near Verdun in 1915. Improving on the French design, the Germans used it extensively.[47] Together, eavesdropping and communications security measures were developed by British forces belatedly, after they discovered that frontline communications were insecure. German Moritz interception stations were eavesdropping on Allied trench communications. This, in turn, spurred improvements in communications security and in monitoring the Germans' voice and Morse code communications by imitating the capabilities the Germans and French had acquired.[48] The BEF equivalent of the Moritz was the IT or IToc (*aye-tok*), an abbreviation for Intelligence, Interceptor, or Intercepting Telephone.[49] Because of its technical focus, the British Army's Directorate of Signals was responsible, rather than the Brigadier-General Intelligence. John Ferris explains how it worked thus:

> Field telephone and telegraph traffic carried via (a) an 'earth return' system (in which the ground was used as the medium of transmission), (b) 'earthed' circuits (where the wire was not insulated from direct contact with the ground), or (c) imperfectly insulated

cables whose current leaked into the ground, could be intercepted in two fashions: through an earth 'pin' (a metal bar placed in the ground) or induction (long 'loops' of wire laid on the front line parallel to the enemy's cables ... 'Pins' and 'loops' intercepted weak electrical currents originating from cables up to several thousand metres away and transmitted them by wire to an IT [Intercepting Telephone] set. Here valve amplifiers boosted the strength of the currents so that operators wearing headphones could overhear the traffic.[50]

The pin and loop interception techniques would later find analogies in the close access techniques in use during the 1980s that relied on emanation or leakage from computer cathode ray tube screens and from emanating crypto devices.

Initial efforts by the British to develop their own eavesdropping equipment and instructions for frontline forces to exercise caution proved inadequate. Most revealing, however, were the messages passed in clear or enciphered messages sent with transmission errors, such as repeating passages in clear (un-enciphered), which led to grave security compromises.[51] Breaking into enciphered enemy communications, and enemy breaking into Allied communications, was facilitated by user errors. Changes of key presented a challenge which was overcome when messages like the following were sent: 'We have not received the new code books yet; please repeat in the old code', or 'Send 0317 in plain text'. A comparison of transmissions would enable the key to be solved.[52]

The Battle of the Somme and its technological consequences

At the Battle of the Somme on 1 July 1916, British orders dictated over a landline to a battalion headquarters near a German-held village were intercepted before the assault. Using Moritz interception stations in their dugouts, the Germans were intercepting messages leaked from telephone lines in the British trenches or transmitted through the ground – at a distance of up to 2740 metres. This helped prevent British forces from achieving tactical surprise.[53] Moritz intercepts warned of the precise time and place of many if not most divisional attacks. Ferris observes that 'tactical surprise was intended to safeguard all of these assaults. Failures in signals security destroyed this shield. While this offensive would have been a debacle in any case, this failure increased its scale.'[54]

The disaster at the Somme appears to have ushered in British Sigint on the Western Front.[55] Australians, along with other members of the BEF, were reminded of the value of capturing equipment such as the Moritz.[56]

Thereafter more attention was paid to communications security training for BEF forces, including Australian signallers. Operation of the IToc machines, however, remained a function of the British Royal Engineers (Signal Service), and a separate trade for linguists employed as Interpreter Operators (Wireless) was designated. The Interpreter Operators were the British Army's first formally recognised intelligence trade group.[57] Recruits came from across the BEF, including those on secondment from Dominion forces, such as Canada and Australia.[58] The number of Intercept Operators working the BEF's listening sets was laid down by the War Office in signal unit establishment tables. Six men per corps were allocated in September 1916, increasing to nine by mid-1917. These levels also applied to the two ANZAC Corps (and subsequently the Australian Corps, which was formed in November 1917 by pooling the Australian divisions from the two ANZAC Corps).[59] By war's end, Britain maintained 11 Wireless Operations Groups in the field, employing 1300 people.[60] These were the forerunners of the Army Special Wireless Sections (and their RAAF equivalent) – the eyes and ears of the national Sigint architecture of the Second World War.

In addition, technical solutions were introduced, notably the 'Fullerphone' for relatively secure frontline communications.[61] In December 1916, Canadian and Australian signals units reported successes in intercepting German trench communications, as well as interference from friendly telephones. Monitoring and reporting of own-troop communications led to disciplinary action, including courts martial, and to a noticeable improvement in telephone discipline along the trenches. In addition, false information was fed to the German Moritz station operators. In one case, orders for a night patrol into no-man's-land were transmitted. The Germans set out to ambush the patrol an hour before the designated time, only to be ambushed themselves by troops lying in wait. By the end of 1916, there were between 20 and 30 intercept stations across the BEF, including with the Australians.[62]

Once the war of movement resumed, and eavesdropping on trenches became more difficult, IToc devices were used to listen in on conversations between German prisoners of war. In the meantime, wireless interception would come to play a more significant role than the IToc system.[63] From mid-1916 onwards, British forces developed the capacity to intercept and exploit German wireless communications. As the war progressed and the capability expanded, Wireless Intercept Sections were formed, renamed in June 1917 as Wireless Observation Groups. These groups worked from static locations, directed by specialist intelligence officers. Their roles grew as battles became less static in 1918.[64]

Lessons in communications security

On the other side of the front, the Germans capitalised on Sigint to achieve remarkable early successes, particularly against the Russians. There, an example of the perils of not 'protecting your own' is Russia's General Samsonov, famous for what was described as one of the greatest blunders in military history. To ensure that his orders reached his corps commanders rapidly and accurately, Samsonov ordered they be transmitted over wireless radio and in plain language. The Germans were listening.[65] Interceptions of daily exchanges in plain language by Russian commanders gave the Germans a crucial advantage at the Battle of Tannenberg in August 1914, and resulted in the capture of some 90 000 Russian prisoners of war by the end of the month.[66] As one scholar observed, 'cryptanalysis apparently entirely escaped the attention of the Russian General Staff'.[67] Recognition of this weakness prompted changes to Russian ciphers, but renewed instances of carelessness betrayed the whole system to their enemies.[68]

On the Western Front, German advances into Belgium and France surprised the combined French and British forces, although timely Allied interception of enemy communications slowed down the German race to the sea.[69] Ironically, while the Germans capitalised on clumsy Russian use of radios, at the Battle of the Marne early in the war, equally lax German use of radios wrecked the German offensive. Even before the war, the French had committed resources to following German radio traffic. This revealed hierarchies, appointments, strengths and intentions, enabling them to anticipate German moves and prevent the enemy achieving a quick outcome in their campaign in the west.[70]

And in the air …

As the war in the air gained pace, spotter aircraft came into prominence, passing wireless radio messages back to artillery units on the ground to direct artillery fire onto sites only visible from the air. France and Britain listened to German Air Force traffic. Zeppelin airships used the wireless radio, reporting what they observed and requesting bearings from German direction-finding stations.[71]

In June 1916, the British Army was given control of the air defence of Great Britain. Until then, responsibility had been untidily split between the Admiralty and the War Office with their respective Sigint organisations, Room 40 and MI1b, both working against aircraft communications. As part of the new arrangements, the areas of MI1b responsible for wireless interception, DF and traffic analysis were split off into a new section called MI1e, with support to air defence as its main operational task.

Working with British and French Sigint units in France and with Room 40, MI1e gave advance warning of air attacks and tracked the raiders' approach using DF. They were helped by the difficulties of night-time navigation. To fix their positions, German aircraft often made transmissions intended to allow their own DF system to locate them and inform them of their whereabouts, which was low-hanging fruit for British Sigint.

MI1e's reporting and information from ground observers and sound location was fed into a plotting and command system – the London Air Defence Area – whose structure and efficiency in 1918 matched that of Fighter Command in 1940. But the defenders' ability to respond was limited by technology. The lack of accurate real-time positional information (DF did not provide pinpoint accuracy) and, until very late in the war, the absence of radio communications with aircraft, meant fighters could not be vectored to an intercept 1940-style, but could only be launched to operate on the general track of raids. By 1918 the British were destroying about 10 per cent of the aircraft launched against them, but far more were lost through navigational errors and crashes on landing.[72]

Above the trenches of Western Europe, the interception and analysis of German Air Force wireless communications traffic provided useful tactical intelligence, disclosing the disposition of enemy aircraft and the artillery's gun emplacements. The intelligence served the same function as radar would in 1940, guiding Allied fighter aircraft towards spotter aircraft. The effect of this breakthrough was a sharp decrease in German spotter aircraft activity, with flying times reduced and a number of spotter aircraft destroyed, crippling the enemy's air power in a war of attrition.[73] The lessons learned from this experience would be profound for the future of Sigint in the air, as later chapters attest.

Australian Flying Corps operations were often prompted by Sigint.[74] In May 1918, for instance, No. 4 Squadron was placed on stand-by for wireless signals, prepared to chase any German machines in their sector. Australian journalist and military historian F.M. Cutlack records

> The sector was divided into three, with a wireless swinging arm (known technically as a 'loop antenna') in the centre of each division. These wireless stations were in direct communication with each other by telephone... The sole duty of the stations was to 'listen in.' 'Listening in' meant that the operator sat with his ear-pieces on, picking out the different aerial sounds of wireless sent by the German aeroplanes to the batteries for which they were spotting. Any one of these stations, or perhaps all three, on picking up a wireless call from a German machine would swing the aerial arm round until the sound on the instrument was loudest, when the arm would be pointing in the direction of the enemy machine sending. The wireless operator would

then ring up a neighbouring station and obtain the direction of its aerial arm. Thus the triangle would be marked up on a map. This operation would take only about five minutes. The stations promptly informed the squadron 'standing by' for the day, and two pilots, waiting for the call, would be allowed seven minutes to find a white arrow on the ground pointing in the direction of the enemy machine heard working ... The pilots would then set out to stalk him.[75]

Wireless intercept teams broke German codes and ciphers, an activity that GHQ Intelligence centralised from the autumn of 1916. The cryptanalytical organisation known as I(e)C was located away from GHQ at Saint-Omer in northeastern France, and by the end of the war numbered 40 personnel, including a dozen women. The process was tedious, intensive and sometimes continued around the clock, but was aided by the capture of German materials, such as a code book in mid-1917 and occasional security lapses by German wireless operators. Still, the eavesdropping system was not flawless; nor did it deliver complete information – notably at critical times. In the lead-up to Germany's March 1918 offensives, for instance, there was a sudden change to German codes, placing the system under considerable strain.[76]

Meanwhile, BEF cryptographic systems were less secure than those used by the United States, France and Germany, but no easier to handle. By 1918, when four cryptographic systems were used for frontline communications traffic, two were cryptographically weak and known to be compromised (the Playfair Cipher and the 'B.A.B. Trench Code'); the other two were considered exceedingly complex to use. German and American cryptographic systems were superior to those of the BEF and often updated to reduce the risk of compromise, but almost all codes were subject to compromise during the war, thanks to frequent document captures and operator errors.[77]

As the campaign on the Western Front progressed, Sigint did not lead to any individually decisive result, given the nature of the attrition campaign, yet it proved invaluable to reconstructing the enemy's order of battle, relying on analysis of a score of snippets of information collected across the front. The cumulative effect made it the British Army's most valuable intelligence source next to combat troops.[78] And by giving the combat troops an advantage, it enhanced their value.

Sigint, deception and diplomatic code-breaking

Sigint also proved crucial for effective deception, enabling the element of surprise which is so important to success in battle. In the lead-up to the March 1918 'Michael' offensive, Germany created phantom armies through signals deception to mislead the Allies about its plans. Indeed, the Michael offensive was arguably almost as successful as the initial German advance had been in 1914. The experience demonstrated that under the right circumstances, signals deception techniques could provide an extraordinary level of surprise.[79]

Thereafter, though, Germany lost the advantage. Until July 1918, German traffic analysis traced BEF dispositions and deployments, but then the Germans 'lost their enemy', as the BEF's security and ability to conceal operations increased. The attack near Amiens on 8 August, for instance, achieved absolute operational surprise concerning the time, place, strength and style of attack. It also broke the nerve of the German Army and its high command, with General Erich von Ludendorff later describing it as the 'Black day of the German Army'. The Amiens attack was a watershed, foreshadowing the future role of Sigint.[80]

Diplomatic code-breaking proved to be of particular benefit, as Germany undertook blatant acts of hostility against neutral countries in the Americas. Britain's enemies, on the other hand, could rarely intercept its traffic and thus exploit its weak cryptography, although two of its allies could, which gave them an advantage in alliance bargaining. Meanwhile, British cryptanalysis became excellent; this was despite initially poor relations between the Navy's Room 40 and Army's MI1b, as well as bottlenecks in the distribution of intelligence.[81]

While the Foreign Office was responsible for managing diplomatic relations, the exploitation of foreign diplomatic codes during the First

World War was undertaken in part by Room 40 and by MI1b. Room 40 worked principally on German naval and a certain amount of diplomatic enemy and neutral traffic, drawing on messages collected by naval intercept stations. MI1b worked mainly on German military and some neutral and even Allied diplomatic traffic, the latter obtained from cable censorship efforts managed by the War Office.[82] The AIF was the beneficiary of this work, alongside other British empire and Allied forces, although the absence of an Australian-controlled Sigint entity during this period would leave Australia dependent on Britain, not just for military and naval Sigint but also, to a certain extent, for diplomatic Sigint.

Diplomatic Sigint and why America entered the war

The greatest demonstration in history of the impact of diplomatic Sigint is the story of the Zimmerman telegram. Despite appeals, and in spite of considerable losses of ships sunk by German U-boats, the United States had steadfastly refused to join the war.[83] Germany's Foreign Minister, Arthur Zimmermann, sought Mexico's help in keeping the American army busy on the Mexican border in exchange for a promise to support the return to Mexico of the territory of New Mexico and Texas. Germany was confident about the security of its cryptographic system. Zimmermann, however, had not taken into account the prowess of British Siginters. While German cryptography had been a focus from the outset, from late 1915 onwards, the British also had succeeded in solving US diplomatic codes, and this gave them considerable advantage.[84]

Room 40 proved to be a remarkably powerful tool in Britain's arsenal. One of its ablest codebreakers, the Old Etonian, publisher and Sigint historian, Nigel de Grey, who had been on the night shift, said to 'Blinker' Hall, the British intelligence chief, 'Do you want to bring the Americans into the war?'[85] De Grey handed Hall an incomplete decrypt of an intercepted telegram from Zimmerman to the Mexican Embassy in Washington for onward transmission to Mexico City.[86] The British concocted a story that the decrypt had been fished out of a wastepaper basket in the

German Embassy in Mexico City – an early example of source protection. Anticipating that the Americans would suspect a forgery, the British got de Grey to decrypt the message again in front of the US Ambassador. Once the news was out, the neutralist sentiment in the United States evaporated. President Woodrow Wilson declared war on Germany and committed the United States to the Allied cause.[87]

As a ruse to maintain German confidence in the telegraphic system and thereby access to its message traffic, the British engaged in some theatrical hand-wringing, publicly praising the Americans for discovering the telegram and castigating themselves for failing.[88] Britain's ruthless exploitation of Sigint matched Germany's exploitation of American goodwill in facilitating German telegraphic communication through the United States to Mexico. But the revelation of Zimmerman's scheme provoked a storm of indignation which directly affected American resolve, drawing in American workers and industry to assist the Allied cause and effectively guaranteeing victory in the war.[89]

Before victory, however, significant battles remained to be fought to bring Germany to an armistice. America's military contribution was small compared to that of Britain and France, but with America's European allies exhausted, it would prove decisive. As Germany's Spring Offensive in early 1918 would soon demonstrate, things could have gone either way that year.

US Sigint prowess and a budding special relationship

The intelligence relationship between the British and the Americans became close, largely due to British openness about their achievements when the first American Expeditionary Force (AEF) officers arrived in 1917.[90] In March 1918, when the German offensive was imminent and British cryptographers were unable to read the new German cipher, the AEF intercepted a German message which was repeated in both the new and old versions of the cipher; thus allowing them to break it. Recognising its significance and urgency, this material was couriered by aircraft immediately to British GHQ.[91]

After the end of the Civil War, American military cryptologic practices had lapsed, but in operations during the Spanish–American War and near the Mexican border in the 1890s the AEF had acquired some proficiency in Sigint. The Radio Intelligence Service was created by the US Military Intelligence (G2) Branch to support strategic intelligence through wireless radio intercept, principally on the US–Mexican border in 1916 and 1917. Fourteen listening-in stations operated near the Texas border 24 hours a day, using radio tractor units for direction finding to identify the source location of wireless radio transmissions across the border in Mexico. The United States' experience tapping Mexican telegraph lines paid dividends on the Western Front.[92] They subsequently established a diplomatic intercept station in 1918 'under cover' at the US Embassy in Mexico City.[93]

With the US declaration of war, US Army Major Ralph Van Deman was appointed to head the Military Intelligence Section of the Army General Staff (designated the Military Intelligence Division in June 1918). While the United States was experienced at traffic analysis and direction finding, its cryptanalytic capabilities in 1917 were under-developed; but they soon picked up speed.

Key to this acceleration, Van Deman recruited a brilliant cryptanalyst, Herbert O. Yardley, who began service in the US Army in July 1917 and soon assembled a staff to conduct in-house cryptanalysis. Yardley has been described as 'an organizational genius, a slick and astute salesman, and a self-taught cryptanalyst who built the first permanent cryptologic organization in the US Army'.[94] By November 1917, he had created the Codes and Cipher Section (MI-8).[95] Thus began the United States' first modern, sustained military cryptanalytic organisation: America's 'Black Chamber'.[96] MI-8 soon discovered spy letters written in invisible ink, solved German diplomatic codes and ciphers, and found specialists to read obscure German shorthand systems.[97]

Yardley would gain notoriety for publishing a book about his experience which became a point of reference for Australian Sigint efforts at the onset of the Second World War, which we will discuss later.

The military intelligence cryptographic unit in the AEF was called the Radio Intelligence Section, or G2-A6. It was otherwise known as the Code and Cipher Section of G2, the Military Intelligence Division of the General Staff. Initially, the AEF relied largely on training by British and French counterparts.[98] In a matter of months, however, the Radio Intelligence Section was managing its own wireless intercept stations at AEF Headquarters in Chaumont (southeast of Paris), Souilly, Neufchâtel and Gondrecourt, south of Saint-Mihiel.[99]

Training for the AEF Sigint work was undertaken by William F. and Elizebeth Smith Friedman, who, as one historian observed, 'were destined to become the most famous pair of cryptologists in American history'.[100] The Friedmans would play a leading role in the development of the United States' Sigint capabilities in the lead-up to and during the Second World War, to the point where the United States emerged with an industrial-scale cryptologic enterprise. Australia would be the beneficiary of this prowess, particularly, as we shall see, from 1942 onwards, when its forces joined those of US Army General Douglas MacArthur in the war in the Pacific.

Before the United States entered the First World War as a belligerent, the Friedmans had been employed by a wealthy benefactor, George Fabyan, who believed a cipher in the *First Folio* by William Shakespeare would prove that the Bard's plays were actually the handiwork of his contemporary Francis Bacon. While they did not succeed in this task, the Friedmans' work for Fabyan prepared them to establish a cryptologic training program for American military cryptologists to work in the AEF's G2-A6. William Friedman would subsequently enlist, seeing out the war as head of G2-A6.[101]

While the US Navy had its own Office of Naval Intelligence (ONI), its main task was observing other countries' building programs and fleet manoeuvres. Given their limitations with regard to Sigint, both the Navy Department and the State Department depended on the War Department's Military Intelligence Division for interception, cryptanalysis and reporting on enemy enciphered communications.[102]

While US cryptologic capabilities expanded quickly during the war, France's work on Sigint at the Quai d'Orsay in Paris also expanded under the head of the French War Ministry's Cryptanalytic Agency, François Cartier. In the lead-up to war, France employed a number of highly competent officers in their cryptanalytic service. As early as 1 October 1914, then Major Cartier and his team of cryptanalytic experts solved the

German cipher key, a double transposition system which used two keys (the same key applied twice), with keywords which were changed into a corresponding numerical key.[103] And as one writer observed, 'even if the French cryptanalytic experts did wonderful work, the numerous blunders made by the Germans considerably facilitated it.'[104]

In addition to interception and interpretation, an additional specialised Sigint role to emerge in the BEF was cryptanalysis. German codes and ciphers were attacked by the intelligence section at the British Army's GHQ. In the spring of 1917, a separate cryptanalysis organisation, designated as 1(e), with additional but smaller 1(e) sections maintained at army level, focused principally on traffic analysis. At least one interpreter operator was employed on cryptanalysis in GHQ from October 1916. During the war that number would grow.[105]

While these technical tasks were undertaken at GHQ and army level, additional intelligence led to demands for support at corps, division and brigade levels; even down to the level of battalions. From 1917, there was also a divisional intelligence officer. Staff duties in subordinate formations, the infantry brigades, were handled in a similar manner, but with smaller staffs. An infantry brigade commander, for instance, had only a brigade major handling 'G' tasks, and a staff captain dealing with 'A' (administration) and 'Q' (quarters, supplies and logistics, as it is now known).[106]

In 1918 Sigint gained further recognition, as a series of breakthroughs saw the return of mobility to the Western Front with a succession of German offensives and Allied counter-offensives. By this time, the skills for effective deception, traffic analysis and code-breaking had been refined.[107]

By 1918, when the campaigning became less static, the Germans used more wireless voice messages. These were difficult to 'live log' except by interpreters who knew shorthand. To compensate, modified 'dictaphones' were introduced, although it appears that only one or two interpreter operators per army were assigned to this work. Consequently, this kind of wireless interception work was not done at corps or division level. Australia's highest command during the war was at corps level, with Sir John Monash on the Western Front and Sir Harry Chauvel commanding Australian and other empire troops in Palestine.[108] This meant that it would be not only

Monash and Chauvel, but their senior commanders and staff officers (who would feature again twenty years later in the Second World War) as well who would return home from the war with only a superficial understanding of the workings and importance of Sigint.

Middle East as a training ground for future Australian Siginters

While few Australians fully appreciated its value, the effect of Sigint was even greater in Palestine and Iraq during 1917, as Britain was able to read Turkish and German radio messages. They struggled to use the material uncovered, but higher formation commanders used it extensively. During the war in Mesopotamia (Iraq), for instance, the British commander of Mesopotamia Field Force, Lieutenant General Sir Stanley Maude, applied indications from code-breaking to steer aircraft reconnaissance onto key targets.[109]

Maude also presided over a well-documented episode featuring tensions between British radio intercepting and code-breaking personnel, signals and staff officers.[110] Australians and New Zealanders were also involved, giving them an early taste of the internal politics and chains of command associated with the delivery of Sigint.

Australian army signallers began operating in Mesopotamia in March 1916, along with a similar unit from New Zealand. At the request of the British Indian government (which oversaw operations in Mesopotamia), this cohort was reinforced in May by a second Australian troop and headquarters unit to form the First (ANZAC) Wireless Signal Squadron.[111] While this was not what we would now recognise as a Sigint unit, signals intelligence would soon become an important function.

As soon as they arrived, the ANZAC squadron's unskilled wireless operators set about conducting traffic analysis, a field in which Australians, as discussed later, came to be recognised as the best. They began copying Turkish radio messages, using press messages from the Germans, British, French and Indians. According to one account, 'after a little while we managed to get down with accuracy the mixed figure and letter groups (which were sent at high speed)' by the Turks. These enciphered messages were unintelligible to the operators, although they began to recognise the enemy stations and call signs.[112]

Eager to exploit further the messages' hidden information, the officer commanding the Australians forwarded this traffic to Maude's headquarters. The British signal staff were reportedly 'somewhat nonplussed' and initially did not consider it worth mentioning to their commander. But when Maude did learn about it, he immediately realised its importance, was 'rather annoyed' that he had not been informed earlier, and asked the War Office in London to send a deciphering expert to his headquarters.[113] At the same time, he ordered that two of the eight Australian signal stations continue to intercept Turkish wireless communications.[114]

The expert sent to Maude's headquarters was Captain Gerard Clauson, an Oxford scholar specialising in oriental languages, including Turkish. By the time Clauson arrived in Mesopotamia and joined the ANZAC squadron outside the town of Kut-el-Amara on the Tigris River, the squadron's wireless operators had improved their interception skills. They now recognised the relative importance of the various messages they were intercepting; even identifying the branches of the Ottoman Army sending each message. By combining their knowledge of enemy call signs with rough-and-ready DF equipment, they had established the locations of many enemy wireless stations such as Dahra Bend (SAR), Samarrah (SMR), Mosul (SBA) and Damascus (DAS).[115]

Within 23 hours of arriving at the headquarters, Clauson cracked the first Turkish code. 'Thereafter, despite daily changes and enciphering of a complicated kind, every enemy message arrived at I (Intelligence) Branch as certainly as if it had been addressed to them.' The two Australian stations in interception work maintained their effort until the Turkish surrender. In the words of one Australian historian, 'the uncanny power which this information, over and above that secured by his other means of intelligence, gave to him [General Maude] is one of the outstanding features of the Mesopotamian campaign.'[116]

The road to this notable achievement, however, was a bumpy one. In the words of Clauson, in a letter of 9 October 1917 to a colleague,

> Coxon [the Commander of the Special Wireless Section in Mesopotamia] is a bloody old ass and f-ing awful nuisance. He has done nothing but stand on his dignity, put up distinctly 2nd rate intercepts and then try and get work taken away from the Anzacs who do thundering well. He could think of nothing except trying to pinch more personnel and material and give a minimum of bearings. Everyone was getting bloody sick of him when your merciful offer to replace him arrived. One station is actually going and the other will in a week, so with Lefroy coming here to give things a start proper and with another fellow vice Coxon we should do well.[117]

In late October, Major Lefroy arrived at Baghdad to take over Coxon's work, and on 4 November, the latter was formally relieved, ostensibly on the grounds of poor health.[118]

The ANZAC Wireless Signal Squadron experience in Mesopotamia stands in contrast to the work of the AIF Signals on the Western Front, where most of the Sigint work was undertaken by BEF elements above the level of the corps headquarters. This experience would be another consequential precursor to the development and fielding of Army Special Wireless units during the Second World War.

Looking back on Australia's First World War experience with Sigint

Sigint continued to play a major role to the end of the war. The Germans had monitored Russian deliberations in the lead-up to the signing of the Treaty of Brest–Litovsk in 1917, which saw Russia bow out of the war. Similarly, though, from 8 to 11 November 1918, the German armistice delegation to the negotiations in the Forest of Compiègne sent numerous enciphered telegrams back to Berlin – all of which were deciphered by the Deuxième Bureau, the intelligence section of the French General Staff. In one of the last despatches, the delegation head was instructed to 'try for milder terms; if not obtainable, sign nevertheless.' This sealed the outcome of the negotiations.[119]

Reflecting on the 1914–18 war, Sigint was a success story for Britain, but its limitations were notable. Wireless intelligence was at the leading edge of the theory and practice of telecommunications, but sometimes mistakes were made. How could commanders trust a source which, on occasion, for instance, located German army stations behind British lines, or the High Seas Fleet inland somewhere in Germany? Overcoming mistrust and the limitations of clumsy and primitive equipment required the skills of some of the best engineers on earth. Sigint was proving extraordinarily powerful, but harnessing its power and applying it in a timely and effective manner involved trial and error.[120]

The modern age of intelligence, it can be argued, began in 1914, as a result of developments in sources, organisation and communication. The telegraph and radio were powerful instruments for collecting, assessing and using intelligence. Imagery and Sigint, in concert with the general staff system, were an effective combination. By 1918, military intelligence systems were sophisticated and used every technique available during the war of 1939–45.[121] Many of these would be foundational for the work of Australia's national Sigint agency after the Second World War.

On the Western Front, the French outperformed the British in military Sigint, where the Americans also did well.[122] On balance, however, Allied tactical Sigint triumphs during the war were trumped by Austrian and German Sigint successes which helped the forces of the smaller Central Powers defeat a larger Russian Army.[123] On the other hand, the effect of the British decryption of the Zimmerman Telegram proved decisive in bringing the United States into the war, hastening Germany's defeat.

With the outbreak of the First World War, the tsarist *cabinet noir* had become Britain's ally. In October 1914, the Russians had given the British the code book captured after the sinking of a German cruiser: the *Signalbuch der Kaiserlichen Marine*. While Australia's seizures of German merchant marine code books were significant at the outset of the war, Britain owed its first success in breaking German naval codes to this Russian windfall.[124]

After the Bolshevik Revolution in 1917, however, the equation changed dramatically. One of the best tsarist codebreakers of British codes, E.C. Fetterlein, escaped to England where he assisted in breaking Soviet codes. He went on to become head of the Russian section at the British interwar cryptanalytical agency, the Government Code and Cypher School (GC&CS).[125] Historians Christopher Andrew and Keith Neilson have observed that 'thanks in large measure to Fetterlein and the accumulated expertise of the Tsarist *cabinet noir,* the most valuable diplomatic intelligence source available to the British intelligence services during the decade after the Bolshevik Revolution was the decrypted traffic of the Soviet government.'[126]

While the Australian land forces may not have developed particularly sophisticated Sigint capabilities during the war (with the exception of the resourceful signallers in Mesopotamia), their maritime counterparts emerged from the war with relatively advanced ship-based capabilities, even if these were not matched with a robust shore-based naval intelligence directorate. What the RAN's participation in this first Sigint war did was to highlight to those involved, in both Australia and Britain, how this small navy could make an important contribution to the collective defence of the empire through specialised and skilled personnel and facilities. Thereafter the RAN was seen as having the potential to provide more than just ships and men.[127] This would be illustrated by RAN involvement in several small-scale initiatives by the British and the Royal Navy in the interwar period, including the training of Japanese linguists, decisions on DF stations, participation in conferences and in the Far East Combined Bureau discussed in the next chapter.

For much of the war, and in most theatres, Australia principally relied on Sigint being performed on its behalf by British intelligence collection, analysis and dissemination mechanisms. Had this not been the case, Australia likely would have been compelled to invest in its own capabilities.

As with the AIF, with its preponderance of infantry and little emphasis on the supporting arms and services, so with intelligence. With the occasional exception, such as the Sigint successes of the RAN early in the war, Australia contributed to the First World War with a limited understanding of the spectrum of capabilities required for modern warfare. Sigint capabilities had developed dramatically in the forces of the United States, United Kingdom, France and others, but little work had been done to transfer these capabilities to the Australians. So, in one sense, while access to British-sourced and American-linked Sigint undoubtedly helped

Australia at the time, it was a mixed blessing. Dependence on British Sigint support meant that there was no need for a national Sigint institution for Australia in the early decades of the twentieth century, and hence none was established. This was consistent with other aspects of Australia's dependence in defence and foreign relations, leading to a lack of self-reliance and national self-confidence.

It would not be until the next major conflagration, as a more independent foreign and defence policy became a priority, that Australia recognised the need for robust and sophisticated Sigint capabilities in direct support of Australian forces and to assist the nation in decision-making.

Nonetheless, the legacy of this wartime experience for the Australian military and naval forces was substantial. While the interwar years would see a precipitous decline in military funding and capability, the legacy of the wartime experience would nonetheless have a lasting effect, as the following chapters explain.

4
AUSTRALIAN SIGINT & THE INTERWAR YEARS

War had driven rapid expansion and increased sophistication of Sigint, particularly for the great powers, and Australia was a beneficiary of this, both directly and indirectly. Yet as the 'war to end all wars' came to a close, many assumed that intelligence, notably Sigint, was no longer required on such a scale. Indeed, in the interwar years, some looked to close down their Sigint organisations, hoping, believing even, that they were no longer necessary.

Sigint would remain important, however, in the years of peace. While the interests of the United States and Great Britain would diverge in the interwar period, Australia remained closely aligned with Britain, although often reluctant to back that alignment with financial commitments.

Australia had become a federation in 1901 and had built up some myths and legends from its war-fighting experience in the First World War and the Anglo–Boer War. But in the 1920s and 1930s, it was not yet a fully independent nation. The interwar experience saw the British empire's self-governing dominions of Canada, Australia, New Zealand and South Africa slowly mature their bureaucratic functions and legislative controls. Nonetheless, Australia and New Zealand still looked to the United Kingdom for instruction on foreign policy and assistance with defence policy.

The Statute of Westminster, which gave the self-governing dominions power to formulate foreign policy, would not be enacted in London until 1932, and was not adopted in the Australian and New Zealand parliaments until a decade later. Australia's outlook on Sigint during this period, therefore, is best examined with a clear understanding of developments in the United Kingdom and, given the influence it would exercise in wartime, in the United States as well.

The Australian government's deference to Britain in foreign and, to a certain extent, defence policy was accompanied by a reluctance to spend on defence and security. Notwithstanding the cutbacks, many lessons of radio propagation and cryptology were absorbed and refined by secret institutions maintained by the governments of the major belligerent powers on both sides. With advances in technology, notably the mechanisation of cryptography, a profound technological change affected Sigint.

Australia has at numerous junctures also sought to compare itself with its 'strategic cousin' in fellow British empire dominion, Canada. Like Australia, Canada was reluctant to spend on defence and security during the interwar years, to the point where, according to historian Wesley Wark, the country reached a state of 'cryptographic innocence ... enjoy[ing] no real tradition of strategic intelligence activity'. 'Modern techniques for intelligence work', Wark continues, 'especially those developed since 1914 in the areas of wireless interception and code-breaking, were virtually unknown'.[1]

Not surprisingly, therefore, the imperative was to cooperate with the United Kingdom (and subsequently the United States) which were more advanced in Sigint and maintained national functions, refined through the war, on which Australia continued to depend. Notwithstanding these constraints, there was a modest but growing awareness among Australian officials of the importance of Sigint to the management of national security affairs.

Britain's Government Code and Cypher School

The existence and functions of Room 40 laid the groundwork for the post-war Government Code and Cypher School (GC&CS), the precursor to the Government Communications Headquarters (GCHQ) of today. In the early days of the Second World War, GC&CS assisted the development of Sigint in Australia, and in the establishment of the direct antecedent of

ASD, the Defence Signals Bureau (renamed the Defence Signals Branch in 1949, Defence Signals Division in 1964, Defence Signals Directorate in 1977 and the Australian Signals Directorate in 2013). Developments in the United Kingdom would have a direct impact on the national Sigint architecture of Australia.

Following the remarkable success of Room 40 and MI1b, in early 1919 the War Cabinet's Secret Service Committee met in London to consider 'reviewing the existing arrangements and organisation of Britain's secret service', including Sigint. Chaired by Foreign Secretary Lord Curzon, the committee recommended the centralisation of cryptographic work within a single organisation. The services objected to civilian control by the Foreign Secretary, but they agreed to establish the GC&CS in November 1919. The director of naval intelligence, Captain (later Admiral Sir Hugh) Sinclair was given the task of setting up GC&CS with a civilian administration, but under the Admiralty. Key personnel from Room 40 moved to GC&CS and their wartime diplomatic Sigint functions were transferred to the new organisation. The first director was Commander Alastair Denniston, RN, formerly of Room 40. With little naval traffic to report on at the time, the initial reluctance to cede control to the Foreign Office was overcome, and in 1922 responsibility for GC&CS was transferred from the Admiralty to the Foreign Office.[2]

Thereafter, the British Army, Royal Navy and later the Royal Air Force (RAF) maintained an active interest in Sigint. This included radio interception from Britain and abroad, with interception and cryptanalytic units stationed on British territory in the Middle East, India and the Far East. The armed services also established sections in GC&CS to gain experience in cryptanalysis and assist with translation. By 1924 a Committee of Coordination would be established and met annually to prioritise and coordinate the activities of the armed services and GC&CS.[3]

The considerable volume of translation work was facilitated by Clause 4 of Britain's *Official Secrets Act* of 1920, which enabled the Home Secretary to sign a warrant requiring that cable companies send originals or copies of all telegrams sent or received in Britain from abroad. Telegrams passed through the British-controlled territories of Hong Kong, Bermuda and Malta also were subject to these provisions. This arrangement continued throughout the interwar period until cable censorship was again instituted, obviating the need to rely on Clause 4.[4]

During the early interwar years, notably between 1919 and 1932, the GC&CS was probably the world's best code-breaking bureau.[5] During this period, the GC&CS had near mastery of the diplomatic code systems of several countries, including Japan, the United States and Italy, and access to the systems of France and the Soviet Union. By the middle of the 1930s, however, GC&CS lost power over Soviet and Italian diplomatic systems, had little luck against Germany, while retaining access to Italian military communications systems. Then in 1939, Britain's success against Japanese systems collapsed. On the other hand, Britain retained access to the diplomatic traffic of the United States, France and smaller powers, augmented by one new triumph: penetration of some Soviet Communist International (Comintern) codes used in Europe and Soviet Army systems in central Asia.[6]

This work was complemented by the collaboration between GC&CS and the Metropolitan Police, as well as with the Security Service (known as MI5), and the Secret Intelligence Service (SIS) (known as MI6). The connection began through the police, who wanted help in searching for illicit transmissions. A station established at Denmark Hill (a few kilometres south of Westminster) provided information not only concerning criminal endeavours, but a range of clandestine networks. Through the 1930s, DF equipment helped locate clandestine Comintern transmitters and uncovered the secret network set up by the German Foreign Office.[7] The Post Office, through its site at Dollis Hill in London, would also become involved.

To coordinate Britain's Sigint efforts, GC&CS had established a 'Cryptography and Interception Committee' in 1924. This committee guided work priorities for the armed services.[8] With the sun never setting on the British empire, demands for Sigint came from around the globe. But as Britain's intelligence historian, F.H. Hinsley observed,

> the three Services could not all have interception stations everywhere and by the 1930s a system had grown up in which the War Office undertook most of the work that was done in the Middle East,

the Navy looked after the Far East and the Air Ministry confined itself to what it could do in the United Kingdom. Even within this general sub-division of responsibility, moreover, inter-Service integration had developed ... Thus from 1937 the naval cryptanalysts at [GC&CS] worked almost entirely on non-naval Japanese cyphers, leaving the Japanese naval cyphers to be worked on at Hong Kong.[9]

It was in this context that the Far East Combined Bureau (FECB) was established in the mid-1930s and, as discussed below, Australian Sigint practitioners would engage with the FECB in the lead-up to and early days of the Second World War.

Britain's diplomatic Sigint setbacks and successes

Britain's acquisition of star tsarist codebreaker Fetterlein paid enormous dividends. For Lord Curzon, Britain's Foreign Secretary from October 1919, reading decrypted messages became almost an obsession.[10] These messages revealed indications of Soviets engaging in subversion and propaganda.[11] What transpired would have long-term repercussions.

Curzon was infuriated by Russian messages decrypted during 1920–21 bilateral trade negotiations. The messages insulted their British partners and included evidence of secret payments to the socialist *Daily Herald* and the Communist Party of Great Britain. While the trade negotiations led to the first de facto recognition of the Soviet regime by a major Western power, the secret payments prompted Cabinet to demand their public disclosure. Copies of intercepts revealing the payments to the *Daily Herald* were given to all national newspapers. In an attempt to protect the Sigint source, the press were asked to say the intercepts had been obtained from a 'neutral country'.[12]

Political pressure grew for further public revelations, with Lord Curzon in the vanguard. After the fall of Lloyd George's government in October 1922, Curzon prevailed. On 23 April 1923, he secured Cabinet approval to prepare a protest note to Moscow, now known as the Curzon

Ultimatum. The Cabinet agreed that the advantages of basing its argument on the actual despatches passed between the Soviet government and its messengers outweighed the disadvantages of the possible disclosure of the Sigint source.[13]

In his protest note, Curzon not only quoted publicly from the intercepts, but taunted the Russians, saying they 'would no doubt recognise' their own communications. In the short term, Curzon's ultimatum was effective. Keen to preserve the trade agreement, the Soviets accepted the conditions, including cessation of the secret payments.[14] They also changed their ciphers, but Fetterlein simply broke them again.[15] The ultimatum had other consequences, however; notably setting a precedent for risking the disclosure of the Sigint source for political reasons. More Sigint revelations would follow.

In 1927, the government of Britain's Prime Minister Stanley Baldwin breached Sigint security even more recklessly. A year after a general strike, amid concerns over secret support from Moscow for the strikers and distribution of subversive propaganda, the government decided to break off diplomatic relations with Moscow. To justify the decision, Baldwin read out four Russian telegrams in the House of Commons which had been decrypted by GC&CS,[16] thus compromising the work of its own intelligence organisation while still failing to prove the Home Secretary's charge that Britain was threatened by an extraordinarily nefarious Soviet 'spy system'. Even with the assistance of the decrypted messages, Baldwin's government was only able to prove the lesser charge that the Soviets were covertly dabbling in British politics.[17]

This time, the net result for British intelligence was most unfortunate: the Soviet Union adopted the theoretically unbreakable one-time pad for its diplomatic traffic. Thereafter, between 1927 and the Second World War, GC&CS decrypted almost no high-grade Soviet communications.[18]

Commander Alastair Denniston, the head of GC&CS, summed up the cryptographic effort on interwar diplomatic messages this way:

> We started in 1919 at the period of bow-and-arrow methods, i.e., alphabetic books; we followed the various developments of security measures adopted in every country; we reached 1939 with a full

knowledge of all the methods evolved, and with the ability to read all diplomatic communications of all powers except those which had been forced, like ... Russia, to adopt [one-time pads].[19]

The effect of the Curzon and Baldwin compromises must have dismayed Denniston.

Machines and one-time pads

The Soviets did not just rely on the public pronouncements of men like Baldwin. In fact, from 1934 onwards, they had an agent, William Wiesband, inside the US Sigint community who worked in the Russia section and leaked information which compromised US efforts.[20]

From the time the Soviet Union introduced one-time pads, revelations of Soviet interception curtailed the ability of GC&CS and US Sigint to continue work. Similar revelations in 1927 about the breaking of German codes during the First World War led to tightened German security measures. But, unlike the Soviet Union, Germany chose to focus on making decipherment more difficult by making their cryptographic machines more complicated. Worst of all, from a British perspective, that same year, the founder of Room 40, Sir Alfred Ewing, gave a public lecture on its activities and the lecture was reported fully in the media.[21]

Meanwhile, the first nation to acquire cipher machines for its armed forces was Germany, starting with the German Navy which adopted an Enigma machine made by Chieffriermachinen AG in 1926. Orders for thousands of additional Enigma machines for use by German authorities would follow.[22] The Germans invested in secret changes to the Enigma machine to make decryption more difficult. The addition of a plugboard added a plethora of additional possible permutations and this was considered sufficient.[23] As events would show, however, while those changes certainly added to the degree of difficulty in deciphering, they did not make it impossible.

In 1927, the US Signal Corps bought an Enigma machine, but decided to develop its own, the M-209, based on the Hagelin design. The Japanese also purchased one and by the mid-1930s had developed their own version, which came to be known by US codebreakers as Purple.²⁴

Meanwhile, the Enigma machine had an indirect effect on the communications security of the British armed forces, which would, in turn, affect Australia. An encrypted telegraphic machine, known as a Typex, was developed by RAF engineers who built on the design concept of the Hagelin-inspired Enigma machine. In the 1930s, an interdepartmental committee examined the range of cipher machine options, and the RAF machine emerged. The Typex was then manufactured for the British forces by the Creed Teleprinter Company. This was coming on line in all three services by 1939 and would come to be widely used in Australia. Its introduction was intended to reduce the risk of enemy exploitation of communications security breaches, although, as with the Enigma machine, vulnerabilities remained, relating particularly to compromise due to sloppy procedures and concerted attack. This is what happened early in the war, enabling Germany to detect Allied shipping and aid their U-boat operations which nearly brought the United Kingdom to its knees in 1940 and 1941.²⁵ Eventually, the Typex would be replaced by a more secure cipher machine for use by British and empire forces, the Rockex. But even then, some vulnerabilities to determined exploitation would remain. Meanwhile, American business spearheaded technological developments independently of Britain; developments which would contribute to shaping Australia's wartime and postwar Sigint capabilities.

US interwar Sigint, Yardley and prohibition

After the end of hostilities in 1918, US–UK Sigint cooperation dwindled. Thereafter, US influence on intelligence matters in Australia was less than that of the United Kingdom; Australia, after all, remained dependent on Britain for trade and defence. It had not yet developed national Sigint institutions beyond basic capabilities in the RAN. Still, the interwar development of US Navy and Army Sigint would prove critical for the Australian national Sigint capabilities in the Second World War. American advances in radio communications were also crucial. Continuous-wave radio was superior to the arcs and alternators designed and used by Marconi and his corporation. The Radio Corporation of America (RCA) and other companies became major international players in continuous-wave radio design, eclipsing Marconi. Marconi responded by discovering some of the properties of short-wave radio communications (enabling signals to bounce off the ionosphere and be heard far away). This led to a revolution in the design and cost of radios, making them far more accessible and portable. The combination of these developments transformed wireless radio in the 1920s and 1930s, to the point where cable was considered to have been superseded by radio.[26] This would have major implications for Sigint.

In the United States, some limited but effective operational Sigint capabilities were maintained and developed. Initially, this was principally through Herbert Yardley's Cipher Bureau (his 'Black Chamber'), which was under the shared jurisdiction of the State and War departments. Capitalising on the cryptological successes of the First World War, the Cipher Bureau solved the codes of 20 nations.[27] One of these was Japan. Knowing in advance how far Japanese negotiators at the Washington Disarmament Conference of 1921 and 1922 would yield if pressed, US negotiators induced Japan to accept less capital-ship tonnage than Japan had wished – the famous 5:5:3 ratio between Britain, the United States and Japan.[28]

Despite such Cipher Bureau successes, financial support dwindled as the 1920s progressed and the memory of war receded. The staff shrank; some members worked only part-time. Then, in November 1928, Herbert Hoover was elected US president. He named Henry L. Stimson, a New York lawyer with no experience or appreciation of US cryptologic endeavours, as his secretary of state. Yardley waited until Stimson had been in office a few months before sending him the solutions of an important series of messages. In the meantime, the US Senate ratified the Kellogg–Briand Pact renouncing war. Now aware of Yardley's work, the high-minded Stimson was shocked, declaring 'gentlemen do not read each other's mail'. Stimson thus dismissed the incalculable advantages offered by Sigint. War was no longer on the horizon, and Stimson strongly disapproved of the long history of eavesdropping by earlier US administrations. He cut off funding to the

Cipher Bureau, destroying the Black Chamber and much of America's intelligence capability.[29] Other arms of government would take an interest in Sigint, but not before Yardley made international news.

Out of a job and unimpressed by Stimson's decision, Yardley wrote a no-holds-barred history of the Cipher Bureau's activities, *The American Black Chamber*. In a manner echoed generations later by the revelations of US contractor turned defector Edward Snowden, Yardley's book scandalised the political and diplomatic world, becoming an international bestseller in 1931.[30] Ironically enough, as war clouds gathered again, and as Australia finally sought to invest in its own cryptological capability, Yardley's book would prove invaluable as a textbook and guide for novice Australian Siginters.

Yardley's hard-hitting writing style and the total novelty of his revelations led to notoriety. In one of the most dramatic moments in cryptology, Yardley revealed how the Cipher Bureau broke Japanese codes during the Washington Naval Treaty negotiations. Reviews of the book were positive, but the damage done reverberated for decades, particularly in Japan, where the book skyrocketed Yardley to infamy and left deep resentment in official Japanese circles.[31] This also galvanised the Japanese wartime cryptographic organisation.[32] While American diplomatic and military cryptology became dormant following Stimson's closure of the Cipher Bureau, the Yardley revelations reinvigorated the cryptological efforts of Japan and Germany.

In the meantime, notwithstanding the demise of the Cipher Bureau, US domestic agencies kept the government's cryptological capabilities alive. Thanks in part to the decision to introduce restrictions on alcohol in the 1920s, rum-running organisations sprang up in defiance of prohibition, with an undercover wireless network comparable in size, technical skill and organisation with the radio operation conducted by enemy agents during the Second World War. The invention of the vacuum tube made small transmitters possible and, by 1928, illegal transmitters were small enough to be concealed on a person's body.[33]

The US Coast Guard monitored and intercepted illicit sources of liquor from British Columbia in Canada, and Mexico, on the West Coast; and Nova Scotia, British Honduras and the West Indies for the East

Coast. Sophisticated Sigint capabilities were brought to bear, including interception, DF, deciphering and translation. In a demonstration of early interagency cooperation, the Coast Guard was assisted by the Federal Bureau of Investigation (FBI), and the radio intercept arms of the Bureau of Customs, the Department of Commerce, the Treasury Department (whose leading cryptanalyst at the time was Elizabeth S. Friedman) and the War Department. MI-8 and the infant Signal Intelligence Service (SIS) of the US War Department provided cryptanalytic assistance, with the SIS using intercepted communications traffic for training purposes.[34] After prohibition was repealed in 1933, smuggling continued, until it was effectively stonewalled by cross-border cooperation between the Royal Canadian Mounted Police (RCMP) and the US Coast Guard. By 1939, organised smuggling had practically disappeared, but the capabilities used to thwart it came in handy, as the international security situation deteriorated in the lead-up to war.[35]

IBM and Purple machines

Meanwhile, as war clouds gathered over Europe and the Pacific, the United States began to reinvest in its naval and military Sigint capabilities. Drawing on this technology and other initiatives, in the 1930s the US Army Signal Corps established the SIS to identify and break Japanese and German codes.[36] In the 1920s and 1930s, William Friedman, Abraham Sinkov (who would later play a key role in the development of Australian Sigint) and Laurance Safford worked to establish and build up American Army and Navy Sigint capability, respectively, thus enabling both services to develop rapidly from early 1942 onwards.[37]

As noted earlier, US cryptological efforts in the First World War used some automated data processing. By the early 1930s, however, the US Navy Sigint unit (OP-20-G) developed cryptological techniques using cards and machinery manufactured by the International Business Machines company (known as IBM). The US Army Sigint unit started working with these techniques a year or two later. In the United Kingdom, the GC&CS also used licensed versions from the mid-1930s. Throughout the 1930s,

IBM made a concerted effort to develop high-performance card-handling machines suitable for data management and accountancy work; technology that would prove remarkably useful during the Second World War.[38]

In 1935, for the first time since Yardley's revelations led to a blackout, the Americans once again broke the Japanese diplomatic code, known to American cryptologists as 'Red', although their efforts were constrained by limited resources due to the competing priorities of work on naval and military codes. In 1939, Japan replaced 'Red' and adopted the more sophisticated Hagelin/Enigma derivative machine system known as 'Purple'. In August 1939, Colonel Spencer Akin took command of the US Army's Signal Intelligence Service, and over the next 18 months led efforts to solve this system. Akin would later appear in Australia as General MacArthur's Chief Signals Officer in charge of the combined Sigint agency (Central Bureau) that would be established in Australia during the Pacific War. Meanwhile, working for Akin, Friedman played a pivotal role in enabling SIS cryptanalysts to 'crack' the Purple machine. This ingenious machine was an analogue of the Japanese Purple. It was invented largely by William Friedman and his team, without access to an original Japanese one. Their remarkable invention enabled a mechanical decryption of the Japanese machine-enciphered code. Aware of the risks of compromise, the Japanese changed the key to the code daily, but further American breakthroughs uncovered a ten-day pattern which simplified the decryption work.[39]

By October 1941, ten analogue Purple machines would be built, one of which would be in use by a US Navy detachment, known as 'CAST' in the Philippines. In addition, the US Army's Station 6 intercepted Japanese radio traffic from the Philippines as well. While Akin, then MacArthur's Chief Signals Officer in Manila, had access to the 'Magic' reports of the analogue Purple machine, surprisingly enough, the Chief Intelligence Officer (G2) was not given access. Akin ensured, however, that MacArthur and his US Navy counterpart had such privileged access. But even then, reports had to be hand-delivered – a process prone to delay due to the physical distance between the interception and reporting sites and the respective headquarters.

As war approached, the focus on Japanese diplomatic codes, at the expense of Japanese army and navy codes became more apparent.[40] As it happened, success against this one machine cipher did not necessarily translate into penetration of all other Japanese ciphers. This was because the Imperial Japanese Army and Navy each employed a range of entirely different cipher systems to those of the Foreign Ministry.[41] This tangle of systems caused confusion and controversy, notably following the intelligence failure over the surprise Japanese attack at Pearl Harbor.

After the commencement of the Pacific War, United States' collaboration with their British counterparts, which included Australia, would blossom.[42] As we shall see, however, there were limits to that collaboration.

Australia's direct involvement in these arrangements began only after Japan's invasion and MacArthur's escape to Australia, along with his Sigint team, in February and March 1942.[43] Meanwhile, Australia's Sigint capability, with few resources and little attention, nevertheless took some small incremental steps forward – until war again seemed imminent.

Australian interwar Sigint and traffic analysis

To understand Australian Sigint between the wars, it is helpful to clarify some terminology. There is a modern tendency to equate Sigint solely or primarily with cryptanalysis, when, in reality, cryptanalysis is only one part of the Sigint system. However, just because the Australian system did not by that time include extensive cryptanalysis or code-breaking does not mean there was no Sigint effort in Australia. In fact, as the previous chapter attests, there had been already a basic system operating throughout the First World War. This was developed further, intermittently and quietly during the interwar period, particularly by the RAN.

The two most important aspects of Sigint, particularly at the tactical level where combat operations took place, were geolocation, through DF, and tactical information about dispositions and possible intentions derived from traffic analysis (TA). DF was vital for telling commanders the location of enemy forces, while TA provided additional information on who was communicating, at what times and frequencies, at what level of organisation, and how active they were. In essence, therefore, TA could provide an enormous amount of 'actionable' intelligence without a single word of a message needing to be understood.

Further, TA provided not only the date and time of the message and the frequency on which it was sent, but also the call signs used for communication and their place in an organisational hierarchy, allowing identification of the units involved. In a fast-moving situation, the greatest utility of TA often would be found close to the source and for a short period. It could also provide the number of messages in the series, allowing accurate collation of the traffic (and verification of missing or incomplete sections) and provide insights into standard preambles and sign-offs to identify the sender. Furthermore, TA revealed the grade of the message, while the level of urgency and activity could be deduced from the speed of reply and total number of messages being sent compared to other equivalent periods. TA also told the user whether the activity was routine or special. In many cases, the only thing TA did not reveal was what the target was actually saying inside the message. The value of TA was considered greatest at the tactical and operational levels, not at the high policy level, where the actual communication was key.

Seen as the foundation from which all cryptanalytical attacks could be launched on a cipher or code,[44] TA was to become one of the strengths of Australian intelligence in the Second World War. Australians were recognised as being very good at it, in fact the best. According to Nigel de Grey, 'traffic analysis played a very active part in the [South West Pacific Area] throughout the war. In this art, the Australian Army held the lead.'[45] But the foundations were laid earlier, including during the interwar period.

During the First World War, Australian land forces had certainly relied on the work of higher (British) headquarters and formations to manage Sigint for them. The RAN, on the other hand, as part of the Admiralty's worldwide intelligence system, had contributed to the Sigint successes of 1914 in operations that led it to take an active interest in cryptanalysis and code-breaking over the following decades.

It should be borne in mind that after the First World War all Australian forces were severely limited by government unwillingness to spend on defence. The challenging financial situation was compounded by a lack of understanding by some key government and military figures of the importance of Sigint, much to the frustration of those who understood its value for Australian defence.

Faced by savage cuts in their budget and operational capability, the ACNB and Britain's Admiralty prevailed upon the Australian government to commission an inquiry into the defence of the Commonwealth by Lord Jellicoe, Britain's Admiral of the Fleet. The purpose of the inquiry was to determine what was needed to ensure Australia's defence and what it would cost to have Britain's Royal Navy maintain a large fleet in the region.

Lord Jellicoe's report

Jellicoe's report, submitted in draft to the Governor-General on 14 August 1919,[46] advocated the establishment of a NID to deal with activities in the Pacific and Indian oceans.[47] Jellicoe also made a number of recommendations for wireless telegraphy (W/T) intercept and DF stations, as well as for the erection of interception facilities as soon as war broke out. His recommendations did not include, however, the training of personnel to operate these stations and act as cryptographers. But an unidentified member of the Australian Naval Staff, commenting on the report, noted that an 'Intelligence W/T organisation has been immeasurably important during [the last] war'.[48]

Although the government was unwilling to pay for a national intelligence system on the scale Jellicoe envisaged, Australia still enjoyed the privileges of being part of the Admiralty's worldwide intelligence organisation, with access to the invaluable interchange of information that organisation provided. In return, Australia through the Australian Navy Office, would supply intelligence to the Admiralty. British and Australian warships were governed by the same scheme of naval intelligence.[49]

Despite government frugality, there were those who remained concerned about Australia's defence and understood the crucial role of Sigint and they did what they could as economically as possible, or, as they say, 'on the smell of an oil rag'.

One of Jellicoe's recommendations, to establish a coast-watching organisation, did indeed come to pass. Costing virtually nothing, it proved to be an undoubted success. It began formally with the appointment of Ancell Gregory, an unpaid volunteer, on 15 March 1920. The secret orders issued on 29 March to the unpaid Honorary Lieutenant A.C. Gregory required him to forward 'certain information' along the strip between North West Cape in the south and Cape Londonderry (the northernmost point of mainland Western Australia), to meet the requirements of naval intelligence. Gregory was to report all information on 'aliens', especially Japanese, and on any unauthorised landings; provide information on fuel, coal and reliable men with specialist qualifications; and supply information on unsurveyed harbours, anchorages and approaches along the coast. Gregory was required to report once a month by registered post to the Department of the Navy in Melbourne.[50] Gregory set to work quickly and Captain Henry Cochrane, the chief proponent of the Coastwatchers, wrote in July 1920 to the First Naval Member, Rear Admiral Sir Percy Grant, declaring that the 'appointment of this officer [Gregory] has already justified itself'.[51]

Australia had been a strong critic of the Anglo–Japanese Alliance since it was signed in January 1902, but London remained in charge of foreign policy. The alliance was affirmed three times, but after the First World War it was seen as an obstacle to negotiations at Versailles and the establishment of the League of Nations. Britain was also mindful of Australian objections and had concerns about Japanese inroads into Britain's sphere of influence in the Far East. In March 1921, conscious of the possible threat from Japan, the Admiralty held a conference at Penang, Malaya (now Malaysia) for the flag officers commanding the East Indies Squadron in Ceylon, the China Station in Hong Kong and the Australian Fleet in Sydney. The purpose was to develop strategic advice from the Admiralty for the 1921 Imperial Conference in London, including future coordination of intelligence collection and wireless communications between the stations in the Far East and the Admiralty.[52] These deliberations preceded Britain's entry into the Washington Naval Conference in November 1921, which, in turn, led to the signing of a treaty limiting naval construction. What emerged, in addition, was a naval defence policy aimed at deterring Japanese aggression, which came to be known as the Singapore Strategy.

Australia agreed to contribute to the development of a W/T network in the Pacific and to contribute to the Admiralty's intelligence system, but, interestingly, refused to endorse the automatic subordination of the Australian Squadron to imperial command at the outbreak of war. In the end, however, limitations of cable and wireless capacity stymied these proposals and Australian and British reluctance to invest in the program meant that it did not go ahead.

The Australian Navy's incremental steps

Again, the Australian Navy took whatever cost-neutral steps it could. In 1921, it distributed Japanese Morse code manuals to all of its naval telegraphic stations and major units, and ordered all ships and stations to practise taking down Japanese Morse code.[53] This was to be a secret training program, undertaken on internal (offline) machines to avoid inadvertent transmission of Japanese Kana code. Wireless telegraphists were encouraged to become proficient in the reading of Kana code. Interception of Japanese transmissions became known as 'Procedure Y', and was conducted from Australian cruisers. The skills acquired would enable telegraphists to listen in to parts of the radio frequency spectrum known to be used by Japan and report on what they heard. By 1922, however, the program was abandoned, with the commander of the Australian Fleet complaining that it was 'quite impossible to give wireless ratings sufficient instruction to enable them to meet the standard set'.[54]

With Australian naval ships conducting transits around the South Pacific and beyond, opportunities were still taken to test and see what they could hear and collect. In March 1922, the ACNB decided to fit all of its light cruisers with DF equipment. In June 1922, early signs of this capability were demonstrated when intercept operators on board HMAS *Sydney* managed to collect Japanese high-frequency (HF) wireless transmissions from the Japanese Mandated Territories north of New Guinea. Again, in 1923, the Navy's Winter Cruise saw a combined Humint and Sigint operation, when two RAAF officers joined HMAS *Adelaide* and HMAS *Brisbane* to survey the islands for suitable airfield locations, notably around the Deboyne and Admiralty islands and around the north coast of New Guinea.⁵⁵

Such operations were most beneficial when coordinated with the work of Britain's GC&CS. They were managed through the Foreign Office rather than the War Office or the Admiralty, which meant that the military arms could not give direct orders to GC&CS. By 1924, however, a Japanese naval section at GC&CS had a dedicated team, supplied by the Admiralty, looking at IJN communications traffic.⁵⁶ Collection from a shore-based establishment in the UK proved inadequate for Far East collection, so British ships on foreign naval stations were tasked to undertake 'Procedure Y' collection.⁵⁷

Eric Nave: a gifted individual who shaped Australian Sigint

According to a British Naval Intelligence report, 'Japanese Naval Sigint work began in 1924 when Paymaster-Lieutenant Eric Nave, a Japanese interpreter, was appointed to the flagship of the China Station to make a start on the job.' Nave, then in the Royal Navy, was nevertheless an Australian. He features prominently in Australia's contribution to Sigint as well as that of Britain.⁵⁹

Frank Birch, the official historian of British Sigint, writes about Nave without naming him, and probably unaware that he was Australian, as follows: 'The most effective revival of naval Sigint during this period – 1919–1935 – may be best traced to the appointment to the flagship of the China station in 1924 [sic]⁶⁰ of an interpreter, assisted by a Chief Petty Officer telegraphist for the study of Japanese communications.'⁶¹

In a letter from Hong Kong dated 1 September 1925, Nave wrote: 'My job is to supervise the interception of Japanese W/T Messages, correct intercepted messages, and issue "Hints to operators", thus training them to the work. Also collection of call signs, wave lengths, etc.'⁶²

An exceptionally talented linguist and cryptanalyst, Nave had originally been 'loaned' to the Royal Navy and posted to Japan for formal language training in 1921. He was among the foremost of those gifted individuals who influenced the history of Australian Sigint. Born in Adelaide in 1899 and employed first as a railway clerk in South Australia, he had joined the Navy in March 1917 and continued to serve after the war's end. He would become a world-famous codebreaker, who, in the opinion of Denniston, the head of GC&CS, 'was possibly the best man we have got on Japanese naval cyphers'.[63]

Nave was, as his biographer described him, a 'man of intelligence'. After seizing the opportunity to study in Japan, his talent was noticed by the British naval staff. Mindful of the shortage of expertise in this area and the growing need, Nave was seconded to the Royal Navy's China Fleet as an interpreter. He went on to join GC&CS in 1928, transferring to the Royal Navy, at Admiralty's request, two years later. He spent much of the next decade in Hong Kong attempting to decipher the Japanese naval code, which he had succeeded in doing for the first time in the mid-1920s. With numerous code upgrades along the way, this would be his focus for the next 20 years.[64]

Early in his career, as a result of decoding one long telegram, Nave was able to reveal why Japan was satisfied with the Washington Naval Treaty ratio of 3 against 5 for the United States, but insisted on maintaining that balance. Their plan was that when the US fleet set off across the Pacific on any future warlike operations, the ships would be harassed by Japanese submarines en route, suffering some losses. On reaching the western Pacific, they would be short of fuel and water and other supplies, their crews voyage-weary and thus the Japanese could be satisfied with their ratio of 3/5. Not surprisingly, this message created a sensation.[65]

Along with a telegraphist as his assistant, Nave set about intercepting Japanese naval wireless telegraphy. This led to more active Sigint collection from the China and Australian stations with material sent to Nave for deciphering.[66] The commander-in-chief of the China Station received instructions to establish a 'Y' station at Hong Kong (with 'Y' procedure being the early cover name for Sigint operations). The station was tasked with collecting wireless radio communications and conducting traffic analysis and low-level decryption of IJN wireless traffic for local analysis, while forwarding encrypted messages to GC&CS. On 12 June 1925, Nave arrived for specialist interpreter duties and W/T work. His RAN supervising officer described him as having been 'employed entirely on Procedure Y and has carried out the work with exceptional keenness and ability'.[67] Nave's work was also being noticed by the lords of the Admiralty, who were 'extremely gratified' by the progress made on the China Station. They 'accordingly approved a favourable notation being made in the records of the Officers of the Royal Navy concerned and ... a similar notation should be made in the records of Paymaster Lieut. T.E. Nave R.A.N. who has rendered valuable services in this regard'.[68]

In another of his RAN personal record reports, Nave was described by his RN commanding officer as an exceptionally able Japanese interpreter. His personal records consistently noting his zeal and energy, as well as tact and 'nice manners'.[69] In 1934, the chief of the British Secret Service, Admiral Sinclair, was heard to say: 'we have agents in all the important cities and ports in the world and yet 90% of the reliable information we get comes from Nave here and a few people like him.'[70]

The Navy capitalises on new technology

Meanwhile, at the Australia Station in 1924, the Navy acquired two Creed Relay high-speed wireless recorders to record messages transmitted in Japanese telegraphic Morse code that was transmitted at a rate considered too fast for manual transcribing.[71] These messages were duly put to use by a select number of naval telegraphists and linguists.

The Creed Relay recorders relied on being close to the target to accurately receive the wireless radio signals emanating from Japanese transmitters. The importance of being close to the target diminished following two developments. The first development, the discovery of the effect of radio wave propagation around the globe, meant that, with more sensitive radio receiving equipment, the signals that bounced back from the ionosphere carried over vast distances, enabling messages transmitted from places like Berlin and Tokyo to return to Earth on Australian soil. The second development was the emergence of the Dictaphone – an American high-speed recorder invented to record music for broadcast, but which could also record messages passed by radio waves. As we've seen, early versions had been used to eavesdrop on German communications in the trenches during the First World War. The Dictaphone comprised a Bell microphone connected to a valve that supplied the amplified signal to a Bell electromagnetic cutting head. So great was the surge in commercial and government interest that Dictaphone opened an office in Australia in 1926.[72]

The introduction of the Dictaphone paid dividends, enabling greater scrutiny of intercepted signals. The messages could be recorded over a far broader range of the radio frequency spectrum (from 200 Hertz to 6000 Hertz – which was 3600 Hertz higher than pre-existing mechanical methods). In addition, Japanese naval traffic could be replayed repeatedly and sent back for further analysis by cryptologists and linguists. This level of control meant that fewer and less skilled intercept operators were required. In addition, the fact that traffic from Japan could be intercepted in Melbourne was a discovery which would have significant ramifications, particularly once war returned. As Japanese messages could be intercepted and exploited from afar, and not just from ships in nearby waters, 'Y' procedure became an even more attractive intelligence technique than it previously had been.[73]

Reports of Sigint successes by British ships equipped with Sigint detachments in the Mediterranean in 1926 convinced the Australian Navy's Director of Signals and Communications, Commander Cresswell that 'it was highly important that [a] similar procedure should be carried out by H.M.A. ships as opportunity occurs'.[74] Efforts to collect information on the Japanese network of stations and supported force elements helped piece by piece to build a picture of Japanese force dispositions and practices in the Japanese Mandated Territories (Caroline, Marshall and Mariana islands) to the north of the Australian Mandated Territory of New Guinea. The New Guinea administrator's steam yacht, SY *Franklin*, was one example. Selected in 1927 as a vessel that would not arouse undue suspicions operating further north, the *Franklin* was fitted with eavesdropping equipment and set course for Japan's Mandated Territories. Despite some technical hitches along the way, the operation was a success, identifying the IJN's submarine net as it stood and, with the help of a Dictaphone, recording a hundred transmissions which were then analysed, exposing a range of procedures, administrative arrangements and call signs, although little sign of sophisticated codes. Cresswell reported that the results achieved by this special W/T party were 'most satisfactory', containing 'valuable information of the operating of Japanese W/T stations in the Pacific'.[75]

Encouraged by the fruits of the SY *Franklin*'s labours, in 1928 the newly commissioned HMAS *Australia* and HMAS *Canberra* undertook special interception collection operations on their Pacific transit to Australia. Nave was among those involved, and the Second Naval Member recommended that he collect information and report on the latter's arrival in England and be responsible for sending correspondence on the subject to the DNI.[76] In his book *Australia's First Spies*, former intelligence officer John Fahey observes that

> Despite the limitations imposed by falling budgets and the lack of linguists and other specialists, enough intelligence was being collected to deduce the IJN's order of battle from [traffic analysis] and the creation of the first net diagrams. In addition, Eric Nave broke the IJN's nine-letter general purpose '43' code, and although the exploitation of this breakthrough was limited by the lack of linguists, it represented a real step forward in the attack on the IJN.[77]

Special interception operations were again launched in 1932 when the IJN Training Squadron, including HIJMS *Asama* and HIJMS *Iwate*, toured Australia in April and May 1932. In a daring, larrikinish venture, possibly inspired by Australian authorities' duping of the master of the German ship the *Hobart* in 1914, the Australian liaison officer aboard *Asama*, Paymaster Lieutenant W.E. McLaughlin, seized an opportunity to copy the frequencies and schedule used by the squadron and provided them to the RAN escort ships for use in monitoring their communications. Considerable technical information was gleaned, including what appeared to be Japanese efforts to map the radio propagation around Australia and the southern oceans – information that Japanese warships and submarines would find useful for subsequent wartime ventures around Australia.[78]

Sigint operations continued from onboard Australian warships into the early 1930s, enabling the capability to mature gradually. When they were not busy monitoring their own communications channels, HMAS *Albatross, Australia* and *Anzac* conducted cruises around the Japanese Mandated Territories and ships' telegraphists eavesdropped on Japanese communications – yet another example of getting the job done with no additional resources.[79] The commanding officer of *Albatross*, Captain H.J. Freakes, submitted a report concerning 12 stations monitored on the voyage, identifying Japanese wireless stations by call signs, location frequencies and associated radio stations.[80] Dedicated four-person teams were assigned for Sigint collection with collected Japanese naval traffic forwarded to the Royal Navy's China Station. But the budgetary constraints of the Great Depression saw these cruises cut back significantly, with HMAS *Albatross* decommissioned in April 1933.[81]

In January 1934, with news of Japanese assertiveness in China and beyond circulating, Britain made clear to the select audience at the Admiralty's Singapore Naval Conference (which included Australian and New Zealand representatives) that the IJN was now regarded as the principal threat in Asia. Recognising the importance of high-frequency direction finding (HFDF), the conference recommended the construction of HFDF stations at British sites in the Far East, as well as in Australia; notably Canberra and Darwin. Australia's reluctance to commit financially to this construction frustrated Britain's Committee of Imperial Defence (CID) and Chiefs of Staff Committee, and meant that the Canberra and Darwin stations would not be built in time for the onset of the war with Japan, with serious consequences: in early 1942, Rabaul would be captured and Darwin would be bombed without the benefit of warning from HFDF stations.[82]

Far East Combined Bureau and Admiral James's memorandum

Another outcome of the 1934 Singapore Naval Conference was the proposal for the establishment of a Naval Intelligence Centre Far East –

a watershed moment for Australian Sigint, When the head of Britain's Secret Intelligence Service (SIS), Admiral Sir Hugh Sinclair, heard of this, he called a meeting with his GC&CS, Admiralty, War Office and Air Ministry counterparts and they agreed to the creation of a more inclusive organisation: the Far East Combined Bureau (FECB).[83] Its purpose was to collect and collate all-source information for the armed services on their Japanese naval, military and air force counterparts, primarily providing warning of hostilities.[84] A building inside the Hong Kong Naval Dockyard on Stonecutters Island, HMS *Tamar*, was selected as the FECB base and the bureau began operating from there in April 1935 with one full-time cryptanalyst and ten W/T sailors plus five RAF airmen. A RAN officer regularly served as deputy of the FECB and Australians from the other two services served there on attachment. The British Army, however, declined to contribute, arguing they would supply people if an expeditionary force were to be sent.[85]

The other game-changing event at this time was a memorandum from Admiral Sir William Milbourne James to Britain's DNI, the Deputy Chief of Naval Staff, on the subject of operational intelligence. The memorandum, issued on 4 December 1936, galvanised British, and by extension Australian, naval intelligence to prepare for war.[86] Admiral James who had been in charge of Room 40 in the latter part of the First World War, apparently coined the term 'operational intelligence' by which he meant Sigint without decryption; that is, traffic analysis. As we've already seen, TA was the area where Australians would be recognised as having the lead. In Room 40, James wrote, traffic analysis had been able 'to keep up a flow of intelligence when there was no cryptography, which was very little different from that which was being issued when there was 100% of cryptography'.[87] Eventually the term 'operational intelligence' was used to designate 'current' or 'tactical' intelligence.[88]

By 1936, James considered that the successes of the First World War would be difficult to replicate. He had been involved in several of the Sigint successes in Room 40, but feared that wider knowledge of Sigint's use since that war and technological advances in the interim, along with the lack of resources committed to Sigint work, meant that there was an added degree of difficulty. He warned that 'we might not be able to rely on intelligence from broken enemy cyphers'.[89] On James's initiative, a nucleus Operational Intelligence Centre was formed in the NID in 1937.[90]

In the end it would transpire that, in fact, information about the existence of the FECB and the proceedings of the Singapore Naval Conference was being passed to the Japanese through well-placed sources – a shorthand typist from the civil government in Singapore and a cartographer from the Singapore Public Works Department were on the Japanese payroll. Their revelations ensured the Japanese had a detailed understanding of the weakness in defence plans and the hollowness of Singapore as a fortified naval base.[91]

Still, the work of the FECB enabled GC&CS to be reasonably fluent in their reading of all main Japanese naval ciphers and to know quite a lot about Japanese army ciphers as used in China. For instance, British accounts indicate that FECB staff in 1939 broke the main Japanese fleet cipher (known as JN-25), which had only been introduced the previous year.[92]

Pacific Sigint, FECB and the Coastwatchers

Another example of Australia operating highly effectively on a shoestring was the establishment of a clandestine Sigint site on the island of Nauru. The site was manned by former Petty Officer Harold Barnes, formerly Leading Telegraphist on HMAS *Albatross* and one of the RAN's most experienced Sigint operators. The clandestine Nauru site is also an example of the willingness of Australians to commit people in order to get the job done when funds were unavailable. Frustrated by lack of promotion, Barnes had left the RAN and was employed as manager of the British Phosphate Commission's telegraphy station on Nauru. There, he produced intercepts for the Navy Office in Melbourne and for the FECB and GC&CS until 1940. The appointment raised some concern among the British about the adequacy of Australian security. In early 1936, Captain Waller of HMS *Tamar* wrote to the DNI in Melbourne about Barnes and his work, flagging the possibility of cooperation, but noting that the DNI, London, was 'concerned with the degree of secrecy which obtains both with regard to the activities with this operator in Nauru and to his communications with Australia'.[93] It is noteworthy that the British were apparently more concerned than the Australians about the security risk of such an experienced Sigint operator working alone in such an exposed location.

Barnes's work at Nauru stimulated plans by the ACNB for the construction of land-based DF stations on Australia's west, east and north coasts. Sites were selected southeast of Fremantle (Jandakot), Sydney (although eventually HMAS *Harman* near Canberra was selected instead) and Darwin (HMAS *Coonawarra*). These were intended to work in conjunction with the Royal Navy's China Station organisation and the FECB. The Board envisaged the Darwin and Singapore stations could monitor waters north of Australia to an estimated distance of nearly 13 000 kilometres, with Darwin and Sydney forming a baseline for the northeastern arm, and Darwin and Fremantle a baseline for the northwest arm. In August 1938 arrangements were also made for the heavy cruisers HMAS *Canberra* and *Australia* to be fitted with DF gear – equipment that would later help them locate and intercept the transmissions of enemy ships, notably commerce raiders.[94]

Imperial cooperation developed as the FECB's capabilities matured. Additional interception and DF stations were established in Malaya, Singapore, North Borneo and on the Cocos Islands. Meanwhile, the Canadians were operating an intercept site at Esquimalt on Canada's west coast. In 1940 they added a DF station, which complemented the work of the stations on the other side of the Pacific Ocean.[95] Similarly, New Zealand expanded its naval Sigint capacity, establishing high-frequency DF stations at Awarua (near the southern tip of the South Island); Musick Point (on the east side of Auckland); Waipapakauri (at the northern end of the North Island); and Suva, Fiji. Other wireless radio intercept stations were added at Blenheim Awarua, Waiouru (south of Taupo, central North Island), and Nairnville (north of Wellington). These stations maintained direct communications with each other and could obtain simultaneous bearings; the work being directed and coordinated by the FECB in Hong Kong, and shared between the Navy Office in Wellington, the Australian Navy headquarters in Melbourne and the Royal Navy in Singapore.[96]

While Japan made intelligence inroads, the FECB started making cryptanalytic advances as well. By 1938, FECB cryptanalysts were working on six Japanese naval codes. Their task was made easier by poor Japanese wireless discipline and repeated errors which enabled significant cribs to be identified (a crib being assumed underlying plaintext to help an analyst understand the method of encipherment) as well as predictable patterns due to the Japanese preference for formality in communications. The introduction of new Japanese naval codes in 1937 and 1939 was undermined by sloppy procedure and unnecessary duplication, combined with occasional mixing of messages transmitted in code and in clear. Nonetheless, Japanese code changes still meant that a substantial part of their communications remained out of reach until a breakthrough came in September 1939.[97]

Throughout this period, the organisation that had been established at the start of the 1920s, the Coastwatchers, had grown considerably. By 1939, there were more than 700 Coastwatchers around the Australian coast and in the islands to the north and northeast. Coastwatchers in the field were issued with wireless radios fitted with specially cut crystals enabling them to operate on a particular frequency in the 6 MHz band known as 'X' frequency.[98]

A notable Coastwatcher was Ruby Boye-Jones, who lived on the island of Vanikoro. Before 1939, this island at the southernmost tip of the Solomon Islands had been best known as the place where the French explorer Jean-François de Galaup de La Pérouse was thought to have been shipwrecked. Boye-Jones's husband Frank managed a Kauri timber concession there, and the base had a powerful radio connecting Vanikoro with Melbourne and surrounding islands. The service was operated for meteorological as well as commercial reasons. On declaration of war, the island's telegraphist left to enlist and Boye-Jones took over. At first, she used voice radio to give weather forecasts and taught herself Morse code to be able to transmit information in bad weather. When the war with Japan moved into the South Pacific, she would be appointed a Coastwatcher. As well as sending meteorological information in difficult weather conditions, Boye-Jones would transmit coded intelligence messages from outlying islands. Her Australian connection would be with the DNI in Melbourne, Commander Rupert 'Cocky' Long.[99]

Ruby Boye-Jones braved death threats from the Japanese and was never paid for her work. She was afforded some official recognition, however; with the naming of an accommodation block of the Australian Defence Force Academy in Canberra in her honour,[100] the awarding of the British empire Medal and the 1939–1945 Pacific Star and War Medal, and appointment as a life member of the WRANS Association.[101]

The Coastwatcher system would prove to be one of the most effective and reliable Humint systems operating on any side between 1939 and 1945, functioning across the Solomon and New Guinea islands.[102] The Coastwatcher network would also be one of the most highly decorated military units in Australian history. It was comprised of 'chancers' living in remote places,[103] although missionaries and cattle station owners, postmasters, police and pilots were also among them. The US Navy commander in the Pacific, Admiral Chester W. Nimitz, later described the work of the Coastwatchers as 'of inestimable value'. Similarly, Admiral William Halsey would later declare that Coastwatcher intelligence saved Guadalcanal which, in turn, 'had saved the Pacific'.[104] As we have seen, it also proved of inestimable value in providing plausible cover for the extraordinary work of the Siginters involved in collecting, direction-finding, decrypting, translating, analysing and reporting on enemy transmissions thought to be secure and indecipherable. While the Coastwatcher organisation matured in the late 1930s and early 1940s, naval Sigint developments occurred more gradually, commencing shortly after the end of the First World War.

Australian interagency Sigint on war's eve

In the early 1930s, the Great Depression forced the Australian government to slash defence spending, but by the middle of the decade a growing sense of foreboding led to a re-examination of priorities. In May 1936, the British Admiralty agreed to an Australian Navy Office proposal for 'the resumption and expansion of W/T Procedure Y'; that is, wireless radio signals interception. Additional funding became available in 1937, and by

May 1938, useful interception was being carried out at Victoria Barracks, Melbourne, on board HMAS *Australia* and at the Nauru Radio Station. The material was also sent to the FECB in Hong Kong for analysis.[105]

In July 1937, a meeting in London was called by the British Director of Military Operations and Intelligence. Representatives of each dominion attended and were informed that the United Kingdom had wireless interception in place and that members of the British Commonwealth should consider 'how best they can gain experience of W/T interception in peace and prepare ... to meet war conditions'. In the case of Australia, it appears that plans were at last in place for the construction of radio intercept stations, as Jellicoe had recommended years earlier.[106]

On the outbreak of war in 1939, however, the Australian Navy found itself underprepared. Desperately short of telegraphist trainees, qualified telegraphists and 'Y' procedure specialists, it also lacked shipboard equipment. Even the high-powered wireless station, HMAS *Harman* in Canberra, was not ready, as the Department of Works needed a further £25,256 to complete it. At Darwin and Rabaul, there was nothing. In 1939, Australia was technically in a worse position than it had been in in 1914.[107]

With little funding and limited connections or influence over decisions about preliminary intelligence arrangements made in London and Washington in the lead-up to war, Sigint in Australia lagged behind. In particular, Australian Sigint initiatives did not benefit from machine assistance to their cryptanalytic effort between the wars. Indeed, it would only be through collaboration with the Americans, from 1942 onwards, that Australian intelligence became familiar with the use of such machines in cryptanalysis. Despite these limitations, however, Australians had made some progress, sometimes ingeniously, in building its Sigint capacity for what lay ahead.

5
SIGINT IN THE SECOND WORLD WAR, 1939-41

At the outbreak of the Second World War, in September 1939, the Second Australian Imperial Force (2nd AIF) was raised, along similar lines to the one raised by that generation's parents in 1914. This time, aircraft would feature more, as would other technology, notably more advanced tanks and more sophisticated Sigint.

At that time, Japan was at war with China, and France's fall to Germany in May 1940 led Germany's Axis partner, Japan, to encroach into French Indochina. Only when Japan struck the Western powers in the Pacific did the full scale of the war become apparent. For the United States, this included Hawaii, the Aleutian Islands and the Philippines. For the United Kingdom, it encompassed Malaya, Singapore and Burma, as well as its self-governing dominions of Australia and New Zealand and a range of smaller colonies in the Pacific. For the Dutch, it meant the Netherlands East Indies.

Previously, Australia, like fellow self-governing dominions Canada, New Zealand and South Africa, had dutifully contributed to the empire's defence a generation earlier. In Australia's case, that included sending five infantry divisions to Europe and contributing forces in multiple contingents in Egypt, Palestine and Mesopotamia (Iraq). The higher order logistics and intelligence, however, was left to Britain to organise.

Again in the Second World War, Australian forces contributed to the empire's defence in Europe, the Mediterranean and Middle East, although this time with more advanced and autonomous capabilities. But it would be in the Pacific where an existential threat and the imperative to collaborate with the United States Armed Forces transformed Australia's capabilities – including local defence industry, broad logistic support capabilities and more robust national intelligence arrangements.

For decades after the war's end, the use of Sigint during the Second World War was spoken about in hushed tones by the cognoscenti. It was only from the late 1960s and early 1970s, notably with the release in 1974 of Frederick W. Winterbotham's book *The Ultra Secret*,[1] that historians started to discuss openly the role of Sigint in the war.

Since then, a range of works have explained various facets of American and British wartime Sigint.[2] Australia's side has been captured by some former practitioners, who wrote of their personal experiences in their twilight years,[3] as well as some more recent works by historians not involved in the war, but with access to some of the subsequently-declassified Sigint material.[4] As noted earlier, in the late 1970s, the author of Britain's *Official History of British Intelligence in the Second World War*, Professor Harry Hinsley, made clear that he had not attempted to cover the war in the Far East 'when this was so much the concern of the United States'.[5] While US historians covered aspects of it from their own perspective, Australia's official war historians, led by Gavin Long, avoided the topic of Sigint. As a result, Australia's story was not told.

Drawing on some previously unexplored and eye-opening archival material, as well as the views of other scholars, this chapter, and the two that follow, explores Australia's wartime experience of Sigint, reflecting on UK and US ties that helped shape the national Sigint capability. Together, they set out how this experience laid the groundwork for the national Sigint agency established after the Second World War – an organisation which would eventually be known as ASD, supported by the service Sigint components. The antecedents described in these chapters are critical to understanding how and why Australia's postwar Sigint architecture emerged as it did.

Early wartime Sigint and national intelligence arrangements

On 3 September 1939, Britain declared war on Germany. Just one hour later, Australian Prime Minister Robert Menzies declared on radio that it was his 'melancholy duty' to inform the Australian people officially that

Britain had declared war, 'and that as a result, Australia is also at war'.[6] At that time, Australia had not yet acceded to the Statute of Westminster; Australian federal legislation appropriating this statute would not be signed into law until the end of September 1942 (Canada had done so a decade earlier). As a result, Australia was legally obliged to follow Britain, and Menzies's turn of phrase reflected the familial closeness of Australia's relations with Britain, as well as his personal inclination. In military affairs, while the Australian armed forces were distinct entities, they too were closely connected with their British counterparts, with Sigint no exception.

Preparations for war, however, were just beginning. Recognising the looming war clouds, the DNI, Commander Rupert 'Cocky' Long, had arranged for representatives of the Air Force and Military Intelligence to consider what actions they should take. Following a series of meetings, by early June 1939 they had persuaded the government to consider establishing a defence security organisation. But with competing priorities, the idea gained little traction.[7]

With the outbreak of war in Europe, and with Japan's aggressive actions in China making the British position in Hong Kong increasingly precarious, the FECB was relocated from Hong Kong to Singapore.[8] Australia had played a supporting role with the FECB in the preceding years, and had a firsthand appreciation of the value of its work.

At this stage, despite the FECB move, Japan featured little in Australia's calculations over the strategic equation. Rather than remaining focused on the threat from Japan, Australia's armed services set about raising forces to support Britain in its war against Germany and Italy. Unlike in 1914, when there were German forces in nearby New Guinea to confront, in 1939, the problems of Europe were real, but seemed far away and less pressing. The five divisions of the militia were legally reserved for the defence of Australia, so to deploy forces abroad in support of the British empire, a second expeditionary force had to be raised. It was 28 September, less than four weeks after Menzies's declaration of war, that the first division for the 2nd AIF was raised.

Initially under the command of Lieutenant General Thomas Blamey, the 6th Division (as well as the 7th and 9th Divisions raised subsequently) would be deployed not to Southeast Asia, but to the Middle East. There, they saw action fighting the Italian and German forces in North Africa, Palestine, Greece and Crete as well as against the Vichy French forces in Lebanon and Syria. Raised in July 1940, the 8th Division would deploy in 1941. Two brigades would be sent in defence of the Singapore Strategy, while the three battalions of the 3rd Brigade went to Ambon, Timor and New Britain, to Australia's north. They would suffer through years of captivity at the hands of the Japanese from February 1942 to the war's end.

Aware that intelligence would be important to the country's defence, the Menzies government was nevertheless unwilling to spend money on it. And at first, national coordination was not straightforward.

National Sigint coordination: first steps

In November 1939, Commander Long wrote to the Chief of Naval Staff, Sir Ragnar Colvin, recommending 'the Authorities At Home' be asked whether, in their opinion, a cryptographic organisation should be set up in Australia and if so, they should be asked to provide the 'essential specialists and material'.[9]

Colvin was lukewarm about the idea, recalling that no cryptographic organisation existed in Australia during the previous war, although 'a table used with a German code (which had been obtained from a German merchant vessel) was broken down'. He questioned the point of such an entity, as 'any local organisation would appear to be a duplication of the U.K. effort in the Far East'. He sought the views of the other two services, cautioning that the cryptographic organisation enterprise 'should not be lightly entered into' and that nothing should be done without the advice and assistance of the GC&CS in London.[10]

Air Vice-Marshal Stanley Goble, Chief of the Air Staff, was against the proposal. He considered that 'we would not be justified in setting up a Cryptographic Organisation here at this stage', although he still wanted the Admiralty's advice.[11] Interestingly, the Chief of the Air Staff's opposition to an Australian cryptographic organisation was reminiscent of the RAF's position regarding the establishment of GC&CS.[12] In view of the closeness of the relationship, it seems that the Australian approach was inherited from them. British Sigint's official historian Frank Birch explains that for the RAF, there was no air traffic to intercept and exploit until the reconstruction of the German Air Force. RAF needs were met by the Sigint work of the Navy in the Far East and of the Army in the Middle East, while in the Air Ministry, 'the higher authorities', having no experience of Sigint in the First World War, were, according to the Director of GC&CS, 'very sceptical about the value of wireless interception and the intelligence obtained therefrom'.[13] Birch writes that in 1940, the RAF's Sigint service still lagged behind the others, and 'still had not shouldered all of its responsibilities'.[14]

Unlike Goble, the Army's Chief of General Staff, Lieutenant General Ernest K. Squires, was firmly in favour of an Australian cryptographic organisation. Squires responded that 'we should have at least a nucleus organisation in Australia ... The work is clearly of a highly skilled nature and the sooner a commencement is made the better.'[15] Again, the approach of the Australian Army reflected that of its British counterpart. As Birch notes, in the period leading up to the Second World War, 'if the RAF was the least Sigint-conscious of the Services ... the Army was probably the most'. He attributes this to the continuity preserved in the Middle East, where No.2 Wireless Company practised traffic analysis, cryptanalysis and interception, collecting raw material for military studies at home.[16]

On the recommendation of the Defence Committee, Menzies wrote a less than enthusiastic letter to the Secretary of State for Dominion Affairs in London, suggesting the possibility of establishing a nucleus organisation be examined.[17] He would be obliged to wait six months for a reply. Commander Denniston, head of GC&CS, eventually suggested that Australia send students to 'a training course for cryptographers' in England, and that they 'assist the Home authorities in the interception of fixed commercial stations in various parts of the world'.[18] But Squires, in particular, was determined Australia should start a cryptographic bureau of its own.

When the militia 3rd Division was mobilised, a divisional signals unit had been raised at Squires' instigation, with a limited signals intercept and analysis capability. Then, in January 1940, a small cipher-breaking group was established at Victoria Barracks, Sydney. Initially this comprised Professor Thomas Gerald Room and Richard J. Lyons of the Mathematics Department of Sydney University. Their introductory textbook was Herbert O. Yardley's *The American Black Chamber*.[19] Later they were joined by classicists from the Greek Department, Professor Arthur Dale Trendall and Athanasius 'Ath' Pryor Treweek. They were referred to as 'men of the professor type'.

In light of what would follow, it is worth considering this cipher-breaking group more closely. One member, Treweek, was a polymath; highly gifted in mathematics and linguistics and with a photographic memory. He had a refined ability to connect the dots, and devoured crossword puzzles with ease. In the 1930s, Treweek foresaw the value of learning Japanese and he applied himself to this in the lead-up to war. A lecturer at Sydney University in ancient Greek, he was also a militia officer in the artillery, eventually rising to the rank of lieutenant colonel. When war came, he was a captain slated to be sent to Singapore, but was moved instead to Melbourne 'for administrative work at Victoria Barracks' – a cover for his real work on breaking the codes. Treweek was known for his sense of humour, developing little puzzles including single letter substitutions to hide observations about unpopular officials. But he was also known as a 'gentle genius'.[20] After the war, he would go on to a Nuffield Fellowship in London, pursuing a PhD on the Greek mathematician Pappus.[21]

During the war, this unusual team – also referred to as the Sydney University Group – succeeded in breaking the low-grade Japanese LA-code used for consular messages.[22] The general staff officer for intelligence gathered this 'nucleus organization' to teach each other skills using Japanese consular cable traffic obtained from the Cable and Wireless Company. There was growing recognition, though, that this initiative was not nearly enough.

Nave and the nucleus of D Special Section

At this point, Eric Nave, by now promoted to the rank of Commander in the Royal Navy, re-enters the story. In many ways, Nave personifies the closeness of the Australia–Britain relationship, with the advantages and pitfalls this could entail. A comment by Nigel de Grey, author of the internal unpublished GC&CS history, conveys the extent of this closeness: 'The author has, with great reluctance, adopted throughout this chapter the convention of distinguishing those members of the British race who came under the Home Government as British, from those members of the British race who came under the Australian government, who are dubbed Australians.'[23] In other words, in the eyes of the British, Australians in general were still, most definitely, 'one of us'.

Without mentioning Nave's name, de Grey writes about him:

In February 1940 a naval officer of the F.E.C.B., in fact possibly the most experienced of their cryptanalysts, succumbed to a tropical disease and proceeded on sick leave to Australia ... it was definitely settled that this officer could not return to the tropics and from then onwards he was lent to the R.A.N. and became the leading spirit in their Cryptanalytical Section.[24]

Having taken up an appointment on the FECB commander-in-chief's staff in October 1937, Nave returned to Melbourne in May 1940 due to ill health.[25] He was then 'lent' back to the RAN, where he took the lead in cryptanalysis. By this time, Nave's standing and prowess as a codebreaker were legendary. He established the so-called 'Naval Group', which initially consisted of himself, another paymaster, and a small clerical staff.

In September 1940, this group would become known as the Special Intelligence Bureau. Given the limited resources, Nave was restricted to studying the codes used in the mandated islands traffic being intercepted by Australian stations and the commercial shipping traffic from Japanese merchant ships. Before Italy joined the war in June 1940, the traffic from Italian merchant ships had proved that commercial traffic could be an important intelligence source. A cryptology element was thus added to a small RAN signals interception operation already set up by commanders Long and Jack Newman, the latter the head of RAN Communications.[26]

Subsequently, the so-called 'Sydney University Group', the army's group of four academics working from Victoria Barracks in Sydney, moved to Melbourne to join what was now known as Nave's Special Intelligence Bureau or Section (the term was used interchangeably). They would form

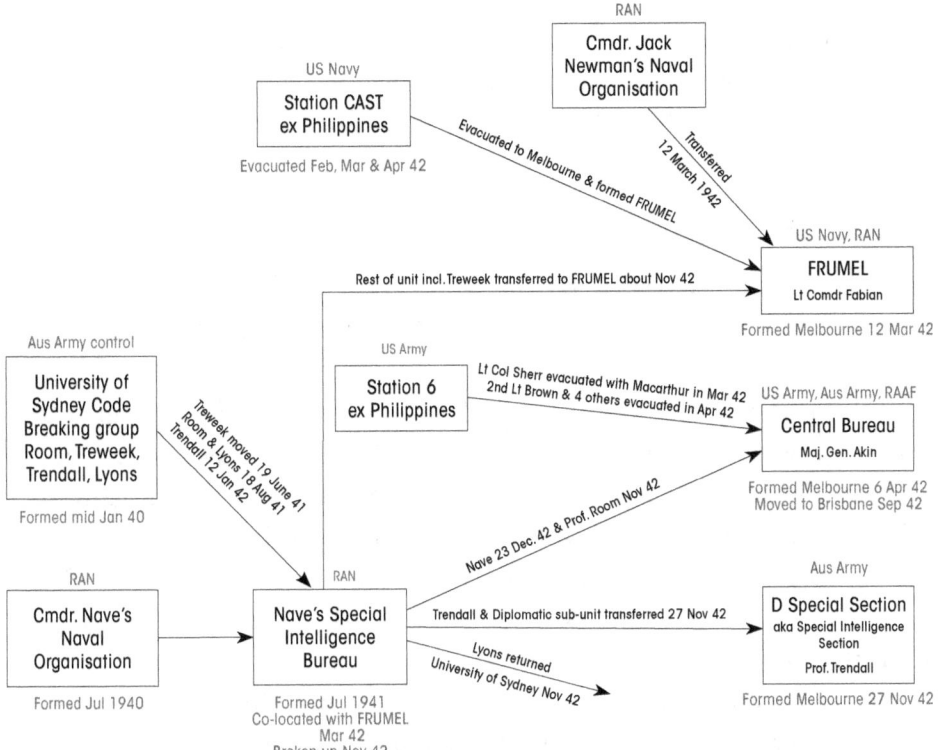

Notes
1 Special Intelligence Bureau was also known as Cryptographic Section, Melbourne Communications Intelligence Unit.
2 F/Lt Roy Booth RAAF was in Combined Operations Analysis Centre in Melbourne before joining Central Bureau.
3 Maj. Spencer B. Akin and Lt. Col. Sherr were members of the 'Bataan Gang' – both evacuated by patrol torpedo boat with Macarthur. Sherr was the interim initial Director of Central Bureau.

Figure 1. Evolution of code-breaking organisations
© Peter Dunn OAM www.ozatwar.com (used with permission)

the nucleus of 'D Special Section', (discussed below) which focused on breaking Japanese diplomatic codes for the rest of the war.

In October 1940, the British responded to Menzies's letter, agreeing that it was inadvisable to establish a large-scale operation that would duplicate the work of GC&CS, but going on to suggest the expansion of the small section under Commander Nave in Melbourne that was working with the FECB in Singapore. The British would assist with training cryptographers and welcomed Australia's assistance in the interception of Japanese fixed commercial stations.[27] The Defence Committee also asked the Chief of Naval Staff to take up the matter of training with his other service colleagues.

Incidents at sea, the value of Sigint and need for improvements

With focus on Japan increasing, deployments to the Mediterranean also saw a heightened focus on supporting naval operations against enemies based there. At sea, the RAN had some Sigint successes and partial successes in operations against Italy, Vichy France and Germany. The latter included the pursuit of the Italian MV *Romolo,* shortly after it left the port of Brisbane on 5 June 1940 and in anticipation of Italy declaring war, which it did five days later. Pursuit was aided by the monitoring of *Romolo*'s radio transmissions and by DF tracking. As it transpired, inexperienced DF personnel gave a succession of misleading bearings, and coordinated plotting of them seems to have been faulty; a mistake which would have been detected and corrected had there been more experienced staff at hand. In the end, the ship was tracked and, on 12 June, scuttled by the Italian crew north of Solomon Islands despite calls for surrender from the Australian ship in pursuit, HMAS *Manoora*.[28]

Although they were limited in how much they could read from available intercepts, the German Navy's code-breaking service, B-Dienst, occasionally supplied the Italians with useful insights. As Mike Carlton notes in *The Scrap Iron Flotilla,* B-Dienst was able to monitor spikes in traffic volumes and extract operationally useful information – including the indication that a flotilla was about to put to sea. The Italian response set the scene for the Battle of Calabria on 9 July 1940. The B-Dienst kept up supplies of fragmentary updates, but not enough to change the outcome of encounters at sea.[29]

At this time, HMAS *Sydney,* commanded by Captain John Collins, RAN, was in the Mediterranean, as part of the British fleet. As it conducted a sweep around Crete and the coast of Greece, it encountered the Italian cruiser *Bartolomeo Colleoni* at Cape Spada. Collins identified the likely path of the Italian warship but, rather than notifying the fleet headquarters, he maintained radio silence. Collins attacked with the knowledge that Italian tactics precluded engagement unless they overwhelmingly outnumbered the enemy. In the ensuing exchange, the Italian warship was stopped and sunk. While active Sigint was not directly involved in this case, the awareness of preferred Italian naval tactics and the maintenance of radio silence proved decisive.[30]

In November the following year, the German armed merchant vessel, HSK *Kormoran*, claiming to be the Dutch ship, *Straat Malakka*, sank HMAS *Sydney*. *Kormoran* was also sunk in the exchange, but the baffling loss of the faster and superior Australian warship is in part explained by German deception and operational security. German raiders operated with a version of the Enigma machine, the code of which had not yet been broken and the short transmissions of which had not been detected. Despite being aware of the risk of such raiders, HMAS *Sydney* disappeared, with the loss of the entire crew.[31]

These three incidents at sea demonstrated the need for, among other things, heightened vigilance, greater operational security and improved Sigint capabilities. They added to the growing awareness in intelligence and command circles in Australia of the need for more active local involvement in the development of Australia's own Sigint capabilities.

By this time, a Central War Room had been established at Victoria Barracks in Melbourne. This room was to be used to coordinate Australia's war effort, supported by a Combined Operations Intelligence Centre (COIC), established in August 1940. Initially, it was staffed by one officer from each of the three armed services, who met daily from mid-October. By January 1941, following the German raids on ships at Nauru, the Navy increased its staffing, with Commander Long appointed COIC director.[32] A joint service organisation, it distributed coordinated intelligence to Australian commanders, performing work akin to that of its US Army counterpart: the military intelligence (G2) organisation. The COIC would be absorbed into General MacArthur's headquarters once it was established in Melbourne in April 1942.[33]

Land force operational experience in the Middle East

While its national Sigint architecture was evolving slowly, the story of Australian army Sigint in the Second World War was emerging; beginning, so it is said, in the foothills of Mount Olympus in March 1941.[34] In May 1940, a 'Special Wireless Section' or radio intercept unit had been raised for the 1st Australian Corps. It went on to serve on the ill-fated expedition to Greece in March 1941, followed by Crete and Syria, before being recalled to Australia in March 1942.[35] The Special Wireless Section was commanded by Captain Jack Ryan, also known as 'Uncle Jack', who had served as a young wireless operator on HMAS *Sydney* when the *Emden* was destroyed in 1914. It was Ryan who chose the banks of a stream under Mount Olympus as the site for the unit.[36] This unit, partnered with its British equivalent, was initiated into traffic analysis and the British also shared with them German Air Force low-grade codes.[37] The experience was dramatic and, once the Germans attacked, short-lived. But it would leave a lasting impression on several Australians who would go on to play major roles in the Pacific War, proving to be the most successful army officers who served in Central Bureau.[38]

Following the evacuation of Greece, the British and Australian intercept units moved to Crete, but by this time the Australians were well trained and able to report their own intercepts.[39] The highly capable and popular Lieutenant Alistair 'Mic' Sandford,[40] a barrister by training, was in charge of the Intelligence Section.[41] He later became an assistant director of Central Bureau,[42] and would be instrumental in securing Nave's services for the Bureau.[43] Sandford had also undertaken Sigint training in the United Kingdom before his arrival in Crete. A prodigious and highly effective networker, his contemporaries in Crete recall two special characteristics of the extroverted, eccentric, and, as we now know, gay, son of Sir Wallace Sandford,[44] a member of the Adelaide establishment:

> His gift for making the lowliest soldier feel that he mattered and his unique way of making a point. This he would do by letting slip some outrageous or grossly exaggerated remark which you would

immediately 'brush off'. However, the remark would stay around in your mind till it occurred to you that there might be some truth in it after all, and at the very least, it kept you thinking. That proved to be a clever way of getting his message over.[45]

There was more to Sandford than his undoubted wit. He was mentioned in despatches in recognition of gallant and distinguished service in the Middle East.[46]

The fall of Crete and the perils of ignoring Sigint

The Australians in Crete, some of whom play key roles later in this story, seem to have been a resilient group, and what they observed in Crete must have been educational. For those in Bletchley Park's Hut 6, tasked with decoding German signals, the fall of Crete was 'the greatest disappointment of the war'. The Hut 6 codebreakers were frequently reading the enemy's communications as quickly as their intended recipients. One message, received two weeks before the invasion of Crete, described in detail the German plans. To insiders, it seemed 'now, we really have got them this time!' The force commander on Crete, New Zealand Major General Bernard Freyberg, was warned that the crucial point of the invasion was to be the airborne attack on the Maleme airfield. He was given the time and every detail of the operation in advance, and considering the appalling difficulty and danger of any airborne invasion in the best of circumstances, the view at Hut 6 was that the attack would be ignominiously thrown back. Sandford was active in passing Sigint. The codebreakers awaited the operation with anxiety, but also with a considerable degree of confidence. In the event, it was a 'damned close-run thing'.[47]

Inexperienced in the use of such Sigint, inclined to focus on sea ports and wary of the unknown, Freyberg chose not to concentrate forces on Maleme. He instead hedged, placing forces in defence of the sea ports.

Ultra was the designation given to the compartmented and closely guarded secret, or ultra-top-secret, reports which pertained to information derived from messages intercepted and decrypted from German Enigma and Lorenz encryption devices. In essence, the Nazis relied on these systems through to the end of the war. Knowledge of what this entailed and the reports emanating from them was restricted to the highest levels of command to the point where operational commanders were discouraged from using it unless they had compelling and plausible alternative cover for their actions. It appears protecting the Ultra secrets may have mattered more than holding Crete.

With these concerns in mind, British General Wavell ordered Freyberg not to mention Ultra to anyone else on Crete and forbade him from making any changes to Allied force deployments on the basis of Ultra information without Wavell's prior approval.[48] While Wavell's orders seem reasonable, it may be that Freyberg felt unduly constrained by them, but this in itself does not justify his failure to use the Ultra information. As the Sigint had forecast, the Germans took Crete from the air, and the Allies lost a great deal of shipping in trying to save the force, extracting the withdrawing troops under frequent aerial attack. From the Allied point of view, the best that could be said about it was that, though their conquest of Crete was ultimately achieved, it was so enormously expensive that the Germans never attempted an airborne invasion on that scale again.[49]

Rommel's uncanny intuition

Australian soldiers had deployed to North Africa before the Greece and Crete campaigns and would be there for some time thereafter as well. Like others, they could not fail to be intrigued by the seemingly supernatural foreknowledge of enemy intentions of Commander of the German Afrika Korps, General Erwin Rommel. German commanders such as Rommel, like their Allied counterparts, often received the credit for successes that were in large part due to timely Sigint reporting. Despite access to Ultra intercepts of German communications, the Allied campaign faltered for a variety of reasons including mistakes by the British command, as well as high-quality German equipment, tactical skill and, not least, access to high-grade intelligence on British plans.[50] Britain's official historian noted

British 'failure to allow for the efficiency of [Rommel's] field intelligence'. Reading British hand cipher field communications 'provided them with at least as much intelligence about Eighth Army's strengths and order of battle as the Eighth Army was obtaining about those of Rommel's forces'.[51] Rommel's hand was strengthened by yet another source.

In Cairo, the US military attaché, Colonel Bonner Fellers, was a conscientious officer and a prolific writer of detailed cables, reporting promptly and frequently to Washington on British war plans in the Middle East. The only catch was that his messages were transmitted over a cipher system that the Germans had broken. In successive battles, Rommel seemed to know exactly how and when to do the right thing to derail British plans. Thanks in large part to Fellers's meticulous and copious reporting, Rommel's performance seemed superhuman. In mid-June 1942, for instance, his forces encircled the North African port of Tobruk, capturing 25 000 Allied prisoners, largely due to having detailed British plans in his hands only hours after they were developed in Cairo.[52]

Accounts by German intelligence acknowledge that for a long while they 'enjoyed superiority in radio interception'.[53] By late June 1942, however, a careless German radio broadcast revealed the significance of the US military attaché's messages, and within 36 hours the messages from Fellers ceased. He was subsequently issued with new, more secure, telegraphic cipher machines.[54] The capture in mid-July of Rommel's Sigint unit, 621 Radio Intercept Company, by soldiers from the Australian 9th Division, was particularly revealing. It provided clear evidence of how sloppy Allied security measures were, and orders issued subsequently called for strict adherence to Allied wireless communications security procedures.[55]

The combination of the unit's capture, and the new security measures for diplomatic and tactical reporting proved to be a turning point. From then on, Rommel was 'forced to grope around in the pitch dark'.[56] Rommel's apparent *Fingerspitzengefühl* – his uncanny knack of knowing exactly when and where to strike, had deserted him. His predicament was compounded by the 'sheer British superiority in tanks, artillery and aircraft for which no amount of tactical skills and self-sacrifice could compensate'.[57]

In the lead-up to the battle of El Alamein of November 1942, the

Australian 6th and 7th AIF divisions had been withdrawn to return to defend Australia against the Japanese thrust southwards. But the 9th Division remained to participate in what would prove to be the battle which marked the major turning point in the North African campaign. The British Eight Army commander, General Bernard Montgomery, approached the battle cautiously, but with an extraordinary advantage his predecessors had lacked: he was armed with Ultra intercepts and faced an adversary blind to the Sigint that Fellers had once so generously, albeit unwittingly, provided.[58] After the Battle of El Alamein (23 October–11 November 1942), as with the Battle of Stalingrad which also came to a climax that November, the tide would turn in the fight against Nazi Germany. At that point, the 9th Division was withdrawn from the field to return home to Australia and prepare to make another contribution – this time in the fight against Imperial Japan.

Support for empire

Like the Army, the RAN responded by participating in operations against the enemies of the British empire – in the Atlantic Ocean, the Caribbean, the Mediterranean, the Indian Ocean, the Persian Gulf and the Red Sea. At war's outset, however, the Navy was stretched, as it consisted of only two heavy cruisers, four light cruisers, five destroyers, three sloops and a variety of support and ancillary craft.[59]

Similarly, much of Australia's air power was committed in support of the empire. The RAAF, which was formed in 1921, emerging from the Australian Flying Corps after the end of the First World War, set about developing basic air power capabilities for the defence of Australia, initially with surplus aircraft from the previous war.[60] While elementary Sigint had been performed during the First World War, including direction finding and simple eavesdropping, there is little evidence that Sigint capabilities featured in RAAF planning in the lead-up to the Second World War. Military aircraft production commenced in the late 1930s, but in insufficient numbers and with planes of inferior quality to those manufactured by Japan.[61]

In the meantime, under an agreement reached in December 1939, Australia committed to the Empire Air Training Scheme or EATS (also known as the British Commonwealth Air Training Plan, or BCATP). Australia pledged 28 000 aircrew in support of Britain's Royal Air Force (RAF), with basic flying instruction at flight training schools in Australia and advanced air training in Canada. By the end of the war, 37 000 Australian airmen had been trained as part of the scheme.[62] Many were simply integrated into RAF units.

War exposes Australians to new technology

Australians, working alongside the Allies, would also be exposed to a range of other new technology. This included the ASV; an air-to-surface vessel radar aid designed to give indication of range of any solid object near the aircraft. Initially, the ASV's scan and range was quite restricted, but increased over time, as refinements were introduced into service. Mostly used for navigation checks near land, particularly in bad weather, the ASV also proved its worth in identifying surfaced German U-boats.[63] German radar countermeasures included a 'Metox' search-receiver aboard U-boats. These items were introduced by mid-1942 and created a demand for further ASV improvements, which, in turn, prompted additional countermeasures in early 1943 that then led to the introduction of a device known as 'H2S'.[64]

The H2S device, which presented a rough map of the ground immediately below the aircraft on a cathode-ray tube display, improved the accuracy of bombing runs and minelaying at sea.[65] The H2S led to the introduction of the 'G-H' system in 1944 which enabled even greater bombing accuracy.[66] While the official historian of the air-war of 1939–45 (writing in the early 1950s) focused on the ASV, H2S and G-H, there is no indication in his account of the place and significance of Sigint. That revelation would have to wait a further two decades. The delay helped protect security of what was still considered sensitive technology, but it would also lead to an erroneous disregard for Sigint amongst the uninitiated.

Australians integrated into RAF units were exposed to the benefits and challenges in radio direction finding (RDF), and radio countermeasures (RCM), or radio direction and ranging – a concept which came to be known simply as radar. Australian EATS participants also were exposed to the working of 'Window' (also known as 'chaff') – an early form of countering enemy electronic intelligence (Elint). This was the radar countermeasures technique used to confuse enemy radar-assisted anti-aircraft artillery batteries by dropping metalised strips along the way over hostile territory and thus flooding enemy radar control screens.[67]

While many EATS graduates would remain in Europe and the Mediterranean theatres, with the outbreak of the Pacific War the majority of RAAF aircrew would complete their training and then return to fly with RAAF units in the South West Pacific Area.[68] Some of the aircraft were fitted with Sigint and Elint equipment, particularly later in the war.

Writing on the early RCM or radar developments in Australia, military historian Craig Bellamy has recently observed that one single nation did not invent radar. Rather it was a case of 'multiple discovery' or 'simultaneous invention' with concurrent developments in the United Kingdom, Japan, Germany and the United States.[69]

Nave's Special Intelligence Bureau

In 1939, Nave had returned to Australia and made himself indispensable to Australia's Sigint efforts. Despite repeated Admiralty efforts to reclaim him, Nave's local supporters managed to keep him in Australia for the duration of the war, initially working with the Special Intelligence Bureau.[70]

Work on Japanese diplomatic ciphers was first begun by Special Intelligence in December 1941 under the auspices of the Navy. In an effort to draw together the disparate Sigint initiatives launched separately by the Australian Navy and Army, the Defence Committee met twice in November 1941 to approve the organisation, establishment and mission for the Special Intelligence Bureau.[71] From that point, the Special Intelligence Bureau comprised Nave and his team of RN and RAN paymasters, as well as Professor Room; and a team paid for by the army, including Treweek (by now a major in the militia), as well as A.A. Mason, R.J. Lyons and Lieutenant Longfield Lloyd.

Once engaged in actual code-breaking, some interesting differences in approach began to emerge between the mathematicians, Room and Lyons, and the classicists, Treweek and Trendall. Treweek explained it this way: 'the mathematicians wanted 100% accuracy ... But often you would get a garble in the text. And it was little things like garbles that gave the show away.'[72]

Trendall, a New Zealander who had been in Australia for two years, was an expert in fourth-century BC Greek vases, with a growing international reputation as a classical scholar. Renowned for his uncanny intuitive judgement when inspecting a new classical item, an obituary after Trendall's death in 1995 recounted that:

> One achievement which gave him great satisfaction is akin to that of the astronomer who predicts the existence of a star before it is discovered. On the basis of characteristics in some Sicilian and other South Italian vases which could not be attributed to the influences of other schools or regions, Trendall postulated that there must be in the region a whole school of vase-painting, none of whose vases had been discovered, and he listed some of the characteristics that would mark this style. In the 1970s a group of vases was found in the Lipari Islands closely corresponding to his prediction.[73]

Trendall's special gifts lent themselves to the art of code-breaking. While the mathematicians worked on theoretical solutions, Trendall considered the intuitive approach to code-breaking to be far superior to the mathematical, analytical approach. As Trendall described his own experience years later: 'You get a feeling for it. Your eye lights upon something, and ... bang.'[74]

As for Treweek, he had the advantage of military experience, having been commissioned in the Field Artillery in 1932 and he remained on the active list throughout the war. As noted earlier, he had also foreseen the war with Japan and begun teaching himself Japanese in the early 1930s before receiving some help from a mysterious Miss Lake.[75] Miss Lake proved to be Margaret Ethel Lake, a teacher of handicrafts at the University of Sydney who had also studied Japanese at the University of Sydney before then living in Japan for over a year, studying both Japanese handicrafts and the language.[76] After the war, Lake published a translation of the account of the capture of Singapore by Colonel Tsuji Masanobu and from this it was evident that she had completely mastered the language. She was born in 1883 but the year of her death and other details of her life are unknown.

Treweek was not the only Australian to have elected to begin learning Japanese; Professor Room had done likewise halfway through 1941. As the Japanologist Peter Kornicki observes, it is difficult to name anybody in Britain or the United States who was so convinced that war was inevitable that they went to all the trouble of learning Japanese.[77]

The first step in providing specialist training in Japanese was actually taken by the Royal Military College at Duntroon, which established a lectureship for the training of interpreters for the Army. The University of Sydney followed, creating a lectureship in Japanese in 1917. Subsequently a three-year course in Japanese was organised. Professor Room was also taught by Miss Lake, who had joined the department some years before as a lecturer.[78] Writing to the Registrar of the University of Sydney, Room observed the 'two terms I have spent under Miss Lake have proved as useful in my present job as the twenty years' mathematics!'[79]

The work of Trendall, Room, Treweek and others echoed developments in Canada, where two mathematics professors from the University of Toronto, Gilbert Robinson and Harold Coxeter, were involved in the establishment in March 1941 of a cryptologic bureau in Canada's Department of External Affairs. Yardley's work had been part of the cryptologic education of Trendall and Room and he would play, at least initially, an even more prominent role in the education of their counterparts in Canada, where he was hired to assist in establishing the bureau.[80] But Yardley's tenure was cut short following an intervention by Commander Alastair Denniston, the head of GC&CS. Denniston insisted on his removal. The Canadians consulted William Friedman in the US Army's Signal Intelligence Service, who confirmed that the United States, like their British counterparts, would not cooperate with the Canadians if Yardley was retained. An experienced 67-year-old British codebreaker, Oliver Strachey, was despatched to take Yardley's place.[81]

Women on Australia's home front

The creation of the Special Intelligence Bureau was crucial, but there were still severe shortages of those with specialist signals skills. Here, Mrs Florence Violet McKenzie came to the rescue. In 1938, McKenzie and several other women had founded the Australian Women's Flying Corps at Sydney's Feminist Club. She taught classes in Morse code for the women in the corps and the demand for such courses shot up. The following year, McKenzie and her husband founded the Women's Emergency Signalling Corps, an organisation whose sole aim was to provide signals training to women. This included Morse code with a Morse key, as well as semaphore (flag signalling) and sending Morse with lamps.[82]

Women would come to Mrs McKenzie's signals school, or 'Sigs' as it became known, in their lunchtimes, after work, or on Saturday mornings. They called Florence 'Mrs Mac'. The whole thing was funded by donations, and Mrs Mac never took any money.[83] She appears to have been an inspired teacher. As one of her students put it: 'all of a sudden a wonderful magical moment came when you forgot about the "dits" and "dahs" and you just actually heard the musical sound of each letter. And then you were really on the way.'[84]

Mrs Mac's establishment of a training program suggests she had more foresight about the future demand for signallers than the Naval Board did. In January 1939, the Board noted that the activities of the W/T Procedure (Special) were increasing, due to the establishment of DF stations, the W/T stations at Canberra and Darwin and the additions to cruiser strength. The Board decided therefore to provide for 18 additional telegraphists (there were only ten at the time) in the 1939–40 estimates.[85]

The Naval Board apparently did not expect any difficulty in finding the additional telegraphists. By March, however, they knew that there was a problem. A meeting on 22 March 1939 noted the serious shortage in the numbers of telegraphist ratings required to operate the new W/T stations at Canberra and Darwin, after all suitable volunteers from members and ex-members of the Royal Australian Navy had been accepted. The Board recommended that further publicity be given in Australia to the Shore Wireless Service in the hope that further volunteers would be forthcoming, and an effort was made to obtain volunteers from the United Kingdom and New Zealand from ex-telegraphist ratings of the Royal Navy.[86]

By October, the situation was becoming desperate. A meeting of the Naval Board considered the minute of 25 September 1939 by the Director of Signals and Communications disclosing a serious shortage of telegraphist ratings.[87]

On 27 December 1940, McKenzie wrote to the minister, offering the services of women signallers from her association and the matter was referred to the Naval Board. The Board approved the proposal in principle, but instructed that the matter be further investigated having regard to the existing amenities and RAAF proposals along similar lines. The Board wanted to know particularly whether an organisation could be established similar to the British Women's Royal Naval Service (WRENS) providing for ranks and the enlistment of ratings, uniforms, discipline, pay, etc. The Board also wanted the question of additional accommodation examined.[88]

In early 1941, Commander Newman, the RAN's Director of Naval Communications, was sent to investigate the potential recruits. He tested several of the women in Morse code, sending and receiving, and was duly impressed.[89] Newman reported to the Naval Board that the women of the Emergency Signalling Corps (who had received only part-time, informal training) were 'almost as good as the men' who were experienced, fully trained naval personnel.[90] Newman's assessment of Mrs Mac's pupils went beyond their signalling skills: he later referred to them as 'the cream of Australian womanhood' – a comment surely meant as a compliment.[91]

Today, the Naval Board might seem to have been excessively concerned with suitable accommodation and facilities for women. After an initial investigation and report on the subject, the Board requested further detail on 17 February.[92] On 20 March, the Board decided that, subject to the report on accommodation being satisfactory at *Harman* (a signals receiving base near Canberra), they would recommend to the minister that female telegraphists be employed there. When the system was working satisfactorily, consideration could be given to extending the scheme. The Board also noted that it was likely that an enquiry would be received from Singapore as to the possibility of female telegraphists being available from Australia.[93]

Finally, at a meeting of the Naval Board on 18 April 1941, Billy Hughes, the former Prime Minister and Minister of Navy at the time, approved the employment of 14 women at *Harman*, on two conditions. First, that 'endeavour was made to minimise publicity'; and second, that efforts to obtain 'suitable male telegraphists' continue.[94] It was not until 25 April 1941 that 14 of Mrs Mac's former pupils (12 telegraphists and two cooks) set off for *Harman*.[95] By the time the Navy officially established the WRANS in October, they were the last to form a women's service, despite having employed women since April.[96]

Reluctance to employ women

Minister for Air McEwen was equally reluctant where the enlistment of women telegraphists for the RAAF was concerned, but in January 1941 he admitted it was unavoidable when all attempts to recruit sufficient male telegraphists had failed.[97] On 9 January 1941, the Advisory War Council agreed that women should be enlisted, but only 'to the minimum number for a minimum period'. In the face of great difficulties and 'marked lack of enthusiasm' on the part of some male officers, the recruiting of airwomen began on 15 March 1941.[98] Many women who enlisted were members of Mrs Mac's corps, having arrived with full proficiency in Morse code.

In July 1941, the Minister for Army, Percy Spender, opened the door to a role for women in the army by announcing that women should be used in roles for which they had the ability. But the War Cabinet initially refused to employ female telegraphists despite the shortage of male counterparts. As in the case of the other services, they eventually would give in. On 13 August 1941, the War Cabinet approved the formation of the Australian Women's Army Service (AWAS) 'to release men for more active roles'. Initially, there were 3600 female signallers.[99] Many of the women trained by McKenzie joined the AWAS and served in the Army Signals Corps.[100]

Why the marked male reluctance to allow women to serve when they wanted to and there was clearly a need for them? Ann Howard suggests that male protectiveness towards women was intensified because there were fewer women than men in Australia at the time. Furthermore, Australian women eager to serve were seen as a nuisance by the War Cabinet. As late as July 1940, War Cabinet ministers were saying they did not need women and that training and money would be better used for men. There was a fear that women in uniform could present difficulties in administration and discipline. Not least, there was a concern that trained women would present problems after the war when returning men wanted their jobs back.[101]

Resistance to admitting women to the military was not, of course, unique to Australia. In the United States, for instance, there was a view – shocking in retrospect – that if the Navy 'could possibly have used dogs or ducks or monkeys, certain of the older admirals would probably have greatly preferred them to women'.[102] But once the decision was made to recruit women codebreakers, the US Navy targeted highly educated women from elite northeast colleges, whereas the Army tended to recruit schoolteachers or candidates from teachers' colleges in the south and Midwest.[103] The US military's decision to tap 'high grade' young women was a chief reason why the United States, entering the Second World War, was able to build an effective code-breaking operation practically overnight.[104]

... as good as the men ... and often better ...

By the time women codebreakers were recruited for Australian service, the British were well aware of their indispensable role in Sigint. According to Captain Sandwith, RN, the WRENS

> are as good as the men. They learn quicker and they're much keener, and I think when we've had them longer, they'll be every bit as good as the men operators. They're not as good as the man who has been doing Y work for ten years because these operators develop a kind of intuition. They can run their receivers through the aether and recognise every station by its characteristics.

> We will never get a Wren to do that under ten years, but they're very good, and, because of the manpower shortage which we are suffering from at home, we could not possibly do without them.[105]

Mrs Mac's trainees were similarly high performers. Using a technique known as 'Tina' (a nickname derived from the Spanish word 'tenia' meaning tape-worm to detect individual idiosyncrasies of another operator's Morse), members of the WRANS in *Harman*, Melbourne and *Moreton* (Brisbane) could recognise who the operator at the other end was by the keying of a message and which enemy operator was sending from an enemy base.[106]

Despite the initial resistance encountered in some quarters, and sometimes very difficult conditions, there is no question about the generally high standard of the women's performance in all three services. In addition to providing trained personnel to the armed forces, Mrs Mac's pupils were also providing training. Members of her Women's Emergency Signalling Corps were instructing Australian service personnel in Morse code.

Not all the women who were to make a significant contribution to Australia's Sigint effort were in uniform. On the staff of D Special Section in Melbourne were Dr Elizabeth Sheppard, formerly a tutor at Women's College at the University of Sydney, and Mavis Tilley, an accomplished linguist.[107]

The role of women in the Allied war effort is not to be underestimated. The Axis powers never mobilised their women to the extent that the Allies did. While there are, of course, many reasons why the Allies prevailed in the Second World War, the recruitment of women was one of the important factors.[108]

It is especially sobering to note, then, that, many years later, the contribution of women was still not fully recognised or rewarded. Justice Hope's sixth report of the 1977 Royal Commission into Intelligence and Security noted that:

> It is only very recently that married women have been able to remain as members of the Services. Therefore, it is only now that women are working up into more senior posts in the ranks of sigint operators.

> But a clear view was held in many quarters that women tend to make very good sigint operators and often to be better at it than men.[109]

The ban on talking about their wartime Sigint work for security reasons applied, of course, also to women. Only recently have we begun to hear some of their stories, which are colourful and sometimes hair-raising. Berenice Wormald (1922–2021), a member of the WAAAF who scandalised people on Manly Beach by appearing in a two-piece swimming costume, was the teletype operator on duty in Townsville when the Japanese Coral Sea Fleet was sighted. As the Japanese codes had been broken, WAAAFs were not allowed to leave barracks without an armed escort with instructions to shoot their charge rather than let her fall into Japanese hands.[110]

British-American Sigint sharing and the onset of the Pacific War

During the First World War, British–American intelligence relations had been close, but in the interwar years they became acrimonious over policy differences in the Middle East and in part due to the Americans' anti-colonial instinct. Later, however, they would rediscover closeness as their interests once again converged. With apprehensions of war, there was a renewed interest in collaboration. The attack on Pearl Harbor on 7 December 1941 coincided with the Japanese assault on the Philippines and British Malaya (across the international date line, where it was already 8 December); but collaboration between Britain and the United States predated the attack by the better part of a year and a half.[111]

Indeed, the interest in closer collaboration between Britain and the United States predated the signing of the Atlantic Charter, the eight-point declaration signed by President Franklin D. Roosevelt and Prime Minister Winston Churchill on 14 August 1941 on board HMS *Prince of Wales* (that would sink off the coast of Malaya a couple of months later). The signing of this charter was a watershed moment when the United Kingdom accepted the rights of all people to choose the form of government under which they would live – a move that pointed to an eventual devolution of power across the empire in the years after the war. In turn, the United States initiated the lend-lease program that would underwrite the British war effort. That rapprochement had been building for some time as UK and US interests converged, and was particularly important in terms of intelligence cooperation.

In December 1937, Japanese forces had attacked an American gunboat, the *Panay*. Despite the objections of US Secretary of State, Cordell Hull, President Roosevelt approved secret staff talks between the two countries' navies concerning the Japanese threat.[112] The US Navy Captain Royal Ingersoll was despatched and the talks resulted in an agreement to a type of operational intelligence exchange between the US Asiatic Fleet and the British China Fleet. They also worked out details of cooperation between their two navies in the event of war with Japan.[113] Constrained by a largely isolationist American public, prior to the war, President Roosevelt was reluctant to provide unrestricted assistance to Britain.[114]

By May 1940, American concern after the fall of France was twofold: that the British might be quickly defeated and US technical and intelligence information compromised; and that the British were single-minded in their focus on bringing America into the war and gaining access to their vast technological and industrial resources.[115] US Navy Rear Admiral Thomas A. Brooks, observed that 'to further their own goals, the British were willing to provide the United States with virtually unlimited access to British secrets – technological as well as intelligence – even without a quid pro quo. Their strategy worked.'[116]

Having broken the German Enigma code, in June 1940 GC&CS approached the US Navy about exchanging technical information. Initially they were rebuffed, but a direct approach to President Roosevelt had the desired effect.[117]

In the summer of 1940, Colonel William 'Wild Bill' Donovan visited London. He would later go on to found the Office of Strategic Services (OSS), the wartime precursor to the Central Intelligence Agency (CIA). Donovan met Prime Minister Churchill and the British high command, as well as Sir Stewart Menzies, the chief of Britain's Secret Intelligence Service, MI6, and the titular head of the GC&CS. Donovan's visit is regarded as the turning point in the UK–US relationship. He went to assess Britain's chance of surviving the war. His sanguine assessment prevailed over the more downbeat prognosis of the US ambassador in London, Joseph P. Kennedy.[118]

Secret staff talks, which went by the cover name of 'Standardization of Arms Talks', followed in late August 1940. It was then that the initial offer of Sigint cooperation was made by US Army Brigadier General G.V. Strong, who accompanied Rear Admiral Robert Ghormley, the head of delegation and the US Navy's former director of War Plans.[119] As one writer has observed, 'the British policy was to hold little back from the Americans, hopeful that their openness would produce the cooperation they sought'.[120]

In September 1940, the US Army's Signal Intelligence Service's William Friedman successfully completed the reverse engineering of the 'Purple' cipher machine. Armed with this breakthrough, the United States had something substantive to offer in exchange for British cooperation.[121] The confluence of technological breakthroughs, growing trust and heightened interest in collaboration provided momentum for an agreement. On 22 October, Ghormley met Sir Stewart Menzies, to begin deliberations over the arrangements for a Sigint technical exchange. Still, there remained a reluctance on the part of Sigint policymakers to move too hastily. The US Navy's Sigint office, OP-20-G, remained wary, as indeed did some on the British side, concerned about the prospect of leaks if too many people knew about what was involved.[122]

But that wariness was overcome and, by December 1940, the United States signed an agreement to share technical information about German, Italian and Japanese codes. Then, in January 1941, a four-man team arrived with a 'Purple' machine at Bletchley Park, where GC&CS had established itself after moving out of London. In exchange, the British provided the plans for the German Enigma machine, showing how it was built and how it worked.[123] Getting to that point had required a substantial effort.

Solving the machine encryption puzzle

In tackling Germany's Enigma code, Britain's cryptographers benefitted enormously from the head start provided by three brilliant Polish cryptologists and distinguished mathematicians: Marian Rejewski, Jerzy Rozycki and Henryk Zygalski. The Poles had been breaking and reading some Enigma ciphers since the early 1930s. In 1939 they shared with GC&CS all their knowledge and experience and gave them an Enigma machine.[124] It was also the Poles who first developed the 'Bomba', the original version of a British and later American electro-mechanical machine used to discover the wheel settings for Enigma keys.

Before the Second World War, an Enigma cipher machine expanded on the prewar design by Boris Hagelin. The German Army purchased several commercially produced Enigma machines which were built from a derivative of Hagelin's patented design, and, after some modification, adopted the Enigma for extensive use.[125] The Enigma machine model developed that concept further and included five variable components: (1) a plugboard which could contain from zero to 13 dual-wired cables; (2) three wired metal and black plastic cylindrical rotors, ordered left to right, with 26 spring-loaded copper input contact points to 26 copper contact points positioned on alternate faces of a disc; (3) 26 serrations around the periphery of the rotors, which allowed the operator to specify the rotors' initial rotational position; (4) a moveable ring on each rotor to control the rotational behaviour of the next rotor to the left by means of a notch; and (5) a non-rotating reflector half-rotor used to fold inputs and outputs back onto the same face of contact points.[126]

Under the general name of *Geheimschreiber* (secret writer) the Germans used two devices for enciphering high-level radio-printer communications. These were generically referred to by the British code word 'Fish'. At army level and above (that is, above corps-level headquarters), the Germans used the 12-rotor Lorenz teletypewriter device (known by the British code word 'Tunny', in accordance with their convention of using fish for various ciphers). The simpler, lighter and more easily deployable three-rotor Enigma was used for army-level message encryption. The British called this 'Sturgeon'.[127]

There were many unknown variables in cracking the codes of the Lorenz/Tunny and Enigma/Sturgeon machines. Marian Rejewski had worked out the necessary mathematical equations to determine the wiring of the Enigma rotors, but the Poles realised that the volume of traffic from these machine-generated cipher messages required a machine-generated response. Reverse engineering these machines would take extraordinary effort. The AVA Radio Manufacturing Company had built the Polish copy of the Enigma and also manufactured the Bomba for the Polish cipher bureau in 1938. The Bomba was built to work through all the possible permutations, relying on indicators. Then, however, the Germans added extra rotors to the Enigma machine, making the reliance on indicators inadequate for the task.[128]

Turing, Welchman and a more advanced Bombe

As war became evermore inevitable, GC&CS began hiring mathematicians, including Alan Turing and Gordon Welchman, from nearby Cambridge University. Capitalising on the Poles' achievements, Turing recognised that a larger, more powerful machine would have worked, not so much with the available indicators, but with assumed text. In other words, cryptologists were aided by using text they assumed was probably included in the message. As a result, the electro-mechanical machine that emerged, the Bombe, was not dependent solely on the indicators. In this machine, rotors and wires would simulate Enigma rotors and pass an electrical current from one rotor to the next, looking for the correct rotor setting, looking not for indicators but assumed text (that is, text considered likely to feature in the intercepted messages).

While Turing worked on the main Enigma machine, Welchman worked on another complication: the Enigma plugboard, which used cables to connect one letter with another. Welchman then designed a board that connected each letter with every other letter. This became known as the 'diagonal board,' as the wires created a pattern of diagonal lines. Combined with Turing's test registers, the number of possible rotor settings decreased from thousands to only a few. Analysts could then readily test a few possible solutions on an Enigma duplicate or analogue.[129]

The Bombes took months to design and build. The British Tabulating Machine (BTM) Company was contracted to build them, and they started arriving in August 1940. In the end, 210 were built in England throughout the war. Each one weighed one tonne, standing over two metres high, 60 centimetres wide and about two metres long. Within each set, the top drum represented the left-most, or slowest, rotor on the Enigma; the middle corresponded to the centre rotor; and the bottom Bombe drum represented the right-most, or fastest Enigma rotor. The British Bombes worked through rotor settings in the opposite direction to that used in the Enigma machines.[130] Further breakthroughs were achieved in February 1940 and May 1941, with the capture at sea of German naval Enigma machines and code books. From then on, German naval codes would prove readable for much of the rest of the war.[131] Australia did not receive Bombes, relying instead on the work on British and American Bombes.

Meanwhile, in the United States, the National Cash Register (NCR) Company in Dayton, Ohio, was contracted to work on and build another, more capable, four-rotor electro-mechanical Bombe design, this one developed by NCR's engineer, Joseph Desch. Each machine weighed over two tonnes, stood over two metres high, three metres long and about 60 centimetres wide. Eventually, the British developed a four-rotor Bombe, but insiders admitted the US machine performed better. By September 1944, 25 NCR Bombes had been made.[132]

The Lorenz

The Enigma, however, was not the only German cipher machine. Another, newer one was ordered by Hitler in 1940: the Lorenz SZ40 and later the SZ42 (SZ stood for *Schlüsselzusatz* or 'special attachment', as it was attached to standard teleprinters). The Lorenz carried higher echelon level messages than the Enigma traffic. The British referred to both the Lorenz machines as 'Tunny', but knowledge of the Lorenz machine and the breaking of its ciphers was only declassified in 2002.[133]

The Lorenz was an expensive machine, which was one of the reasons why so few were used: in the region of 200. Designed to work with a regular teleprinter, the Lorenz machines, like Enigma, created an enciphered message. This involved a series of sequenced turns of revolving rotors, but on an altogether grander scale than Enigma. By comparison to Enigma's three (later four and, in some cases, five) rotors, the Lorenz machine had 12, giving it 1.6 million billion possible start positions. When employed as an attachment to a teleprinter, it needed to encipher five-bit coded letters, employing 'Baudot code' (or five-key binary combinations of 'x's and dots representing each letter of the alphabet) addition. This appeared to the Germans to be unbreakable; but the attempt to produce a machine that would generate a random sequence succeeded only in creating a pseudo-random one, which is not the same thing. The mathematicians at Bletchley Park – Bill Tutte, in particular – found that the Lorenz cipher concealed unseen probabilities in the apparent randomness, which would subsequently succumb to a statistical method of attack.[134]

The breakthrough came on 30 August 1941, originating in a stroke of luck: a German operator sent a message some 3976 characters long and, apparently fearing that the first transmission hadn't succeeded, retyped the whole message again, re-using the original starting setting for the rotors. He also abbreviated some plaintext words this time round, and seemingly introduced some typos. These variations gave the cryptanalysts a way in. After intensive analysis, from 1 July 1942, using hand cryptanalytic techniques, Bletchley Park was able to read 500 messages in the subsequent three months.[135]

The Colossus

Max Newman, who joined GC&CS straight from Cambridge in 1942, quickly grew frustrated with the cumbersome business of hand Tunny decryption. He was convinced that more streamlined, mechanised systems might be devised. Supplied with a team, over the following months Newman worked with technicians (most notably Tommy Flowers) at the Post Office Research Station at Dollis Hill to develop a solution. Together they created the 'Heath Robinson' device (taking its name from the celebrated cartoonist William Heath Robinson, who depicted humorous devices of such kind). Slow and temperamental, the device only provided half a solution, producing the sort of simplified message the qualified cryptanalysts could produce. Junior staff still had to complete the decryptions by hand, but it was a vast improvement on what had gone before.[136]

In the second half of 1943, Bletchley Park's computational capacity was transformed by the development of 'Colossus' – the world's first programmable computer, standing two metres high and over a metre wide. Although its logic was essentially the same as Heath Robinson's, Colossus's technology represented a step change. In Heath Robinson, Tunny decryptions were produced by the means of perforated tapes that had to be kept synchronised at all times, and so represented an electro-mechanical approach. The new Colossus machine developed by Flowers and his team relied instead on an electronic analogue to the Lorenz, with the work done by over a thousand thermionic valves (vacuum tubes), switching the current on and off. Colossus basically did what a human operative would have done, trying to decipher the input enciphered message by testing a succession of different imagined Lorenz wheel settings, applying 'trial and error'. But it did it faster than any cryptanalyst could have ever dreamt of doing it by hand.[137]

While Australia did not operate Enigma, Bombe, Lorenz or Colossus machines during the war, Australian forces undoubtedly were the beneficiaries of their discoveries and the Ultra reports which helped steer the Allies towards victory. One example is particularly noteworthy: Sigint intercepts from Lorenz machines passed by the British to the Soviet Union helped provide a crucial advantage by outlining troop and tank numbers, dispositions and intentions. Their crucial insights into German plans for the largest tank battle of the war, the Battle of Kursk in April 1943, would see Nazi forces decisively defeated, thanks in part to the removal of the element of surprise.[138] In the Pacific theatre, however, the focus would be more on Japan's Purple machine. For this, the technological breakthroughs and inventions of the Americans would prove crucial. Purple, unlike Enigma, however, was not available at lower levels of command in the Japanese forces, so there was a greater reliance by the Japanese on code books and manual, as opposed to machine-generated, encoding and decoding. This made the work of Central Bureau different to the counterparts' Sigint efforts focused on Nazi Germany.

Sinkov's mission to Britain

From late January to late March 1941, ABC-1 Talks were intended to 'determine the best methods by which the armed forces of the United States and the British Commonwealth ... could defeat Germany and the powers allied with her, should the United States be compelled to resort to war'.[139] This involved efforts to devise strategy and to work out command, control, communications and intelligence issues. Roosevelt approved the recommendations from the talks, including the lend-lease arrangements to supply Britain with additional shipping and other materials, a determination to establish the Atlantic-first strategy, and the establishment of joint missions in Washington, DC and London.[140] These arrangements would have significant knock-on effects for how Australia was able to engage with them both as the war progressed, adding impetus for Australia to engage the United States directly on its own terms.

Concurrently, initiatives were underway at the working level to facilitate Sigint exchange. Scheduled to lead a four-man delegation to Britain, William Friedman suffered a nervous breakdown. Captain Abraham 'Abe' Sinkov, who was later to lead the code-breaking effort at Central Bureau in Brisbane (and became prominent in the NSA after the war), was selected instead. The Sinkov mission arrived in Britain on 7 February and spent most of its time at the GC&CS at Bletchley Park. The US delegation gave the British two Purple machines and information on various codes. Britain did not have a spare Bombe machine to offer and was initially reluctant to share its findings, but with Stewart Menzies's and Winston Churchill's blessing, GC&CS in return provided detailed plans, explaining sensitive aspects of their Sigint dispositions and capabilities.[141] Some on the US side were concerned the British were holding back, but Sinkov reported that 'no door was closed to us' and the resulting exchange led to 'a saving of several years of labor on the part of a fairly large staff'.[142]

The Sinkov mission proved highly effective, in large part because of the depth and scope of the sharing of highly sensitive information agreed upon. In effect, the mission's main result was to foster an atmosphere of trust to encourage greater cooperation on intelligence matters over the course of

1941, particularly prior to Pearl Harbor and with regard to what was called the Far Eastern theatre.[143]

Looking back from the perspective of the 1950s, William Friedman wrote about the Sinkov mission thus:

> In the exchange of the Purple machine and informational details concerning the Purple system for specific technical data on certain German and Italian cryptosystems ... both the U.S. and the U.K. gained advantages of inestimable value... More complete U.S.-U.K. collaboration in cryptanalytic operations after the U.S. entry into World War II as one of the belligerents... The value of this collaboration can hardly be overestimated ...[144]

In hindsight, it is clear that this important and delicate work in 1941 initiated a level of trusting collaboration that would withstand the ebbs and flows of war, and set the stage for an unprecedented degree of international cooperation on intelligence matters in the postwar years. This was the foundation which would stand the test of time.

Reflecting the developing US–UK closeness, by May 1941 a Joint Intelligence Committee was established as part of the Joint Staff Mission, and the UK Naval Intelligence Department stood up a new branch, NID-18. This new branch was to liaise with the US Navy's Office of Naval Intelligence, notably including its Sigint section, OP-20-G.[145]

During this period, Japan's ambassador to Berlin, Oshima Hiroshi, was a frequent and detailed correspondent back to Tokyo, and the Purple decrypts of his communications gave important and helpful insights into German plans.[146] The greater cooperation encouraged by the Sinkov mission meant that insights from this information about German intentions were being shared.

Japan attacks

In the 'Far East', to Australia's north, in a move that anticipated the results of the Sinkov mission, the FECB in Singapore established liaison arrangements with the US Army's Station 6 and the US Navy CAST Sigint collection site in the Philippines. In mid-March 1941, the British DNI, Rear Admiral Godfrey, had authorised a Sigint exchange between FECB and Station CAST, including partial solutions of the Japanese transport code and a Japanese air force cipher.[147] The FECB saw Japanese movements and accurately estimated their strength, but got the targets and timing wrong. As a result, as John Ferris observes, the Japanese attack on Malaya was a tactical surprise. 'The failure', Ferris writes, 'stemmed from errors about who made decisions in Tokyo, and the logic which impelled them'. Those errors were compounded by being fooled by Japanese deception and secrecy, which included the continuation of normal diplomacy, hiding the intent to strike, the manipulation of normal procedures, including radio traffic, and the use of well-placed deceptive rumours and press leaks.[148] Sigint was helpful, to a point, but not foolproof and, in this instance, with Japan mindful of British eavesdropping, proved a double-edged sword.

As the situation became more ominous, discussions on a limited basis had also been conducted by British and Australian officials with the small Dutch East Indies (Indonesia) code-breaking team known as *Kamer 14* (Dutch for 'Room 14'). The fall of Singapore and the Philippines would end this arrangement, but the trust built in collaborating with each other, between the British and Americans, as well as the Australians, set a strong precedent for cooperation throughout the rest of the war in the Pacific and thereafter.[149]

The 'Winds' message and a conspiracy theory

In Eric Nave's opinion, the most important piece of decoded information he provided during his time with the Special Intelligence Bureau (later referred to as D Special Section) in Melbourne in November and early December 1941 was one that showed that Japan was about to declare war. Nave recalled that on 19 November 1941 a circular message was sent out from Tokyo which the Special Intelligence Bureau received in Melbourne, 'it being a clear indication that Japan was on the point of declaring war'. Japan was advising all overseas posts that in the event of a national emergency involving the breaking of diplomatic relations and the consequent interruption of overseas communications, a warning would be broadcast in the middle of successive Japanese shortwave broadcasts.[150] This would be repeated at the end as follows:

The breaking of international relations

-- With America – *Hibashi no Kaze ame* – [translated as] 'East wind, rain'

-- With Russia – *Kita no Kaze humori* – 'North wind, cloudy'; and

-- With Britain – *Nishi no Kaze hare* – 'West wind, fine'.[151]

The same warning message, sent on 27 November 1941, was intercepted in Washington in another code. Nave and American and British codebreakers intercepted and read the message at about the same time.[152] The message had been passed to the intelligence staffs of the US Navy and War departments.[153] It was also passed up that day (across the international dateline, on 28 November) to Australia's Defence Department Secretary, Frederick Shedden, for the contents 'to be brought to the attention of the Prime Minister', John Curtin.[154] Even after the 'hidden word' message had been sent, authorities in Tokyo continued to send instructions to their diplomats about the destruction of sensitive (including cryptographic) material. Late on 7 December, Tokyo finally sent the 'Winds Execute' message, several hours after the commencement of Japanese offensive operations across Asia and the Pacific. As American Sigint historians Robert Hanyok and David Mowry observe, 'the abundance of source material did not always lead to a clear understanding'.[155]

Nave's team decoded another message, sent on 2 December from Tokyo to the local consulate (and, as it turned out, to all other Japanese consulates), ordering the destruction of code books and burning of other important documents. There was no doubt that an attack by Japan was imminent. Allied commanders also knew that a large Japanese fleet was sailing down the east coast of Asia,[156] but no one knew exactly when or where the attack would take place. This 'Winds' message was to be at the centre of the controversy over the lack of warning of the attack on Pearl Harbor.[157] There would be eight separate investigations into this from late 1941 through to mid-1946.[158]

These facts did not prevent an 89-year-old Nave from being approached, 40 years after the event, by a former UK MI6 officer, James Rusbridger, who suggested they collaborate on a book about Second World War signals. In a 1991 book entitled *Betrayal at Pearl Harbor: How Churchill Lured Roosevelt into War*, Rusbridger published controversial allegations.[159] Rusbridger used the book and Nave's reputation to promote a conspiracy theory about Pearl Harbor: that Churchill had advance knowledge of the attack and withheld it from the Americans in order to draw the United States into the war.[160] The allegations were, in fact, a fabrication and, once he realised what Rusbridger had done, Nave was appalled by the claims in the book, which tarnished his reputation. Eager to set the record straight, Nave publicly disowned the claims about Churchill, even appearing on Japanese television to do so.[161]

One reason why, at first glance, Rusbridger's conspiracy theory hypothesis seemed so compelling was that IJN codes, notably IJN-25, had indeed been broken earlier by both British and American cryptographers. A surprise change of codes, however, at the start of December 1941, coupled with strict adherence to radio silence by the Japanese fleet, left them in catch-up mode and in the dark during the crucial days leading up to 7/8 December.

In fact, the 'Winds' message could not have given specific early warning of the attack, as it was intended only to inform Japanese diplomats that this was how they would find out that war had broken out. Furthermore, as Hanyok and Mowry point out:

> American cryptologists were hostage to the misperception that because the Purple cipher machine was *the* high-level cryptographic system for Japan's diplomatic traffic, therefore it would carry all intelligence of the highest importance about Japan's intentions. But the Purple device was just one diplomatic cryptographic system,

and the information it protected did not include any data about the impending operations of Japan's military and naval forces. The latter exclusion was deliberate: the Japanese War and Navy Ministries effectively restricted knowledge, especially of the strike against Pearl Harbor, throughout their offices and the *Gaimusho* [Foreign Ministry]. In fact, even large elements of the Imperial Japanese Navy were unaware of the Hawaii operations (*Hawai sakusen*)![162]

Hanyok and Mowry go on to observe that 'events had demonstrated that the Winds Execute message had failed to be either a sort of actionable intelligence or a useful warning. That the actual message was heard several hours after hostilities and applied only to Japan's relations with Great Britain further illustrates that the message was irrelevant.'[163]

Over the following couple of months, the humiliation would be almost total, both for US forces and those of the British empire – including Australia. The defeat of British imperial forces, including the Australian 8th Division, in Singapore on 15 February 1942, and the bombing of Darwin on 19 February, were a turning point for Australia.

Admittedly, even if it had been armed with sensitive information on Japanese capabilities and intentions, Australia – even alongside British and US forces in the region – did not have sufficient aircraft, personnel or equipment deployed to stem the Japanese advance. Good intelligence would have made a difference, but probably would not have been enough. Still, this searing experience demonstrated a profound lack of preparation in response to the threat, and a surprising level of ignorance of Japanese military capabilities and intentions. The experience highlighted the need to dramatically increase the effort to find ways to better understand the enemy. The intelligence efforts up to this point, while highly sophisticated in some respects, were in other ways still embryonic and demonstrably not complex enough to match the threat, and certainly not the highly effective end-to-end system it was to become. Thereafter they would be increased to an industrial scale.

In view of the importance of the 'Winds' message and diplomatic traffic generally, in February 1942 Nave set up within his organisation a small diplomatic section, with two senior members of the British Consular

Service as translators.[164] This would include the 'men of the professor type' sponsored by the army, who had in the meantime moved from Sydney to Melbourne.

At the end of December 1941, a short-lived American–British–Dutch–Australian (ABDA) Command had been established, based in the Netherlands East Indies (now Indonesia).[165] The British Commander in Chief Far East, General Archibald Wavell, who had only recently arrived at Singapore, was ordered to Java to lead ABDA Command on 15 January. His reporting lines for command and control involved London and Washington, but excluded Australia. Not surprisingly, Australian Prime Minister John Curtin found this unsatisfactory and protested.[166] As it turned out, the arrangements would be fleeting. Wavell's command quickly unravelled in the face of relentless Japanese advances across the archipelago. Wavell found his way to India and the remaining Allied forces there surrendered on 12 March 1942.[167]

An early European example of the role of cryptology in history was the Babington Plot to replace the Protestant Queen Elizabeth I on the throne of England by the Roman Catholic Mary Queen of Scots. Secret correspondence using a monoalphabetic cipher was intercepted, deciphered and manipulated by Elizabeth's 'spymaster' Sir Francis Walsingham, leading to Mary's execution.
Alamy

Top left: Signalman John Varcoe, awarded the Distinguished Service Medal and immortalised as the sailor in the statue at the Cenotaph Memorial in Martin Place, Sydney, served on HMAS *Parramatta* in the Mediterranean from 1917 to 1919. At that time, Australian forces had no separate Sigint organisations. When necessary, signallers monitored enemy, as well as own-force transmissions. Signalmen like Varcoe were early practitioners of Australian naval Sigint.
AWM ART02996

Top middle: Sigint played a crucial role in HMAS *Sydney*'s defeat of the German commerce raider SMS *Emden* in 1914. After a British wireless station picked up the *Emden*'s signal, the *Sydney* was sent a visual signal, to avoid breaking radio silence, ordering it to pursue the *Emden* and destroy it. Jack Ryan, a young wireless operator on the HMAS *Sydney*, would later command the highly effective Australian Army Special Wireless Group through the Pacific campaign as part of Central Bureau.
Alamy

Top right: The Battle of Jutland of May 1916. Had they been consulted, the Admiralty's Room 40 would have warned the Royal Navy that the German Fleet may have left port. They were not consulted, however, and the Royal Navy's Admiral Beatty was unpleasantly surprised to meet the entire German Fleet before Admiral Jellicoe's ships came to his aid from Scapa Flow. By ignoring Sigint, the British missed this opportunity to defeat the German Fleet.
2B2BKN3

Right: Commander R.B.M. 'Cocky' Long, RAN, Director of Naval Intelligence, was quick to recognise what the outbreak of war meant for Australia. He wrote to the Chief of Naval Staff, Sir Ragnar Colvin, recommending 'the Authorities at Home' be asked whether, in their opinion, a cryptographic organisation should be set up in Australia.
AWM 107006

Above: Unlike the heads of the other two services, the Army's Chief of General Staff, Lieutenant-General E.K. Squires, was firmly in favour of establishing an Australian cryptographic organisation. Squires instigated the inclusion of a divisional signals unit with a limited signals intercept and analysis capability when the militia 3rd Division was mobilised. He was also responsible for the establishment, in January 1940, of a small, cipher-breaking group of 'men of the professor type' from Sydney University.
AWM 00283 – L, AWM ART28533-1 – R

Above right: The German Army purchased several commercially produced Enigma machines which were built from a derivative of a design by Boris Hagelin, patented before the First World War. The Enigma machine had five variable components: (1) a plugboard which could contain from zero to 13 dual-wired cables; (2) three wired metal and black plastic cylindrical rotors, ordered left to right, with 26 spring-loaded copper input contact points to 26 copper contact points positioned on alternate faces of a disc; (3) 26 serrations around the periphery of the rotors, which allowed the operator to specify the rotors' initial rotational position; (4) a moveable ring on each rotor to control the rotational behaviour of the next rotor to the left by means of a notch; and (5) a non-rotating reflector half-rotor used to fold inputs and outputs back onto the same face of contact points.
Photo by author

Polish codebreakers of the Enigma: (from left) Marian Rejewski, Henryk Zygalski and Jerzy Różycki. The first breakthrough against Germany's Enigma code was made not in Bletchley Park, but in Warsaw, by these three formidable codebreakers. They had been breaking and reading some Enigma ciphers since the early 1930s and were the first to develop the 'Bomba', the original version of a British and later American electro-mechanical machine used to discover the wheel settings for Enigma keys. In 1939, they shared their knowledge and experience with GC&CS and gave them an Enigma machine.
MODWT0 (5)

Building on the work of the Poles, Bletchley's brilliant mathematician Alan Turing developed his own 'Bombe', capable of breaking the more complex wartime Enigma codes. A new technique that made the Bombe more powerful was the use of 'cribs' – assumed or known parts of the message – as a starting point.
Alamy 2BDX60B

The Bombe was the electro-mechanical machine developed in response to the challenge presented by the Enigma machine. In this machine, rotors and wires would simulate Enigma rotors and pass an electrical current from one rotor to the next, looking for the correct rotor setting.
Photo by author

Eric Nave was Australia's own world-famous codebreaker and outstanding Japanese linguist. He spent much of his career on loan to the Royal Navy but was instrumental in forging an Australian signals intelligence capability.
Courtesy of the Sea Power Studies Centre and Ian Pfennigwerth

Dr Abraham 'Abe' Sinkov was in the initial delegation to the UK in 1941 to negotiate Sigint early US–UK cooperation. In July 1942, Major Abe Sinkov arrived in Melbourne to become the American contingent commander of Central Bureau, under Major General Spencer B. Akin. In effect, Sinkov was left in charge and demonstrated strong organisational and leadership skills in addition to his mathematical prowess.
NSA, Public Domain

Commander Jack Newman's Office, Fleet Radio Unit Melbourne (FRUMEL). Newman was something of an internal politician, judging by his dealings with Australia's famous naval codebreaker Eric Nave, which did the latter's career no good, and his curious role in blocking a possible transfer of FECB to Australia at a time when it would have been advantageous to both Australia and Britain. On the other hand, he was a staunch and effective supporter of women in Australian Sigint. Messages intercepted and decrypted by FRUMEL analysts during the Battle of Midway bear Newman's signature.
NAA 7648043

Right Mrs Florence Violet McKenzie, or 'Mrs Mac', instructs two trainees as signallers and telegraphists. Mrs McKenzie founded the Women's Emergency Signalling Corps (WESC). Identified, left to right, Mrs McKenzie, Pat McInnes and Esme Kura Murrell. Mrs McKenzie was the firt woman to register under the *Electricians Licensing Act* of 1929 and at the time was the only qualified woman electrician in New South Wales. An electrical engineer and the first female proprietor of an electrical store in Australia, Mrs McKenzie also found the time to correspond with Einstein. Foreseeing the wartime demand for signallers, she opened a school to train women in Morse code and radio. She also overcame male reluctance to employ women. Her pupils made a major contribution in all three services, at FRUMEL and Central Bureau. A number served at the intercept sites of the wireless group in Australia, particularly Bonegilla where Japanese signals from Europe were intercepted. In the photo above, three uniformed members of the WESC at Circular Quay. Left to right: Unknown, Pat McInnes and Esme Kura Murrell.
AWM P01262_001, AWM P02583_001

Mrs Mac's trainees were high performers, like these senior WRANS officers.
NAA768045

WRANS officers and petty officers.
NAA768048

Above: Those working in the 'Code Room' could recognise who the operator at the other end was by the keying of a message. They could recognise which enemy operator was sending from an enemy base.
NAA7648055

Right: 'Training the trainer': The legacy of 'Mrs Mac' extended well beyond those women she trained personally. Mrs Mac's pupils are shown here instructing service personnel in Morse code.
AWM 017675

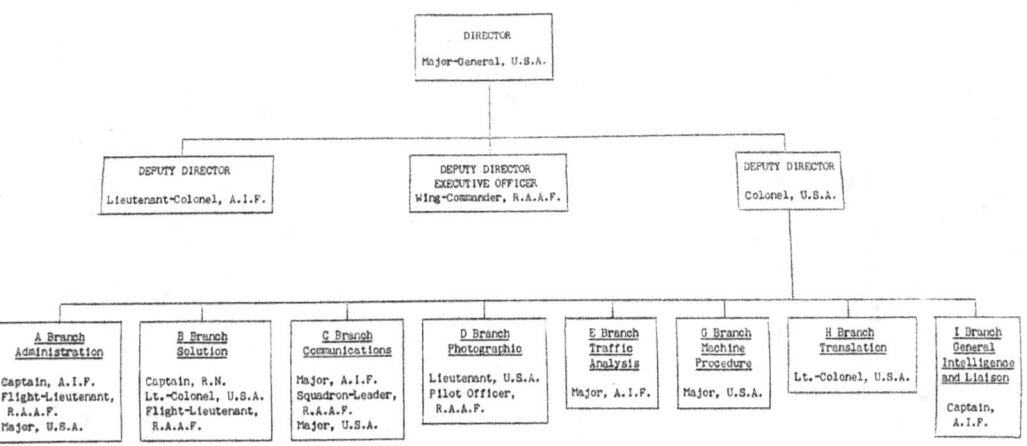

Diagram showing the organisation of Central Bureau: After escaping the Philippines, General MacArthur, now the Commander-in-Chief of the South West Pacific Area (SWPA), arrived in Melbourne. General Headquarters was a combined tri-service organisation which included the US Army, the US Army Air Force, the Australian Army and the RAAF. Central Bureau came under the Signals Branch of HQSWPA, headed by Major General Spencer B. Akin.
UK TNA – Author photo

Group Portrait of senior staff of Central Bureau, including Lieutenant Colonel A.W. Sandford, Deputy Director (back row, far right); Major General Spencer B. Akin, US Army Signals Corps, Director (front row, centre); and Wing Commander H. Roy Booth, RAAF, Deputy Director (front row, right). Largely due to the personalities of the Australian and American senior officers, working relations at Central Bureau were notably harmonious. By contrast, there was longstanding antagonism between the US Army and US Navy, with Sigint a major bone of contention. This had an impact on the operations of Central Bureau, despite the best efforts of the Australians and Americans in charge.
AWM P01443.010

WAAAF Members of Central Bureau with Wing Commander Roy Booth, by then Deputy Director of the Bureau, pictured at centre. Booth was one of the Australians who had worked at FECB, before taking command of the first 'Y' station in Darwin. He went on to play a major role in Central Bureau and in the deployment of RAAF Wireless Units during the war. Many of the WAAAFs were skilled Kana Morse code intercept operators who had intercepted Japanese signal messages from June 1942 through to 1944.
AWM P01443.043

Right: Headquarters of the Central Bureau 1942–45 at 21 Henry Street, Ascot in Brisbane. Central Bureau's Cipher Officer was housed in a garage with no air conditioning and a corrugated iron roof behind 21 Henry Street. The office was the communications hub for other Sigint centres such as Bletchley Park, Arlington Hall and FRUMEL as well as other Allied command centres.
AWM P00473.011

Alastair Wallace 'Mic' Sandford (pictured at the beginning of his career) was the intelligence officer of the Australian Army's Special Wireless Section, which served in Greece, then Crete and Syria before being recalled to Australia in 1942. He later became an assistant director of Central Bureau. Sandford established a cryptographic channel, between the Australian Army and GC&CS, independent of the United States, which was maintained until the end of the Pacific War. Along with Lieutenant Colonel Robert Little, Deputy Director Military Intelligence, he worked to preserve as much as possible of Nave's Diplomatic Sub-Unit which survived as 'D' Special Section, with the alternative name of Special Intelligence Section, in the immediate aftermath of the war. This proved to be the jewel in the crown, the nucleus of the Australian national signals intelligence capability of today.
Courtesy State Library of South Australia

A photo of Sandford later in his career, looking like the sophisticated senior intelligence officer and skillful diplomat he had become.
Courtesy State Library of South Australia

Officers of Central Bureau at Archerfield Aerodrome seeing off Colonel Sandford who is bound for Manila: (from left to right) Unknown; Wing Commander H. Roy Booth, RAAF, Deputy Director Central Bureau; Colonel Alastair Wallace 'Mic' Sandford, AIF, Deputy Director, Central Bureau; Squadron Leader W. 'Bill' J. Clarke, RAAF, Chief Signals Officer Central Bureau; Lieutenant Colonel Jack Ryan, Commanding Officer of the Australian Special Wireless Group. Ryan was a young wireless operator on HMAS *Sydney* when the SMS *Emden* was destroyed in 1914. From 1940, he was the Commanding Officer of No. 4 Wireless Section which served in Greece, Crete (where he was joined by Sandford, then an intelligence officer) and later Syria, before returning to Australia in March 1942.
AWM P01443.045

Above: Athanasius Treweek after the Second World War.
Family photo, used with permission

Right: Major Athanasius Treweek, an ancient Greek and Latin scholar as well as a mathematician, had taken it upon himself to learn Japanese in the lead-up to the war. Active in the militia, he avoided being sent to Singapore and was instead transferred to Victoria Barracks, Melbourne for 'administrative duties', where his code-breaking skills were a valuable asset.
Family photo, used with permission

HMAS *Australia* under attack during the Battle of the Coral Sea, 7 May 1942. The Battle of the Coral Sea was a tactical victory for the Japanese and a strategic one for the Allies. In effect, it was a dry run for the decisive Allied victory at Midway.
AWM 044238

The Japanese Aircraft Carrier *Shokaku* (30 000 tons) under attack, Battle of the Coral Sea, 8 May 1942.
AWM 148953

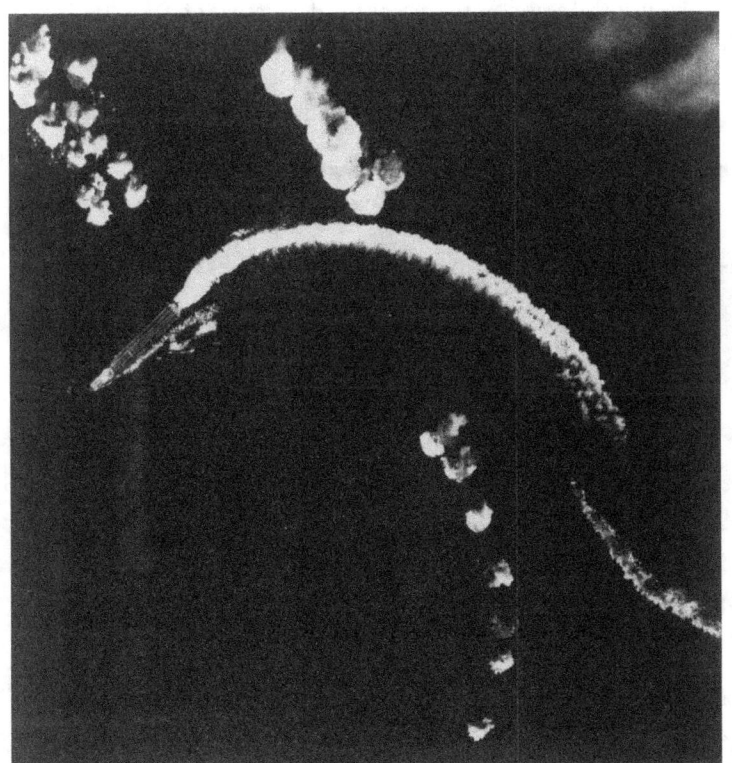

Aerial photograph of a burning Japanese aircraft carrier, June 1942. The Battle of Midway is touted as an American victory, but cryptanalysts of the Fleet Radio Unit Melbourne (FRUMEL) made a significant, but largely unrecognised contribution. Thanks to FRUMEL-sourced information, submarines knew when and where to prey on Japanese ships.
AWM P02018.114

The American Sigaba machine ensured the protection in transmission of the most sensitive material as well as the means to refer matters of higher policy back to headquarters for decisions.
Photo by author

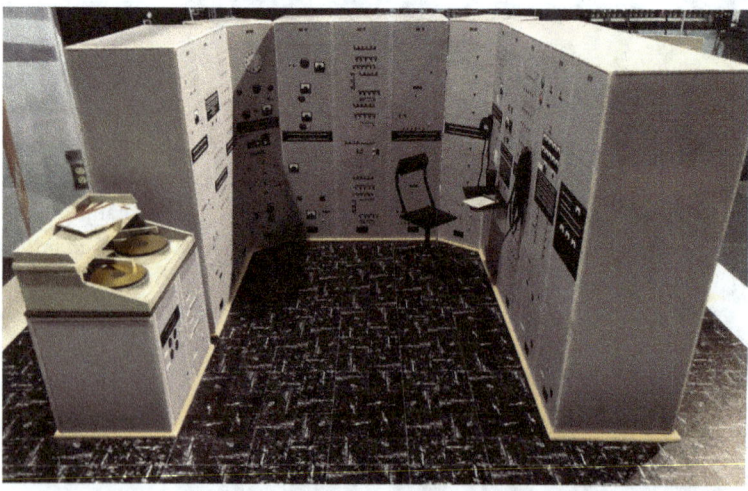

The Sigsaly was an enciphered voice communications telephone invented in 1942 with the assistance of British codebreaker and mathematician Alan Turing. It transformed voice into a digital data signal that, in turn, could be encrypted. The Australian terminal enabled MacArthur to speak securely with his counterparts in Hawaii and Washington and proved vital in protecting sensitive discussions and conferences about wartime arrangements.
Photo by author

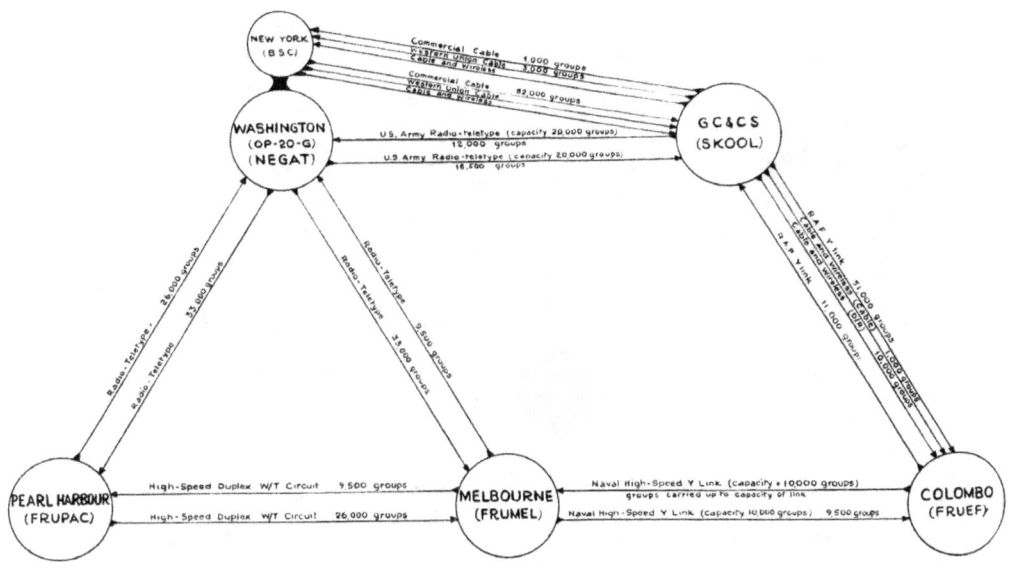

Diagram of the BRUSA circuit in operation in July 1944. The foundation for this was the BRUSA Agreement of 17 May 1943, which formalised and extended collaboration between British and American cryptanalysts and enabled the rapid industrialisation of Sigint in wartime, including the British dominions as well as the two signatories. BRUSA laid the groundwork for the Five Eyes partnership, of which Australia is an active member today.
UK TNA – Photo by author

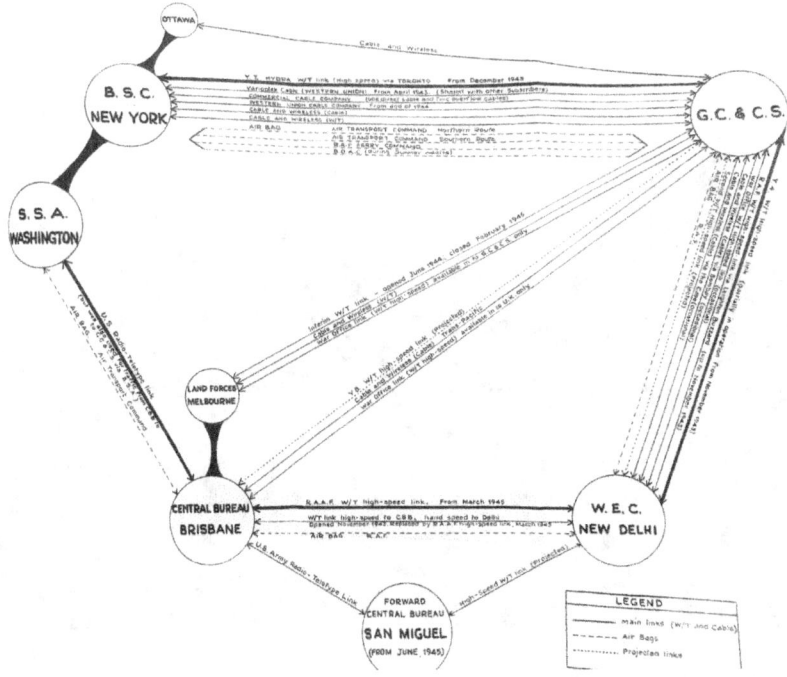

Diagram showing channels for dissemination of Ultra in SWPA as at 15 March 1945. In view of the key position of Central Bureau in this network, the consternation over Ultra leaks in Australia discovered by the British in December 1944 is not surprising.
UK TNA – Photo by author

One of two RAN Hawker Siddeley HS-748 aircraft used for electronic warfare training. The planned installation of electronic countermeasures (ECM) equipment was delayed due to the high cost. Both aircraft were flown to the United States to have the equipment installed and returned in 1981. The aircraft were then operated by a crew of seven: two pilots, one observer/navigator, a tactical coordinator, and three ECM operators.
RAN Seapower Centre, used with permission

Cabarlah, Queensland, direction finding station and the base of 101 Wireless Regiment (later 7th Signal Regiment), the Army's primary Sigint unit, commenced operation on 3 February 1947. From Cabarlah, the Regiment conducted strategic Sigint collection against 'certain Asian targets' under the technical control of the then Defence Signals Bureau.
Courtesy Bob Hartley

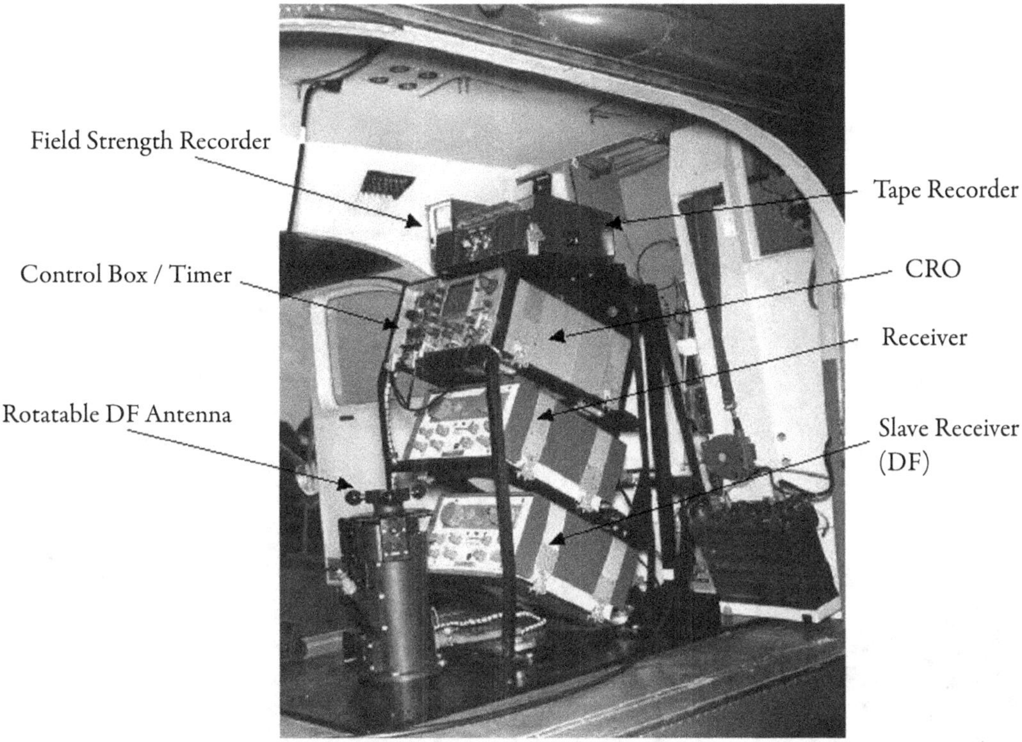

Field Strength Recorder

Control Box / Timer

Rotatable DF Antenna

Tape Recorder

CRO

Receiver

Slave Receiver (DF)

Above: Cessna 180 with Australian designed ARDF equipment. An antenna pod under the fuselage, was lowered once the Cessna was airborne. Cessna aircraft with the equipment on board would be flown up to six times a day to eavesdrop on enemy radio traffic. The aim was to pinpoint enemy headquarters and unit locations for taskforce intelligence and operations staff involved in planning. 547 Signal Troop's ARDF operators, all volunteers, flew at least once a day at low altitudes along straight paths within machine gun range over enemy territory.
Courtesy Bob Hartley

Left: Justice Robert Marsden Hope foresaw increasing public interest in the activities of intelligence agencies, and in 1977 proposed rules to ensure that DSD kept strictly to its foreign intelligence mission. Even Hope, however, did not foresee the advent of cyber and its implications.
NAA A12386 EO 12

Joint Facilities at Pine Gap: The 'Agreement with the Government of the United States of America Relating to the Establishment of the Joint Defence Space Research Facility' at Pine Gap in the Northern Territory was signed in Canberra on 12 September 1966. The facilities epitomise the 'ties that bind' Australia and the United States, but there have always been critics, whose warnings tap into the imprint left by Australian Sigint history, including fears about being still the junior partner, about being dependent and a lack of national self-confidence. It is incumbent on policymakers and intelligence practitioners today to weigh up carefully the pros and cons of Australia's investment in the intelligence infrastructure shared between Australia and the United States. To date, the bipartisan view has favoured maintaining and even deepening those ties.
Courtesy of Kristian Laemmie-Ruff

Right Commander James Armstrong with his son, Tristan. After 35 years in the Royal Navy, Armstrong transferred to the RAN in 1979. The new building housing RANTEWSS was named after him in recognition of his pivotal leadership.
Family photo, used with permission

Below The 'Melbourne component of 7th Signal Regiment' paraded in August 1979 for the final day of duty at Albert Park Barracks. Established in 1942, with assistance from the US Marine Corps, the Barracks were used continuously through to 1979. The detachment commander, Major Clive Williams is pictured saluting the flag being lowered. The detachment relocated to new purpose-built facilities inside Victoria Barracks on St Kilda Road, where it stayed until DSD relocated to Canberra in the early 1990s.
Photo supplied by Clive Williams

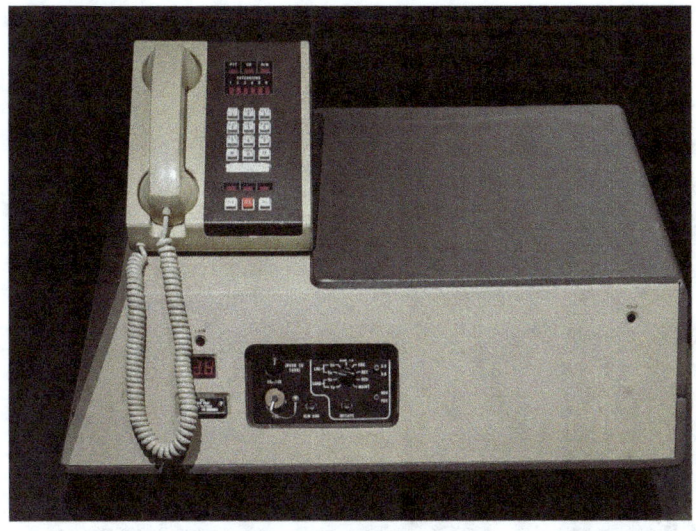

STU-II (KY71) was developed by the United States' National Security Agency in the 1970s and enabled a speaker's voice to be encrypted so as to be unintelligible to an eavesdropper.
Photo by author

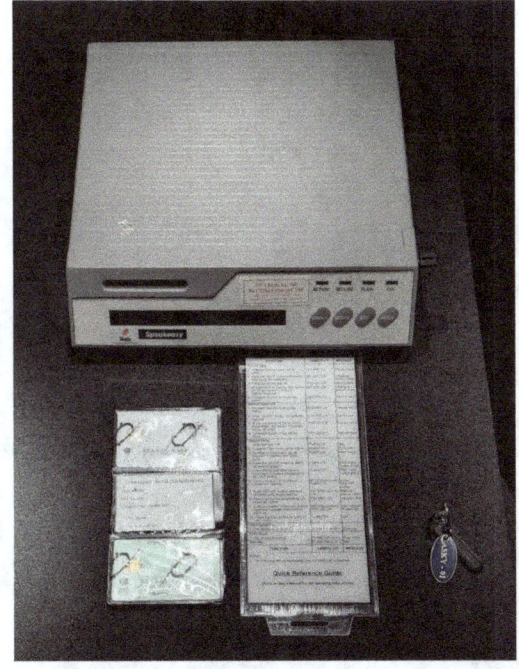

The Speakeasy: DSD and Telstra developed this first Australian-designed secure telephone. It was superseded by secure mobile telephony in 2002. Each Speakeasy was set up and periodically re-keyed, using two cards specific to the device. These were produced by DSD and securely distributed to the Speakeasy user.
Photo by author

Deployed Sigint Support Facility.
Photo by author

Cray computer.
Photo by author

6
SIGINT ARRANGEMENTS & THE WAR IN THE PACIFIC

Japan's entry into the war in the Pacific dramatically altered the dynamics of Australia's Sigint arrangements. Up to this point, they had been maturing gradually, but from this time on they would be transformed. A major contributing factor to this upheaval was that, for the first time, Australia effectively handed over the command of its armed forces based in Australia to a foreign commander, General Douglas MacArthur.

Initial reconfigurations

Following the surprise attacks on Pearl Harbor, Malaya and the Philippines, the British and American forces were caught off guard and ill-prepared. MacArthur's command in the Philippines proved no match for the experienced and battle-hardened Japanese. MacArthur was ordered to leave his fortified headquarters on Corregidor Island in Manila Bay for Australia. From there, he would be charged with defending Australia and mustering forces to mount a counterattack against Japan. This chapter considers the Sigint arrangements Australia put in place to work collaboratively, particularly with the US forces in the Pacific, following the Japanese attack on Pearl Harbor and the Japanese thrust southwards. It also addresses the formalised UK–US Sigint arrangements and how these shaped the Sigint arrangements that contributed to Australia's war effort.

With the defeats at the hands of the Japanese across Southeast Asia and the Pacific, what was left in Allied hands would eventually be divided

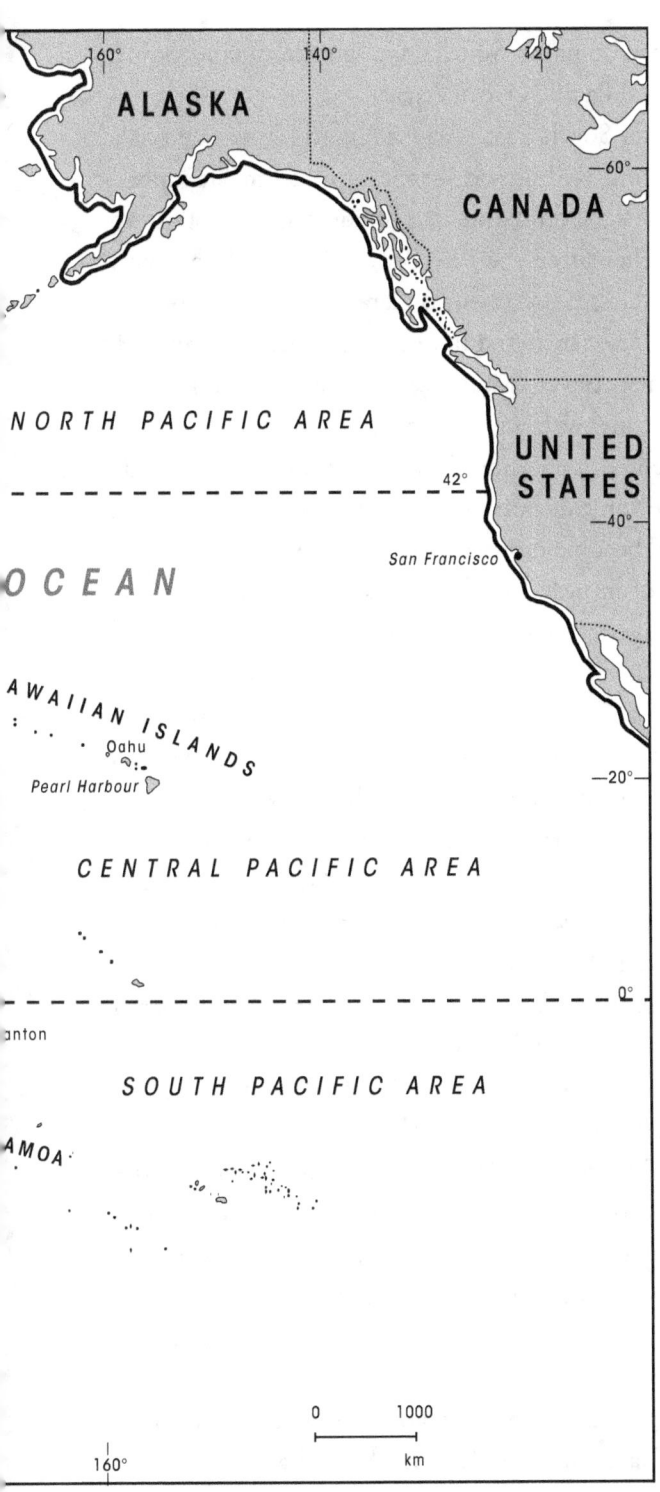

Figure 2.
Pacific areas map, 1942

geographically into three broad domains: South East Asia Command, South West Pacific Area (SWPA) and Pacific Ocean Area.

Britain was responsible for South East Asia Command, covering India and Burma. The FECB, having withdrawn just in time from Singapore, remained with South East Asia Command to provide Sigint support to British empire forces there. Thereafter, the FECB would work closely with the misleadingly named 'Wireless Experimental Centre' (WEC) in British India. Australia, however, was not included in South East Asia Command. With this line drawn on the map, ties to Britain were suddenly less important for Australia's Sigint operations, although connections with the United Kingdom would remain essential for standards, interoperability and a wide range of support measures. From this point on, and for the duration of the war, the United States would become more important for Australia.

In office only a couple of months, after the loss of confidence by the House of Representatives in Robert Menzies, and the 40-day interregnum of Arthur Fadden as prime minister, Australian Prime Minister John Curtin faced extraordinary challenges. This former First World War pacifist found himself leading a nation in what seemed to be an existential war. America's General MacArthur, a politically conservative Republican, became the key partner of the Australian Labor wartime leader. The commander-in-chief of Australia's military forces, Blamey (promoted to General following his return from command in the Middle East), and the secretary of the Department of Defence, Frederick Shedden, were effectively relegated to a second-tier status. Eager to accommodate MacArthur, Curtin was prepared to make significant concessions, including giving the American general *carte blanche* for the command and control of the forces. This included the necessary intelligence arrangements to prosecute the war.

At the beginning of the Pacific War, Australia was not yet privy to the sensitive Sigint arrangements concerning Ultra and Magic – the codeword used for the decryption of Japan's Enigma equivalent, the Purple Machine – that had been agreed between the United States and United Kingdom. Australia had no special liaison units (SLUs) or special communications links for carrying Ultra reports.

Meanwhile, there was justifiable concern about Australian security. The British were concerned that security within the Australian government and its institutions was lax. Ministers disregarded security measures, with Evatt and Curtin in particular providing briefings on sensitive matters to the media. Then there was the cavalier approach to security within the Australian military itself. Only the Sigint organisations and the senior levels of naval and military intelligence were seen to be handling sensitive security matters appropriately. Later in the war, there were revelations of leaks from Australia via Chinese Nationalist reporting and via the Soviet legation in Harbin, within Japanese-occupied China.

British officials understood that neither they nor the United States could prevent Australia from intercepting Japanese diplomatic traffic if Australia decided to do so. The answer was pragmatic: accept that keeping the Japanese ignorant of Sigint successes was vital, and this meant bringing Australia into the tent, because once inside, 'they could force [the] Australians into complying with the security requirements' that were a strictly maintained precondition for those with access to Ultra material.[1]

Naval Sigint Sharing, and the fate of the FECB

By the time of the Pacific War, naval Sigint had developed considerably over the decades since the First World War. New weapons technologies, and new detection and intelligence capabilities such as radar, high-frequency direction finding and Sigint necessitated major tactical changes. The implications would become clear in the first months of 1942.[2] This period was as chaotic for Sigint as it was for other aspects of war planning and operations. With the surrender of Singapore on 15 February 1942 and the American surrender at Bataan on 9 April, British and American Sigint specialists who had managed to escape arrived in Melbourne. The majority of the FECB had withdrawn to Ceylon, but a small number had also made it to Melbourne. Americans came in far larger numbers, with up to 75 arriving by air and submarine. These men would form the American core of the US Navy's Fleet Radio Unit Melbourne (FRUMEL) and MacArthur's Central Bureau. A coordinated and integrated effort seemed likely and would have been sensible to organise, but it was not to be.

The organisation that emerged following the fall of Singapore was very nearly quite a different one. On being forced by the Japanese threat to evacuate Singapore, Melbourne was the clear first choice for the FECB base, as it would offer a safe place to continue its activities. On 22 December 1941, the commander-in-chief of Britain's Eastern Fleet sent the following signal to the ACNB in Melbourne:

> In view of anticipated scale of air attack I am considering removal of Naval Y and Naval Special Intelligence Sections from Singapore by first opportunity. Request your early observations on whether this organisation could be set up in Australia. Party would consist of about 20 officers, 55 operators of which 30 are WRNS, 14 women assistants and four clerks. Total personnel, including wives, roughly 50 men and 60 women plus some children. Equipment would include 20 English [sic] receivers.[3]

On Tuesday 23 December, Lieutenant Commander Jack Newman, the head of the Australian naval W/T Intelligence Unit responded – evidently without consulting Eric Nave or the DNI, Commander Long. His reply was surprisingly inhospitable: 'No spare receiving station exists at present, but Naval Board would make every effort to hasten erection and installation. Office accommodation can be arranged.'[4] Reading this as a less than enthusiastic welcome, the commander-in-chief of the Eastern Fleet turned to the commander-in-chief East Indies. The reply on Christmas Day was brief and both forthcoming and welcoming: 'Accommodation available immediately. Request officer may be sent first opportunity to arrange details.'[5] Commenting on the FECB's request and Newman's response, Nave recalled:

> I only saw these messages after despatch; this was a great pity, had they signalled the Defence Department accommodation would have been provided. Now this highly experienced unit would have to evacuate the Singapore location and would largely lose its effectiveness ... it was a tragedy that experience built up over so many years was now dissipated.[6]

Nave was right. After FECB's move to Colombo, on 15 April 1942, again threatened by Japanese advances, a section of it moved to Kilindini in Kenya. During the next six months after the transfer to Kilindini in June–July 1942, (British) naval intelligence in the Far East 'sank to its nadir and lay there'.[7]

Newman's motivation in this matter of the FECB's transfer is questionable. The FECB could have been accommodated, and the receivers required could have been found. Australia would have benefitted from their experience and numbers. It is possible that Newman feared his 'patch' would be taken over by FECB and seized this opportunity to prevent it.

Evidently, there was no love lost between Nave and Newman. In his unpublished autobiography, Nave recounts that in the period immediately before war with Japan, when Nave was on loan from the Royal Navy, attached to its Australian counterpart, Newman, then Director of Communications, suggested that he and Nave 'keep the D.N.I. out of these matters and handle them ourselves'. Nave felt he could not agree to this, as it was contrary to practice both in London and the Far East. Nave recalls that 'it coloured my relations with him from that time on. He lost no opportunity trying to undermine my position, but I was far too busy to allow this to intrude into my work, and as I was the only experienced body available, I had to do practically everything myself.'[8]

US Navy Sigint team arrival and FRUMEL

With the fall of Corregidor imminent (it fell on 6 May 1942), the US Navy Sigint team arrived in Australia, and it was decided that the Australian Navy element of the Special Intelligence Bureau in Melbourne should join these recently arrived US Navy personnel to form a new unit, Fleet Radio Unit Melbourne, or FRUMEL.[9]

The equipment salvaged from Corregidor was, by Australian standards, quite extensive, and included portable direction-finding equipment and valuable Kana typewriters. They had no Hollerith-derived IBM card-punching equipment, however, and the Purple machine that OP-20-G had sent needed multiple faults fixed before it could operate with any reliability. Ralph E. Cook, a US Navy electrical engineer and former IBM employee, was one of those sent to Australia. He began automating as much of the cryptanalytic process as he could, and he kept the IBM equipment running. His main problem was obtaining more IBM equipment, and it would be late 1942 before any more arrived from the United States. Even then, the equipment was missing the card feeder and printing mechanism for the tabulator. The tabulating section at FRUMEL had to hand-punch cards for three months until the missing equipment arrived.

A key factor in Sigint arrangements from this point on for the rest of the war was that US Navy and US Army chains of command were kept quite separate, and this also applied to Sigint. Tensions between the US Army and US Navy from the interwar years had not abated, and this certainly affected US Navy Sigint personnel in Melbourne.

In light of these strained dynamics, the Allies had agreed that the Pacific Ocean areas, including New Zealand and part of the Solomon Islands (including Guadalcanal), were to be assigned to the US Navy under Admiral Nimitz based in Hawaii. The Hawaii-based Fleet Radio Unit Pacific (FRUPAC) was the principal Sigint organisation supporting Nimitz; but, from 1942 onwards, FRUMEL was established as a combined US–Australian naval organisation under command of a US Navy officer, Lieutenant Commander (subsequently promoted to Captain) Rudolph Fabian; a career military officer who would run FRUMEL 'in rigid military fashion'.[10]

Fabian had worked at the US Navy Station CAST before it was bombed out of the Cavite Naval Yard and forced to move to the tunnels on Corregidor Island. On 5 February, Fabian and 16 others were transported to Java by submarine to assist the Dutch cryptanalysts at *Kamer 14* in Bandung on the island of Java. Then, as the Japanese closed in on Java, they were taken by a second submarine to Fremantle, before then travelling across to Melbourne.[11]

In Melbourne, and thus within MacArthur's area of operations, FRUMEL worked with FRUPAC to meet Nimitz's intelligence requirements. MacArthur was not given command of FRUMEL. On one level, that was understandable, as FRUMEL was intended to serve naval requirements more than MacArthur's land-based priorities. Yet in a move that astounds practitioners, FRUMEL was not even authorised to liaise directly with its US Army Sigint counterparts in Central Bureau.[12] Central Bureau and FRUMEL did not exchange technical cryptanalytic material. Instead, as the NSA historian Edward Drea has observed, 'they worked on separate cryptanalytic material as if the other did not exist'. As a concession, Fabian would provide regular update 'information only' briefs, but they had to be in person with MacArthur.[13] The firm distinction between the remits of Central Bureau and FRUMEL appear to reflect the tense relations between MacArthur and his naval counterpart, Nimitz, in Hawaii.

While FRUMEL operated under draconian restrictions, it did provide insights to the commander of the US Navy's 7th Fleet, to MacArthur in person, and to the US submarine command based in Western Australia. Notwithstanding the strict distinction between the commands, FRUMEL's contributions would transform the military equation in the Pacific in the battles of the Coral Sea, Midway and beyond.[14]

Housed at the Monterey Apartments in Melbourne, FRUMEL soon had teletype machines in operation 24 hours per day, receiving message traffic from its collection sites scattered around the country. IBM Hollerith machines had been used by the US Sigint sites at Corregidor, but as it was impossible to load them onto the submarines bound for Australia, they were left to be dumped at sea. Instead, new machines were ordered from the United States for use by FRUMEL. There, they were set up in the garages to the Monterey Apartments, about 1.6 kilometres from Navy Headquarters in Melbourne.[15]

Moorabin, in suburban Melbourne, was FRUMEL's principal intercept site, and remained so until relocated to Albert Park in 1944. Forward intercept units were also established in Townsville, Queensland; Adelaide River in the Northern Territory; Exmouth in Western Australia; and, for a few months in 1944, at Cooktown in northern Queensland.[16] By the end of the war, the US Navy had 775 receivers in the Pacific theatres, of which 58 were in Australia.[17]

The Holden Agreement

After the United States joined the war in response to the attack on Pearl Harbor on 7 December 1941, US pressure for exploitation of Enigma and building its own version of the powerful mechanical aid known as 'Bombes' increased. Finally, in response to a signal of 10 September 1942 that the US Navy had decided to commence Bombe construction, Commander Travis, then head of GC&CS, visited Washington. In the resulting 'swap', Britain took responsibility for the supply of German and other European codes and ciphers, while the United States took the lead on Japanese codes and ciphers. This was the basis for the Holden Agreement,[18] which was decided between the two parties, at a senior level and without regard for Nave or Australia's newly created Special Intelligence Bureau in Melbourne, and came after FRUMEL and Central Bureau were already taking form and commencing operations.

The 2 October 1942 Holden Agreement was named after Captain C.F. Holden, USN, who was Director of USN Communications and the United States signatory. The Holden Agreement was the latest in a series of negotiations between the United States and the United Kingdom in which Australia had little say. Remarkably, it also set out to dispose of Nave's Australian naval Sigint unit and Nave himself in particular, casting doubt on the future of the D Special Section.[19]

To understand the Holden Agreement and its implications, it is worth looking back to the beginning of collaboration between the British and the Americans in anti-Japanese Sigint in February 1941. This related to collaboration between the FECB and the corresponding US Army Radio Intelligence Unit in Manila which was disrupted by the fall of Singapore and Corregidor.[20]

In the framework of this rather informal collaboration, the Americans gave the British the Japanese diplomatic cipher machine and (in the Far East) some minor Japanese naval systems. In exchange, they received, from the FECB, all past and present knowledge about, and recoveries from, JN-25 – the principal Japanese naval code; and from GC&CS, the full range of British investigation of diplomatic systems and of all German service ciphers except Enigma. Frank Birch, British Sigint historian and drafter of the Holden Agreement, writes that this was an excellent deal for the United States.[21] It was also not a bad one for the British, as according to the British naval historian, Nigel de Grey, it eventually made possible 'a resurgence of anti-Japanese Sigint'.[22]

What the United States really wanted, though, was complete knowledge of Enigma. The methods of solution of the Enigma machine were disclosed to them, but they were pledged not to pass on this information to any but a very few specified officials. In addition, they were not to ask to be shown any resulting intelligence. Part of the justification for this caution was that the Americans were not yet in the war.[23] Another reason was that, with the growing threat of war with Japan, the British wanted the Americans to concentrate on the Japanese problem. However, the United States had other ideas and during a visit to Washington in August 1941, Commander Denniston found that the US Navy Department was investigating the Enigma problem independently.

Denniston then proposed an exchange between officials in February 1942 on cooperation. Discussions were held in Washington between the Americans, British and Canadians, from 6 to 13 April 1942, followed by a general conference. Three key recommendations emerged: (1) Canada was to be linked with the US network for interception in North America of Japanese communications; (2) with regard to Japanese naval material intercepted in Washington, London and Ottawa, all JN-25 intercepts were to be forwarded to Washington and all flag officer cipher and naval attaché traffic was to be forwarded to both London and Washington; and (3) a permanent committee was to be set up to coordinate the activities of the respective intelligence organisations (although this did not happen until 1944).[24]

The Holden Agreement provided that the US Navy would take primary responsibility for IJN traffic decryption. The agreement specified:

> 1.(a) The British to disband the British-Australian naval unit at Melbourne and turn over to the U.S. unit there such personnel as the U.S. may desire, except Commander Nave, who is to be recalled. Requests by the U.S. for any particular individuals from Kilindini or Melbourne will be entertained by the British. The future status

of the diplomatic party at Melbourne will depend upon wishes of the Australian Government and the senior naval and military authorities in that area, which the Admiralty will ascertain.[25]

The struggle over Nave

FRUMEL was established as a combined naval organisation with a US and an Australian commander. The US section under Fabian, the Australian section under Nave and the Australian W/T intelligence team under Commander Newman were housed together in Monterey Apartments. It is perhaps an understatement to say that FRUMEL witnessed, for a time, 'some clash of nationalities and personalities'.[26] Nave, for one, was involved, and ended up being blamed for the tensions; almost certainly unfairly.

Nave's fate seems to have been sealed by a damning report written by a British visitor, Lieutenant Commander E. Colegrave. The report lays the blame entirely on Nave's inability to 'hit it off' with either Newman or Fabian and, indeed, his alleged inability to cooperate with anyone.[27] An insight from one of Fabian's men, however, casts a different light on the clash of personalities and what happened next:

> [Fabian] said that Nave was handling [communications intelligence] insecurely. Fabian was known to be a fanatic about Comint security (he had discussed shooting all 60 men in the unit in Corregidor to avoid their capture and interrogation by the Japanese) but I think it could have been a way of getting rid of a rival who was the logical candidate to put in charge of a combined US and Australian Comint unit. Nave was senior, older and more experienced at solving Japanese naval codes and a better Japanese linguist.[28]

Nave's internal political difficulties were certainly not all of his own making. In addition to the personalities involved, it appears they stemmed from Australia's position as junior partner to the British and the Americans and a lack of assertiveness on Australia's part in that role.

Nevertheless, Nave was made to shoulder the blame, with the Holden Agreement singling him out by name for recall. This must have indicated that Fabian at the time had the high-level bureaucratic backing that Nave lacked. What stands out in this message is the description of Nave's unit, known as the Special Intelligence Bureau (or SIB), as 'British–Australian' when, in fact, it was a solely Australian unit. The Australian authorities apparently did not consider it worth their while to take issue with their unit being disbanded and turned over to the United States by the British.

Understandably, Nave wanted to know the reason for the decision to recall him, and pointed out to his superiors in GC&CS that:

> In [the] past 12 months we have tackled and solved most lower grade naval codes, intercepted and passed on particulars to other posts interested. We have read much of this traffic and supplied the intelligence to the United States Navy here. Almost daily this information, which is mostly of urgent but local operational value, is being passed on by them to the authorities concerned and they have on many occasions remarked its value.[29]

The struggle over Nave with the Admiralty, who wanted him back at GC&CS, went to the highest level, as the following message from the Admiralty indicates: 'Their Lordships regret that the decision to recall Paymaster Commander Nave to duties in UK must stand'.[30] Despite their Lordships' edict, Nave, who did not wish to relocate to the UK for unspecified research duties, was to find a berth in Central Bureau through the good offices of the Australian Army. A minute from Lieutenant Colonel 'Mic' Sandford advising that General MacArthur supported Nave's retention suggests Sandford helped secure this valuable backing.[31] This United States support would be abruptly withdrawn, however, 'due to the British Navy's need for his services'.[32]

Nevertheless, with some skilful lobbying and fast footwork by Sandford, Lieutenant Colonel Robert Little (the Deputy DMI) and Lieutenant General Smart (Australia's senior military representative in London), the Admiralty eventually agreed to Nave's loan to the Australian Army in exchange for the return of two Foreign Office staff who had been attached to the SIB, and a visit by a 'Special Intelligence Officer' (none other than Sandford) to confer with GC&CS and 'explain our set up'.[33]

The BRUSA Agreement

In due course, the collaboration outlined in the Holden Agreement would be deepened and formalised in the agreement between Britain and the USA known as BRUSA. Signed on 17 May 1943 in Washington, DC, by Commander Sir Edward Travis, who succeeded Denniston as head of GC&CS, and by Major General George Strong, the head of the US Army Intelligence Branch (G2), this arrangement greatly extended collaboration between British and American cryptanalysts. The term BRUSA was coined by 25-year-old Harry Hinsley, who would later write a multi-volume history of British Intelligence in the Second World War and who, later in the war, would visit Central Bureau as Travis's assistant. Reflecting US inter-service rivalries, the US Army was at first reluctant for its British counterparts to inform US Navy interlocutors of the BRUSA Agreement.[34] With the passage of time, however, the arrangement would eventually become more widely known. In effect, it laid much of the groundwork for the terms and conditions that would be agreed after the conclusion of hostilities in what is now known informally as the 'Five Eyes' partnership, of which Australia is an active contributing member.[35]

While the Holden and BRUSA agreements set out the terms and conditions of deeper Sigint collaboration, they also provided the framework for a wide range of procedures and processes, some of which were already in place, to be implemented and standardised. These arrangements would enable the rapid industrialisation of Sigint in time of war, not only between the two signatories but also the British dominions. They would also epitomise a deepening relationship. While a number of these measures might be seen by the uninitiated as nitpicking and unimportant, they were essential; not only for the protection of the intelligence from the most sensitive of sources, the potentially war-winning Ultra decrypts, but also for regulating a complex and growing system. Many of these measures would prove to be the makings of the lingua franca for a new global enterprise and would be readily recognisable to those in the business today.

Diplomatic (D) Special Section

While the work of FRUMEL and, later, Central Bureau fitted into the overall scheme outlined as part of the BRUSA, they focused on the Sigint directly related to winning the battles. Beyond that, though, there was also some other important, but highly sensitive Sigint work undertaken by a little-known entity known as the Diplomatic Special Section, or as it was more commonly (and more opaquely) known, D Special Section, which was set up by Nave. Knowledge of this section's work would remain a closely guarded secret for two decades after the Sigint stories of the Army and Navy became public, with the Australian government refusing to admit that it engaged in the interception of diplomatic communications, even in wartime.[36]

The 52nd Australian Army Special Wireless Section (52 ASWS) of the Australian Army Special Wireless Group (ASWG) at Bonegilla and later at Mornington in the state of Victoria intercepted Japanese diplomatic communications which were later decoded by D Special Section. The 85 men and women of 52 ASWS operated 24 hours a day. When reception was poor on certain frequencies, monitoring was conducted from other ASWG stations in Darwin and Kalinga, Brisbane. Intercept was also received from the New Zealand Army special wireless section at Nairnville Park, near Wellington, and a Royal New Zealand Navy (RNZN) station. The New Zealand stations extended D Special Section's coverage another 2500 kilometres further east. The New Zealand intercept was forwarded to Australia on service lines to Central Bureau and FRUMEL. All diplomatic traffic was sent to D Special Section. In Melbourne, the intercept was again sorted and those messages that could be exploited in Melbourne were retained with the rest sent on to the Government Communications Bureau at Berkley Street in London – the diplomatic section of GC&CS.[37]

Japanese diplomatic traffic from their posts in Asia and elsewhere, including Nazi Germany and neutral European states, yielded valuable insights and intelligence. From Southeast Asia came reports on requirements for raw materials for the war effort, delivery means and the effectiveness of allied interdiction of the sea lanes. A message of considerable local interest from the Japanese representative in Dili revealed that the enemy was reading the Australian guerrilla code in Timor.[38] From Europe came some of the first reports following the attempt on Hitler's life, including one from the Japanese minister in Stockholm to the Foreign Ministry in Tokyo. The message was intercepted by 52 ASWS at Bonegilla in July 1944 and passed to D Special Section in Melbourne for decoding. The intercepted message explained that a coup d'état was planned by an elite German Army group known as the *Adelklassen* (nobility classes), who wanted peace with Russia. The Gestapo had already discovered that there was a plot and had begun to act. This forced the conspirators to speed up the execution of their plan, which led to its failure.[39]

Japanese diplomatic reporting on war planning and operations, while stock-in-trade for the section, were not the only Sigint reports offering rare insights. The most productive source was the W/T circuit from Tokyo to the temporary Soviet capital at Kuibyshev (now Samara), some 870 kilometres east-south-east of Moscow with the call sign RTZ. From February to July 1943, Bonegilla intercepted 1421 messages from Kuibyshev to Tokyo and 1212 from Tokyo to Kuibyshev. By May 1944, RTZ messages collected from the site at Mornington accounted for 27 per cent of the messages from stations outside Japan.[40] The realisation was dawning on the Allies that, because of the vagaries of high-frequency communications, many of the messages intercepted in Australia and New Zealand were unobtainable from sites in Great Britain and North America. D Special Section was proving its worth in the emerging global Allied Sigint enterprise.

As a result of the Holden Agreement, the SIB was dissolved, and on 5 November 1942, the Chief of Naval Staff wrote that the 'United States Officer-in-Charge has recommended that the Diplomatic Section should cease to function within the Combined Bureau (FRUMEL in this case) and that this work should be carried out in Washington or London'. The Chief of Naval Staff added that he concurred with this proposal.[41] However, after the intervention of the army's Deputy DMI, the skilful bureaucrat Lieutenant Colonel Robert Little, the decision to abandon the diplomatic effort was overturned. Little had advised that 'it would be a retrograde step to disband the Section', and so they decided they should continue to function under Army auspices in future.[42]

Retaining the diplomatic Sigint capability in Melbourne under the control of Australia's DMI had the advantage of keeping it out of the clutches of MacArthur. This small group of expert diplomatic codebreakers informed decisions of the War Cabinet and was, in effect, the jewel in the crown of Australian Sigint.

Although D Special Section would lose Nave as well as professors Room and Treweek to Central Bureau, it did continue to receive small numbers of trained linguists from the British Consular Service, as well as other highly qualified civilians and Australian soldiers with the requisite aptitude and academic credentials. The Army's senior officer at Central Bureau through the latter war years, Lieutenant Colonel Sandford, was a strong supporter of D Special Section, loaning both military cryptanalysts and linguists when the unit was under heavy pressure.[43]

Staffing decisions were often at the whim of Professor Trendall. One linguist, John Charles Davies, fell out with Trendall, and was sent away to Central Bureau. Another highly qualified potential recruit, John Thomas Laird, never made it to D Special Section but was despatched directly to Central Bureau.[44]

Beyond managing the increasing Japanese diplomatic traffic, there were other demands on these soldiers and civilians, who never exceeded 33 in number throughout the war. In addition to the punishing regime of working 16-hour days, seven days a week, just to keep up with the traffic flow, a number of key staff also had other duties. The intensity of the work did take a toll. For example, a Mr H.R. Sawbridge, who helped make Japanese language propaganda broadcasts, had a nervous breakdown, presumably from undertaking both his regular work and the additional broadcasts. The colleague who replaced him also collapsed under the strain of work.[45]

The case to maintain the diplomatic effort turned on the experience and skills of the linguists in Melbourne, especially the British cohort from Singapore and the quality of intercept from the Army's Australian Special Wireless Group site at Bonegilla. With the Japanese Foreign Ministry only using two high-grade ciphers at the time, the task for D Special Section was to solve the one known as Fuji. Traffic in the other, known as JAA (Purple), was forwarded to GC&CS for processing on their replica Purple machine.[46]

Australian arrangements for protecting 'Y' material, and continuing security concerns

Prior to these agreements Australia, in its own way, had set out procedures for the protection of 'Y' material. In July 1942, with the authority of the CGS, the DMI, Colonel John Rogers, introduced procedures to protect the transmission of intercepted enemy wireless communications, as well as stipulating how such information could be used below army division level on the battlefield (that is, at brigade and battalion level).[47] More detailed instructions were issued in January the following year by the Australian Army's Signal Officer-in-Chief (SOIC), Major General Colin Simpson, setting out cipher procedures for the handling of 'Y' material.[48] Both of these directions, and at least one other similar instruction from Deputy DMI, Lieutenant Colonel Little, reflected the wishes of the 'Y' Board in London, with Little informing the DMI, the SOIC, Central Bureau, and the D Special Section of enhanced requirements.[49]

Personnel security was among the considerations for the protection of Ultra and other Sigint. In December 1942, Australia's senior military representative in London, Lieutenant General E.K. Smart, was notified by the British Deputy DMI of their requirements for the security vetting for all who might be engaged in Sigint work, as well as those who would have access to such intelligence. While this was in the context of Australians needing to be 'screened' (the term in use in Australia at that time) in order to have access to British intelligence or engaging in Sigint work with the British, it was also suggested in the language of the day that Australia itself might like to consider such arrangements.[50] Smart informed the CGS, Lieutenant General John Northcott, of the British requirements and vetting processes, with the Australian Deputy DMI acknowledging the requirements, aspects of which do not seem to have been implemented promptly or comprehensively.

Sir Edward Travis, the Director of GC&CS, continued to be concerned that the D Special Section, now operating under the auspices of the Australian DMI from Victoria Barracks in Melbourne, among other entities, was not complying with instructions. London was also seeking reassurance that Australia was prepared to accept British regulations regarding Ultra and Sigint. Australia was reminded, in an oblique reference to the Holden Agreement, that British Indian forces, the US Navy and potentially the other US forces (Army and Army Air Force) had accepted these regulations.[51]

The head of the British Army and RAF Liaison Staff in Australia, Major General R.H. Dewing, perhaps for the first and certainly not the last time, reinforced GC&CS's requirements and need for reassurances of correct Australian implementation. There was some confusion on the Australian side, with separate and, at times, conflicting instructions being sent to cipher staff. Lieutenant Colonel Sandford, who was often at the centre of Sigint negotiations, was on an important extended visit to GC&CS at the time. He was asked to seek clarification; the staff of the Australian DMI provided reassurances that arrangements were in place to ensure implementation of the new instructions.[52] Within this new series of instructions were definitions of what constituted 'Special' Intelligence and what did not. There were also directions on the use of code words, standardised terminology and the use of ciphers for transmitting Sigint.[53] Aspects requiring clarification seem not to have been cleared up until Sandford reported to DMI Rogers on his findings in London in late October 1943. Even at this time, the US Army still had not promulgated specific regulations for the handling of Ultra and 'Y' intelligence.[54] Central Bureau and Australia were perched uncomfortably on the fault line of, at times, conflicting British and United States approaches.

Revised empire-wide instructions for Sigint and Ultra

In 1943, the London Signal Intelligence Board (LSIB) succeeded the 'Y' Intelligence Board, reflecting a wider reorganisation following the BRUSA Agreement. In October, the LSIB issued a series of wide-ranging instructions designed to replace the instructions issued incrementally over the previous year and a half. These were intended to mandate the procedures for the management, dissemination and control of Sigint and particularly Ultra material across the British empire. These prescriptive instructions were far-reaching and served both to ensure the protection of Ultra and to manage and standardise an increasingly complex worldwide enterprise. Procedures for matters such as the 'sanitisation' of Sigint (that is, for release of sensitive sourced information at lower levels of classification and the protection of its source) were canvassed, as well as liaison arrangements. While giving explicit direction on procedures, the instructions allowed for the creation of local Sigint boards. Despite this, it is clear that these local boards were to be subordinate to London.[55]

Australia, unlike Canada, did not establish a local Sigint board. Instead, Central Bureau and particularly Lieutenant Colonel Sandford filled that role. In this peculiarly Australian approach, it would seem that Central Bureau increased its authority over the D Special Section in Melbourne, becoming the lead for all matters excluding local administration. As ever, Sandford, who always seems to have worked collaboratively with the Assistant DMI, Lieutenant Colonel Little in Melbourne, set the tone of the relationship with GC&CS. Sandford was someone in whom the British had confidence.[56]

Throughout 1944 further measures were taken to ensure the smooth running of the British empire components of the now global enterprise. In February, GC&CS mandated a number of procedures to improve message security and address practices that were the cause of confusion and a risk to security.[57] This signal was followed in March by a broader joint British and United States agreement on the definition and use of security classifications which was disseminated by MacArthur's General Headquarters South West Pacific Area. The agreement also included the requirement to appoint a Top Secret Control Officer at each level of command to be responsible for security arrangements to protect the most highly classified intelligence material, particularly Sigint. These requirements and guidelines bear an uncanny resemblance to the systems that would emerge after the war and which remain in place today across governments in the so-called Five Eyes countries.[58]

If these previous arrangements were designed to enhance the security of classified information, including Sigint, it was the creation and implementation, especially in the Pacific, of the GC&CS's Special Liaison Units (SLUs) and the appointment of Special Liaison Officers (SLOs) that would be the final measures designed to protect Ultra and Magic material. While SLUs had been used by the British as early as 1940 to support the BEF in France, their employment was later refined in the Mediterranean theatre, extending to support Admiral Louis Mountbatten's South East Asia Command. The SLOs and their small SLUs not only provided the communication links for the receipt and dissemination of Ultra and Magic material but were also responsible for the 'indoctrination' (or in-briefing) of senior commanders and their staff into the 'compartments' (of closely controlled, ultra-sensitive information). Moreover, in addition to ensuring that the existence of Ultra and Magic was never divulged, the SLOs had the authority to ensure that senior commanders did not immediately act upon the information provided without first discussing the matter with the SLO and gaining his approval.[59] The highly secure communications using the British Typex and later, in the case of the Americans, their Sigaba machine, ensured the protection in transmission of the most sensitive material, as well as the means to refer matters of higher policy back to headquarters for decisions.

The appointment in Australia of the experienced Squadron Leader Sidney F. Burley, RAF, initially as the leader of SLU 9, and later SLUs 7 and 8, was an important one.[60] Initially located at MacArthur's headquarters in central Brisbane, Burley and his team then moved to Central Bureau's main facility at 21 Henry Street, Ascot, a few kilometres to the northeast. By this time, in early 1945, Central Bureau was significantly smaller at Ascot, with a substantial advanced echelon having been established at Hollandia, Dutch New Guinea, and later a forward presence at Leyte in the Philippines. As Allied forces advanced towards Japan, so too did the SLU, by now numbering three under Burley's command with sub-elements in a number of locations. By mid-1945, SLU elements were colocated with their supported headquarters as follows: 1st Australian Army (Lae), the RAAF's 1st Tactical Air Force (Morotai), RAAF Forward Echelon (Morotai), Advanced Headquarters Land Headquarters (Morotai), Central Bureau Advanced Echelon (Hollandia), RAAF Northwest Area (Darwin), RAAF Command (Brisbane), Land Headquarters Rear (Brisbane) and Central Bureau and General Headquarters Rear (Brisbane). Furthermore, the SLO and his SLUs were linked to supporting commands, Sigint production centres and key decision-makers in Delhi, Hawaii, Washington and London.[61]

Sigint arrangements and MacArthur's command

In Australia in 1942, the Allied forces were faced with a situation of strategic defence. Japanese naval, land and air forces had reached the chain of islands to the north of Australia. MacArthur could therefore expect attacks on the northern territories, most likely – and, as it in fact transpired – by air units from the newly won island bases. The first deployment of Sigint units was planned to counter this threat. The scheme was to place the field sections in the best possible positions for air-to-ground coverage and to connect them – by landline where possible, but otherwise by W/T (radio) – direct to Central Bureau.[62]

After escaping the Philippines, General MacArthur, now the commander-in-chief, SWPA, and his General Headquarters arrived in Melbourne without a dedicated Sigint unit. In the Philippines, MacArthur had had at his disposal only a very small group whose major function was interception, for Washington, of mostly diplomatic material.[63] Major General Spencer B. Akin, who had been in charge of signals for MacArthur in Manila, investigated what Melbourne had to offer; this included Jack Newman's naval signals unit and the nucleus of an interception organisation for military radio traffic, Jack Ryan's Army Special Wireless Section, including Lieutenant Colonel Sandford, which had returned from the Middle East a few weeks earlier.

On 8 March 1942, Sandford wrote to the chief staff officer of the First Australian Army Corps, recommending the establishment of a special intelligence group and its incorporation in the Special Wireless Group.[64] On 1 April 1942, Akin and Sandford had a meeting (of which there is no record) in Colonel Arnold's office (Signals), which was then located in MacRobertson Girls' School in Melbourne.[65] On 5 April, it was agreed that 'Y' sections should come under the control of GHQ and that Number 5 W/T Section and associated intelligence personnel remain at GHQ so that suitable personnel could be trained and a central cryptographic bureau established.[66] The RAAF had already established an interception site in Darwin in 1941, and had sent Flight Lieutenant Henry Roy Booth to Singapore to examine RAF methods and Sigint organisation. He returned armed with insights on FECB and RAF arrangements which contributed to the deliberations.[67]

A note dated 5 April from the Australian DMI, Colonel Caleb Roberts, set out the arguments for the establishment of a Central Bureau: the need to 'feed' wireless groups with broken ciphers, frequencies and so on, as the Central Bureau GHQ Middle East had done. Furthermore, the US Army and RAAF needed a Central Bureau in Australia because although there was already a naval 'Y' (wireless intercept) Centre, Japanese naval and army traffic were totally different.[68]

Following these discussions, Central Bureau was formed in Melbourne on 6 April 1942 as a combined US Army (including US Army Air Force), Australian Army and RAAF organisation.[69] The Bureau was under the direction of General Akin. As General Headquarters was a combined tri-service organisation, there was an expectation that Central Bureau would be the same. But the longstanding tension between the US Army and the US Navy (with Sigint resources a major point of contention) would have an impact on the operations of Central Bureau. This in spite of the best efforts of the Australian and American senior officers in charge,[70] who did not mince their words: 'the need for collaboration with other cryptanalytic services was emphasized by the lack of proper cooperation between Central Bureau and the Naval "Y" service'.[71]

In the absence of such cooperation, a division of labour emerged. Central Bureau was to concentrate on all military communications, while

FRUMEL concentrated on naval communications. After some deliberation, the FRUMEL Director graciously indicated that 'they were quite happy to relinquish any interest in air-ground work and would like Central Bureau to take over the entire function'.[72]

Sigint was central to the Allied intelligence collaboration, but a range of other intelligence-related bodies were created during this period as well. These are captured in figure 3. Of note, significant institutions included the Allied Translator and Interpreter Service, the Allied Geographical Section and the Allied Intelligence Bureau (concerned with special warfare, covert operations, propaganda and human intelligence collection operations). But none were more closely managed from MacArthur's command than Central Bureau.[73]

Central Bureau

Central Bureau also had a division of labour. The Americans were responsible for cryptanalytic tasks, while the Australians focused on traffic analysis. The Australians had little cryptanalytic experience, but their troops had gained considerable experience in traffic analysis fighting alongside other British empire forces in the Middle East. In contrast to naval Sigint personnel, there was a 'remarkable level of harmony and cooperation', considering the workforce comprised at various times 'representatives from fourteen services and many nationalities including American, Australian, British, Canadian, New Zealander, French and Filipino'.[74]

When MacArthur moved his headquarters to Brisbane in July 1942, Central Bureau went with him.[75] There was still a shortage, though, of Australians with experience in Japanese code-breaking. After the prolonged bureaucratic struggle with the British Admiralty and lobbying by Lieutenant Colonel Sandford, Deputy DMI Little and Lieutenant General Smart in London, Eric Nave was again 'loaned' by the Royal Navy, this time to the Australian Army. His appointment to Central Bureau in Brisbane was most timely and, as a bonus, he brought Professor Room with him from SIB Melbourne.[76]

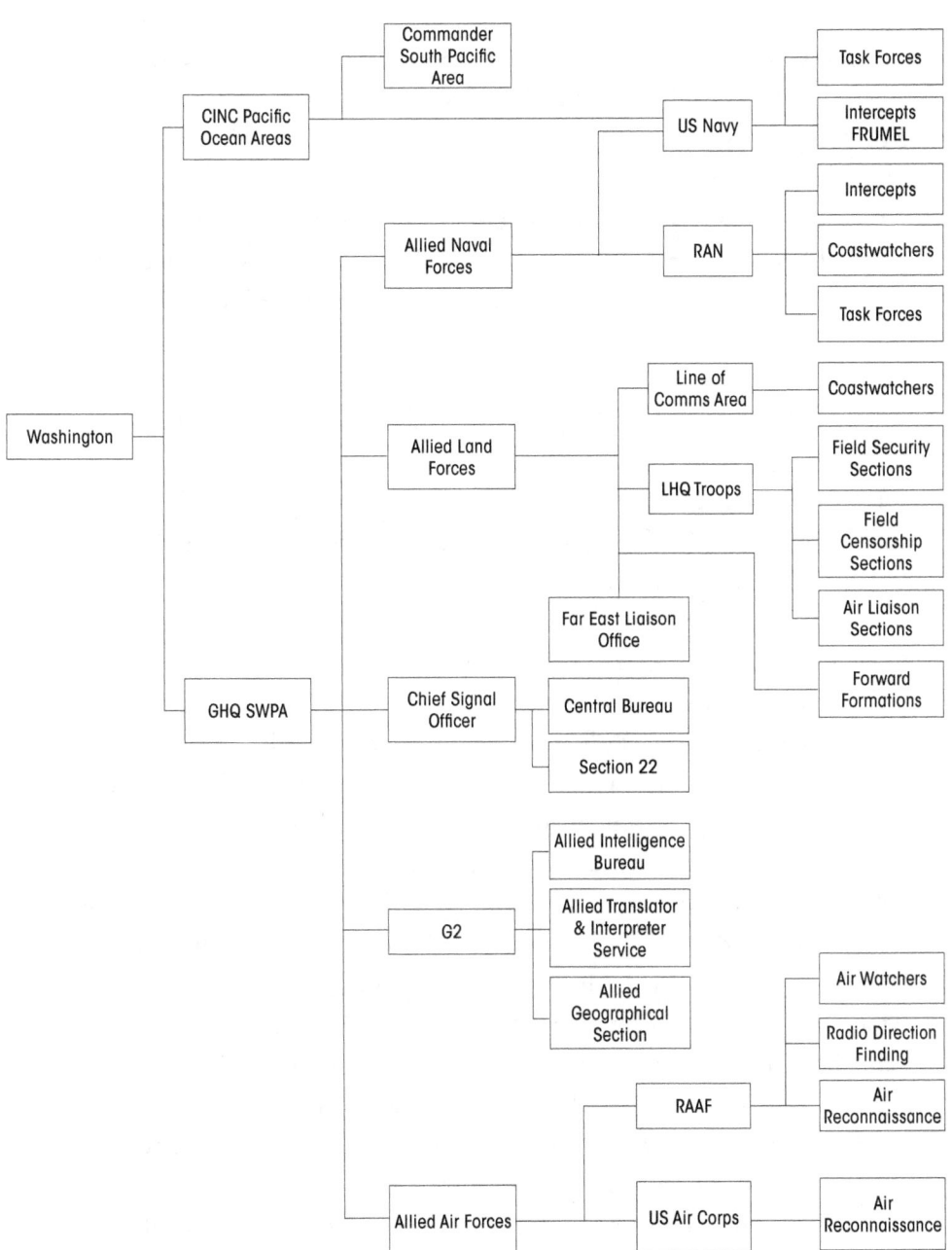

Figure 3. Allied intelligence organisation in SWPA, May 1943
SOURCE Letter Brig. Rogers to DMI London, India, Middle East, Ottawa, Wellington, 10 May 1943, Adv. HQ AMF G Int. Sect. War Diary AWM 1/2/2, cited in David Horner, *High Command: Australia's Struggle for an Independent War Strategy, 1939-1945* (Sydney: Allen & Unwin, 1982), p. 241.

As we have seen, one of the conditions for allowing Nave to stay in Australia was a visit by an Australian 'Special Intelligence Officer' (namely Sandford) to confer with GC&CS and 'explain our set up'.[77] This visit of four weeks in London and one in Washington was a significant coup for Army Sigint, as well as for Sandford himself.[78] It established a cryptographic channel between the Australian Army and GC&CS, independent of the United States, which was maintained until the end of the Pacific War.[79]

The organisation of Central Bureau

Organised along functional, not service or national lines, initially, Central Bureau consisted of only four sections: traffic analysis, cryptanalysis, translation and administration. Over time, however, the number of sections grew to include 'solution', communications, photography, machine (i.e., tabulating) procedure, and general intelligence and liaison. These were organised under three assistant directors, one US Army, one Australian Army and one RAAF. Traffic analysis and general intelligence and liaison sections were commanded by an Australian Army officer. The Americans were ahead of the Australians on high echelon cryptanalysis and in the use of machine tabulation, and concentrated their efforts on these. Of the 1450 employed by Central Bureau by August 1945, 750 were from the United States (including 159 women), 235 Australian Army (including 49 women) and 459 RAAF (96 of them women) – approximately half American and half Australian. In the field, however, the Australian contribution was larger, with 915 Americans and 1639 Australians (along with 333 Canadians), 4339 altogether.[80] Nigel de Grey notes that in June 1945, as many as 126 929 Japanese signals were intercepted and handled by Central Bureau, although not all were decrypted.[81]

The unique circumstances and unusual arrangements meant that the inter-Allied and inter-service functions and channels of command over Central Bureau would not be formalised until January 1943. A joint fund was established, and orders for equipment were placed in the United States, while additional intercept operators and interpretive staff of the wireless intercept field units were recruited and trained. The training involved low-echelon traffic analysis and deciphering, with intercepted traffic being passed back from the field units to Central Bureau for further analysis.[82]

By 1945, the intelligence produced by Central Bureau was disseminated far and wide. Daily, monthly and other periodic reports (on technical developments or matters of cryptographic interest) were distributed, with 24 copies sent out daily. Of these, seven were sent to recipients in the South West Pacific Area, including:

G-2 (Major General Willoughby and his staff),
US Navy (FRUMEL),
Chief Signal Officer (Major General Akin),
DMI (two copies),
Allied Air Forces (two copies).

A further seven copies were sent overseas to:

Washington (Signal Security Agency),
Canada (NDO),
India – Wireless Experimental Centre (WEC) and 'YIB', and
London (GC&CS).

A further ten copies were distributed around Central Bureau for internal use.[83] Given its growing geographic spread, that number of copies was justified.

Central Bureau staff and personalities

In its formative year, Central Bureau drew in veterans of the Australian Special Wireless Group headed by Sandford, members of the British signal detachment evacuated from Singapore, the US Army Station 6 team (recently arrived from the Philippines) along with the 121st (later redesignated 126th) Signal Radio Intelligence Company, and the US Army

Signal Corps 837th Contingent, under Major Abe Sinkov – an experienced cryptologist and lead negotiator in the US–UK Sigint arrangements.[84]

Central Bureau was fortunate to have Sinkov, who was inclined to share and work collaboratively and respect his Australian counterparts in Central Bureau (and after the war in his position at NSA). This stood in marked contrast to the obstructionist approach of his US Navy counterpart Fabian at FRUMEL. As a later Deputy Director of DSD, Major General Steve Meekin observed, 'in part, it is out of the Central Bureau experience and the respect gained there that helped so much in facilitating trusted engagement and collaboration after the war'.[85]

In his letter recommending Sinkov's nomination as an officer in the Order of the British Empire (OBE) and awards of Mention in Despatches for three others, it is somewhat surprising to learn that by 30 August 1945 Sandford was expressing certain reservations about Nave – the man who he had championed earlier to be appointed to Central Bureau. Sandford wrote:

> I have recommended nothing in the case of Capt NAVE since I thought I should like to have your advice on this point. Technically, Capt NAVE is by far the most brilliant officer in the Unit, but he is so lacking in initiative and appreciation of changing operational requirements of our forces that his efforts must be consistently guided by Maj CLARKE or myself. One cannot forget however that it was he who 'broke' the first Japanese cypher made by any of the military units during this war, and on that account alone he should perhaps be specially considered. Can you give me your advice on this.[86]

Sandford's assessment is at odds with what we know of Nave earlier in his career, and also later when he helped to establish ASIO. But Sandford's view is consistent with those of others who believed Nave's best code-breaking days were behind him.[87] It may be that the harsh and unjust treatment of Nave in the wash-up of the Holden Agreement, particularly in view of his remarkable service record, induced a form of post-traumatic stress which may have affected his work for a time. Nave would eventually be awarded on OBE in 1972.[88] His death in 1993 was reported in British as well as Australian newspapers and in the *New York Times*.

The evolving role and partnerships of Central Bureau

Initially, in keeping with its character of a field organisation, Central Bureau's primary mission was solving field problems and exploiting low-echelon material.[89] Subsequently, it provided MacArthur with Sigint through cryptanalysis, and translation.[90] Its work also included preparation and dissemination of reports from Japanese sources, as well as planning and coordinating intercept activities. The Central Bureau's 'Radiogoniometric' and Identification unit identified and located Japanese headquarters, units, craft and aircraft, as well as movements and intentions, and meteorological data.[91] A photographic section was established to reproduce documents, microfilm traffic for distribution abroad, as well as developing incoming microfilm and photographing special events.[92]

Central Bureau was also a communications security monitoring body, maintaining Allied communications security throughout the SWPA. It worked closely on code-breaking with the US Army's Signal Intelligence Service in Arlington Hall near Washington, DC, and exchanged intelligence with US Navy and British authorities in neighbouring theatre commands, including Britain's South East Asia Command and the US Navy's Pacific Ocean areas.[93]

In the early days of Central Bureau, the Australian Army provided most of the deployed signals intercept teams. They dealt with air-to-ground traffic, mostly naval-air communications collected from northern Australia and the Australian Mandated Territory of New Guinea. The intercept stations provided early warning of enemy air attack, and their traffic analysis was most effective. The Japanese never achieved the same level of signals security as the Germans, so traffic analysis played a major part in the Allied campaign planning in the SWPA, which MacArthur commanded throughout the war. In this, the Australian Army had the lead; two years in the Mediterranean had set them on the right road and honed their skills. The Americans had not practised the art in peacetime and neither had the RAAF. It was not long before Central Bureau decrypts began to contribute.[94]

According to senior officers in charge of Central Bureau, the relationship between radio intelligence and cryptanalysis emphasised the need for continuity and collaboration. For example, it was frequently possible from traffic analysis to obtain hints and, at times, definite confirmation, concerning multi-part messages, addressees, general context, and other items of significance in cryptanalysis.[95]

The Sigint mission involved many parts working together, including disparate elements across the SWPA and beyond. As Edward Drea observed, 'converting Morse Code dots and dashes into military intelligence ... began at one of the scattered intercept sites'. Intercept operators copied coded Japanese radio broadcasts and transmitted the re-enciphered results to Central Bureau. Cryptanalysts then attempted to decipher the code. If unable to do so, they would glean technical data from the message. In both cases, the technical findings would be transmitted to the US Army Signal Intelligence Service Headquarters at Arlington Hall. At times of peak workload, Central Bureau would draw on Willoughby's Allied Translator and Interpreter Section (ATIS) or call for help from FRUMEL to 'ensure complete coverage of decrypted Japanese radio messages'.[96]

The ATIS was set up by Major General Richard K. Sutherland, MacArthur's chief of staff, under MacArthur's headquarters, with Australian and US personnel employed as translators and interpreters. Initially staffed by Australian enlisted personnel, it soon came to be predominantly staffed with *Nisei* – Americans of Japanese descent.[97] The lack of officers trained in cryptanalysis meant that ATIS translation of such messages was unsatisfactory and sometimes even incorrect. With a massive volume of work, ATIS experienced delays which often rendered their work worthless when time was of the essence. As a result, Central Bureau reached an arrangement by which all 'documents of cryptological importance' were passed directly to Central Bureau from ATIS, leaving the Bureau's translation section to handle these documents.[98]

Once translated, the decryptions were passed via Sinkov, who determined which items a courier would hand-carry from Central Bureau to

Spencer Akin at HQ SWPA. The Signal Corps' responsibility ended when Akin passed the translated version of the Japanese message to Sutherland and MacArthur, who would then decide what to do with the information. Generally speaking, Central Bureau's cryptanalysts did not hear about the results of their work. Occasionally, Sinkov would pass on word of major successes, such as torpedoed Japanese ships, but generally the work was all-consuming and completely compartmentalised.[99]

As for the analysis of Central Bureau's Ultra, this responsibility fell to the G2 Major General Willoughby, who received, rather than produced, the special Sigint reports. The G2 and his staff then integrated the Ultra reports with their other intelligence material sourced from aerial reconnaissance, agent reports, captured Japanese documents, prisoner of war interrogations, and so on. This was then used to prepare theatre estimates and intelligence appreciations for MacArthur.[100]

Central Bureau produced detailed reports on a wide range of subjects. These included Japanese intentions for granting independence to the Dutch East Indies, the functions of Japanese agencies, the movements of Japanese hospital ships, surveys of Japanese medical reports, detailed analyses of the Japanese headquarters in Rabaul, intelligence activities of Japanese forces in China, the activities of Australian secret intelligence and special operations sections, Japanese relations with Portuguese Timor, and Allied prisoners of war.[101] These reports give a sense of the wealth of information available to Central Bureau which, as the end of the war approached, was quite extraordinary.

The Australians of Central Bureau built the organisation from the bottom upwards, using what they had to the best advantage while allowing the rest to grow on the strength of what had been accomplished. Despite pitfalls and mistakes along the way, in the words of Nigel de Grey, 'it is impossible not to admire Australian enterprise and initiative in the speed with which they got into action'.[102] Central Bureau had 'grown up on its own',[103] having little early contact with other centres and yet 'it controlled its field organisation, ultimately not only Australian but also American, even more closely than was done in the last phases by the Sigint Directorate in South East Asia Command where GSI staffs were introduced'.[104]

With growing demand for training, there was an informal arrangement whereby each section was responsible for training its own staff. By mid-1943, however, an intelligence school was formed to provide basic training in Sigint methods, with emphasis on the air-ground problem. Over 300 RAAF and US personnel undertook this training.[105]

According to American principles for the organisation of military staff, the Central Bureau came under the Signals Branch of Headquarters SWPA, headed by Major General Akin. From this perspective, Central Bureau was of no concern to the Intelligence (G2) Branch, except as a source of supply. This was in contrast to the British principle, broadly practised by the Australians, of the indivisibility of the process of Sigint. The contrasting styles required a cooperative approach from the Australians.[106]

By April 1944, the Director GC&CS reached 'a categorical undertaking' that all Sigint results, whether US or UK sourced, should be open to the Australians, 'thus gaining recognition by the [US] War Department of the equal status of Australia to Great Britain and America in the partnership'. But the 'factor above all others' which made these principles and organisation possible and effective was the promulgation of the 'Regulations for handling and dissemination of high echelon Signal Intelligence in the S.W.P.A. and S.E.A.C.'. These regulations ensured that the British in India and South East Asia Command, the Australians, forces in the SWPA, as well as those in the United Kingdom and the United States and operating elsewhere all 'had the same security code in respect of Japanese high echelon Sigint'. This meant that they could each administer the regulations through their specially appointed security officers. De Grey observed that 'this also meant that the flow of intelligence between all the several areas could be with the "swept channels" so to speak free and unfettered'.[107] This remarkably trusting arrangement laid the foundation for relationships that would endure long after the end of the war.

Central Bureau's technology

MacArthur's appointment as commander-in-chief, SWPA, meant that his headquarters and supporting elements would soon become the beneficiaries of significant Allied technological breakthroughs. These breakthroughs drew on the cumulative experimentation, invention, and trial and error of cryptologists in several countries; notably Poland, the United Kingdom and the United States.

The concept of machines that would undertake computations on our behalf – that is, computers – was an innovation that underwent significant trial and experimentation during the war. This happened to a large extent with Sigint. Much of it related to finding ways to automate procedures that were a rigmarole and extremely time consuming for people to undertake, but which, with what now look like relatively simple electro-mechanical arrangements, could be compiled relatively quickly with simple, binary, 'yes/no', or 'on/off' computations. Such computations could be used for enciphering and deciphering as well as for statistical analysis – which would prove so useful in identifying the probability of particular code settings. These binary computational electro-mechanical arrangements would be the early harbingers of electronic computers – machines which rely on the same kind of binary code of ones and zeroes but without the mechanical moving parts that would consume much space in the early machines used during the war. This was the origin of the network of computers that would form the basis for the micro-computers that would become the building blocks of the cyber age a generation or more later.

The Sigaba

One example of the machines at MacArthur's headquarters was the Sigaba cipher machine, incorporating 15 rotors, each with 26 alphabetic contacts (five cipher rotors, five control rotors and five index rotors set daily). The Sigaba was developed in the mid-1930s by the US Army Signal Intelligence Service, with the assistance of William Friedman and his assistant Frank Rowlett. They took Edward Hebern's design of an electro-mechanical rotor cipher machine, which employed crypto rotor wheels like those in the German Enigma cipher machine. Recognising the 'intrinsic cryptologic vulnerability that was mathematically exploitable' in such an electro-mechanical system, Friedman reduced the risk by introducing a key tape with holes punched in, which would permit feeler contacts to turn an electrical current on or off, causing a cipher machine's rotors to stop. In this way, with each key stroke, randomly placed holes in a five-group key tape would produce an apparently random stepping for one or more of the cipher rotors. The cipher test would be printed on a small tape. In turn, the decryption process worked in reverse.[108]

The Rockex

As early as April 1932, Friedman had demonstrated the randomly stepping rotors to his key cryptanalysts, one of whom was Sinkov (later the senior American officer in Central Bureau). Following some modifications to the key tape, they shared the concept with US Navy counterparts. It was this device that US forces brought to the war in 1941.[109] It was a device unmatched by the Germans, Italians or Japanese, and superior to the British Enigma-like Typex cipher machine and later the Rockex cipher machines, with which the Australian armed forces were equipped. The Rockex, incidentally, had a number of similar features to the German Lorenz machines. The US Army's first batch of 459 machines was fielded in June 1941.[110] MacArthur's staff brought this equipment with them to Australia in March 1942. It would be used throughout the remainder of the war for high-level communications, most notably between US President Franklin Roosevelt and Britain's Prime Minister Churchill.[111]

The Sigsaly

Another piece of advanced technology that aided the war effort was an enciphered voice communications telephone known as the Sigsaly. Mindful that earlier devices were probably not secure, the Bell Telephone Laboratories, under the direction of A.B. Clark, with the assistance of mathematician and leading British codebreaker Alan Turing, collaborated on a device that would transform voice into a digital data signal that, in turn, could be encrypted. The Sigsaly appeared in 1942 and was deployed globally the following year. With over 40 racks of equipment and weighing over 50 tonnes, the Sigsaly was not portable. Nonetheless, eventually a dozen Sigsaly terminals were established around the world, including in London, Washington, Algiers, Hawaii, Oakland and Australia. As the war progressed, they were also set up in Paris, Guam, Frankfurt and Berlin. The Australian terminal supported MacArthur, enabling him to speak directly with his counterparts in Hawaii and Washington, and proved vital in protecting sensitive discussions and conferences about wartime arrangements.[112] The Sigsaly set a precedent for encrypted voice communications that would be refined in the decades ahead, becoming an integral part of the technological capability of the Allied intelligence partners.

Central Bureau's need for more, and more advanced, equipment quickly became evident. In early June 1942, it set about acquiring a dozen acoustic recorders, several 'shavers', at least 16 'transcribers' and 80 'undulators', as well as 2000 'cylindrical records' 'as an initial purchase'. In addition, it purchased Radio Corporation of America (RCA) direction finders (£600 per set) and two 'type DFP 5' portable Marconi ultra-high-frequency direction finders at a cost of approximately £500 each (including spare tubes).[113] More such orders would follow in the months ahead.[114]

IBMs and the independence of the Australian Central Bureau component

Central Bureau also benefitted from the use of what the British called Holleriths and the Americans called IBMs – including IBM punch-card tabulators and other IBM machines – in a range of tasks. Australia, being part of the British empire, fell under the contractual arrangement whereby British Tabulating Machines (BTM) managed the sales and maintenance of IBM-designed machines. Historian David Dufty has observed that 'when they were configured correctly for a code that had been broken and for which the codebook was known', the procedure was quite straightforward: 'the punch card operators could set up cards with the numbers from an incoming message, and the machine would read the card, strip the additive, decode the groups, and print out a Katakana version of the text'. A thin piece of cardboard with holes punched in it, the 'punch card', was the primary method of providing input.[115]

The original batch of IBMs used 110 volts, but Australia operates electricity at 240 volts. After some 'shocking' encounters installing such

American electrical equipment, subsequent shipments brought 220-volt equipment. Maintaining two voltages required transformers and agile-minded technical and engineering staff.[116]

The IBM machines streamlined otherwise massive and time-consuming tasks. They did this by stripping additives off the codes rapidly and in bulk, and then looking for repeating patterns in the layer of code uncovered beneath the additives. Repetition could mean the same message had been encoded twice with different additives.[117] At first, simple tasks were managed, like sorting words, particularly frequently recurring words. Sorting words meant creating punch cards for every word in every message. This was clearly a laborious task. Beyond sorting words, the IBM tabulators could do other helpful tasks, including sorting by call sign; this meant that messages from a particular enemy unit could be grouped together. Alternatively, a pile of messages could be sorted by indicator values, such as messages with the same enciphering additives.[118] (A more detailed explanation of the component functions is at Appendix A.)

From 1942 to 1944, American officers directed and supervised the use of tabulating machines. By mid-1944, however, the Australians started to explore investing in their own equipment to support analysis. The aim was for the 'Australian contingent' to be able to operate as a self-sufficient, independent capability should the need arise.

By mid-March 1945, with MacArthur's focus on the campaign in the Philippines, efforts were made for Australia to acquire its own machines. In a minute to Sir Frederick Shedden, the Secretary of the Department of Defence, the case was outlined thus:

> The [Commander in Chief] considers it essential for operational reasons that the Central Bureau (Australian Component) be entirely self-contained and independent of the US resources (i.e. machines and personnel) on loan to them. This arises from the fact that should the Aust Component and the US Component of Central Bureau be required to operate independently at any time, it would be impossible for the Aust Component to do so for considerable time since the operating personnel require a high degree of training and the machines are only procurable from USA.[119]

The comprehensive list of IBM requirements specified by Central Bureau showed that the Australians were seeking a substantial capability, and were familiar with the capabilities of each machine in the suite proposed for acquisition.[120]

> The IB Machines required have been developed during the war to meet the requirements of a special intelligence activity in top secret category. Such machines are quite different in principle from those used for records, statistical etc. purposes and are not obtainable from manufacturers other than the IBM Company.[121]

Outside the walls of Central Bureau and its tightly-held circle of secret 'special wireless' collection units, few understood the request. There were questions as to why, given resource constraints, such specifically US-sourced equipment was necessary: 'Unfortunately machines of other makes, even if available, cannot be considered because the need for coordinating the work in different theatres necessitates the use of machines common to all'.[122]

As with nearly all future Australian Sigint proposals for cryptanalytic machines, the need to be compatible with allies was a critical factor in the equipment items selected for acquisition. And this would not be the last time that unindoctrinated personnel would question sole-sourcing proposals without competitive tenders for sophisticated Sigint equipment.

It was also clear, however, that the Australian component in Central Bureau had, either deliberately or otherwise, not become involved directly in the operation and maintenance of the IBM machines at either Central Bureau or FRUMEL. In early January 1945, the DMI acknowledged 'it will be necessary to arrange for a nucleus of about six trained personnel to operate and service the machines. It is understood that such personnel can be obtained in Canada only. None are available in Australia.'[123]

The acquisition of the IBM machines was seen as important for Australia after, as well as during, the war. As the DMI, Brigadier John Rogers, explained, 'the provision of these machines would make the Australian Component of Central Bureau independent, should it become necessary during the war. It would also be of considerable benefit to the Australian Services in post-war activities.'[124]

Unfortunately, although the proposal was approved by General Blamey, the Secretary of the Army, the Minister for the Army and the Treasurer, it moved so slowly through Army channels, despite being stamped 'urgent', that it was not finalised before the war's end. By then, the view was that the 'functions which will probably be allotted to the [Australian Military Forces] and RAAF components in the near future do not require machinery of this nature'.[125] Despite this disappointing result, efforts to acquire and retain such machines actually had not failed. It is something of a mystery that somehow the Hollerith IBMs or BTMs, which were used throughout the rest of the war, were somehow transferred at war's end to the RAN and subsequently to the Australian Sigint organisation established after the war known as the Defence Signals Bureau.[126] It may be that the close personal relationships forged during wartime between the American and Australian personnel were somehow involved.

The role of women in Sigint and Allied relations

By September 1944, when the decision was made to move Central Bureau from Brisbane to Hollandia in the Dutch East Indies (now Jayapura in Indonesia), the directors would face a problem. MacArthur wanted to move the entire bureau – the codebreakers, the translators, everything. But there were some obstacles to doing this. There was a policy problem regarding the hundreds of women working at Central Bureau as Australian Women's Army Service (AWAS) and Women's Australian Auxiliary Air Force (WAAAF) personnel. Legally, they were not allowed to serve at Hollandia because of the Australian government's policy of limiting the total number of women who could serve overseas to 500. That was not the limit for Central Bureau – it was the limit for the entire Australian military forces.[127]

The traffic analysis department had a large number of women in its ranks who would have to stay behind, so Akin asked for help from his contacts in the United States in meeting the shortfall in personnel. They agreed to send several hundred women from the United States, members of the Women's Army Corps, known as WACs.[128]

Sandford, meanwhile, assiduously lobbied the Australian government to change its mind about restricting women from overseas service.[129] With the submissions in, the War Cabinet discussed the matter again on 18 May 1945. Clearly, the Americans wanted the Australian government to let the women go to the Philippines, and the army and air force agreed. The Cabinet had been informed by Central Bureau, as well as by their own military chiefs, that the women were a vital part of the war effort, and their advisers were telling them that 500 women should be allowed to go. The politicians considered all this, and yet they refused. Public outrage over the assault and massacre of Australian nurses by Japanese soldiers on Bangka Island off the coast of Sumatra in 1942 would have been a factor. The Secretary of the War Cabinet, Frederick Shedden, issued the following ruling:

> The proposal that members of the Australian Women's Services, at present on the strength of the Allied Central Bureau, be transferred to Manila with this establishment was not approved. A definite limit has been laid down by the Government on the number of women to be permitted to serve outside Australia and War Cabinet did not consider that this should be increased.[130]

While women were not to be deployed overseas, that did not preclude them from participating in the work of the special wireless units of the three armed services stationed across Australia.

Service special wireless interception units

Central Bureau oversaw a network of army and air force interception and direction finding Sigint sites across Australia and further afield. FRUMEL did the same for navy sites. Their combined efforts ensured broad coverage of the intentions and activities of Japanese forces. This level of coverage was possible because Central Bureau had complete command of its field organisation, including the American units.[131]

The connection with the service elements was close and tightly held, with few uninitiated to the world of Sigint knowing which units were where and doing what. For Central Bureau, the plan involved sending units to forward locations to provide field commanders with local, timely, and actionable intelligence. Several types of field sections were employed – including those belonging to the Australian Army, the RAAF, the US Army and later the Canadian Army. These units provided air raid warning intelligence, and intercepted low-echelon traffic and tactical intelligence. Intercepted raw material inaudible from rear areas considered potentially useful was passed up the chain 'for cryptographic solution' by Central Bureau, where it was exploited.[132]

To manage the training requirements of these disparate elements, the Central Bureau Training Group (the Intelligence School) was established for personnel assigned to Australian and Allied units.[133]

Headquarters ASWG, commanded by the First World War veteran Lieutenant Colonel Jack Ryan, was initially in Bonegilla, Victoria, and remained there from May 1942 to April 1943, when it moved to Brisbane until disbandment at the end of the war. In August 1942, 270 AWAS personnel were posted in. From then on, the ASWG included hundreds of AWAS personnel.[134] The ASWG played a prominent role supporting both deployed forces and the work of Central Bureau. Its skill and reputation built on the work undertaken in support of deployed forces in the Mediterranean in the early years of the war.

The RAAF had not played such a prominent role early on but from mid-1941 were eager to catch up. Initial training of RAAF personnel was undertaken on a 'special intelligence' course at Victoria Barracks in Melbourne in July 1941. After the onset of the Pacific War, the RAAF became more actively involved in special wireless activities, and the trainees were among the first personnel assigned to duties with the No. 1 Wireless Unit. The unit was established as an intercept station in Townsville, in north Queensland, on 25 April 1942, under the stewardship of Wing Commander Roy Booth. It focused on interception of Japanese naval and military traffic, and, like the ASWG, worked as part of Central Bureau. Initially, it comprised seven RAAF, one Australian Army and four United States Army personnel. In the end, over half a dozen RAAF wireless units were established and operated in Townsville and Darwin, across the north of Australia, in Papua and New Guinea and beyond.[135]

Members of the WAAAF served as well, as did members of the US Women's Army Corps (WAC), who arrived later in the war. Indeed, women served at all of the Central Bureau locations including field sections across Australia. Women made a key contribution to Central Bureau's operational efficiency; when the organisation moved from Brisbane to San Miguel, on Luzon in the Philippines without the Australian Women's Services, their absence was keenly felt.[136]

When it came to advancing from the New Guinea campaign and launching forces into the Philippines, MacArthur made sure that the RAAF Special Wireless Units were included as part of his force, but not the land or maritime equivalents. The land forces mission was steered towards places including what is now the Indonesian island of Borneo. Evidently MacArthur placed considerable trust in the RAAF special wireless teams.

Radio countermeasures and radio direction finding (radar)

Technological advances during the war allowed ships and aircraft to be used for collection, not only of Comint but non-communications intelligence known as electronic intelligence or Elint. This included radio direction finding transmissions, or radar. The Navy was the first service to show interest in radar, looking to install receivers in ships to pick up enemy radar transmissions for early warning.[137]

Radio direction finding (RDF, or radar) special investigations meetings were held from May 1942 onwards. Convened in Melbourne, these meetings involved RAN and RAAF officers and scientists from the Council for Scientific and Industrial Research (CSIR) Radiophysics Laboratory at the University of Sydney. Within a few months, a radio countermeasures (RCM) organisation was operating, and RCM operators were being trained at HMAS *Rushcutter* in Sydney's Rushcutters Bay. It also became clear that the maintenance skills of ship's telegraphists would not stretch to the advanced electronics involved in radar. Wireless mechanics, therefore, were trained specifically in electronics for six months at what would become the Royal Melbourne Institute of Technology (RMIT), followed by equipment training at the RAN shore establishments of HMAS *Rushcutter* and HMAS *Watson* in Sydney, and at HMAS *Harman* and the Belconnen transmitting station in Canberra.[138] Australian-initiated RCM missions were flown out of Port Moresby as early as September 1942.[139]

Recognising the value of radar and radar countermeasures, MacArthur's GHQ SWPA sought to control them. With a surge in US forces arriving in theatre, momentum built for a US-led organisation to take the lead, while capitalising on the work already undertaken by the Australians.

Section 22 and close US-Australian collaboration

On 5 July 1943, a radar and radio countermeasure division was established, under the operational control of GHQ SWPA in Brisbane.[140] The division was under the command of the Chief Signal Officer at GHQ SWPA, Major General Akin, and was innocuously renamed Section 22 in November 1943.[141] This new unit absorbed the Australian RCM initiatives which had commenced in mid-1942.[142]

In April 1944, the USAAF commenced operational training of RAAF personnel before the re-equipment of RAAF units with four-engined B-24D Liberator bomber aircraft. Australians were assigned to crew positions on USAAF combat missions.[143] These aircraft, along with PBY Catalina flying boats, were equipped with 'Yagi'[144] 'rake' homing transmitting arrays, SCR-717 and SCR-717B microwave search radars.[145]

The term 'Ferret' was coined to describe aircraft which were involved in radar countermeasures. Ferrets were flown over enemy territory to identify and analyse electronic signals, usually radar transmissions, for intelligence purposes. The earliest airborne Ferret operations from Australia occurred from 3 to 5 September 1942. A USAAF B-17 Flying Fortress aircraft equipped with ARC-1 radio receivers flew from Port Moresby to Milne Bay, then across the Trobriand Islands to a Japanese naval base in the Shortland Islands.[146]

In November 1942, a radio countermeasures specialist, Lieutenant H.A. Hallett RNVR, arrived in Sydney from Britain, with a number of British naval ratings experienced in the new art of radio direction finding, and a treasure trove of related radar detection equipment. Hallett was involved in establishing the sensitive and classified multi-national, multi-service organisation, Section 22. The section was formed in Brisbane, reputedly taking its name from the door of the office from which it worked, and investigated Japanese radar stations in New Guinea, Solomon Islands and the East Indies.[147] One person involved, Charles Darby, recalled being part of a

> group of eight [Wireless Operator/Air Gunners] ... trained at HMAS *Rushcutter*, the Royal Australian Navy shore base in Sydney, during February and March 1943. Our instructor was a British radar 'boffin', Lt Hallett RN (we never knew his Christian name). In early April we were all posted to operational units, four being assigned to RAAF Catalina squadrons operating out of Cairns in North Queensland.[148]

By February 1943, RAAF Catalina and Hudson aircraft, as well as USAAF B-17Es and Liberator B-24Ds were equipped with this airborne radar equipment and actively involved. Section 22 personnel were drawn from the RAAF, the RNZN, the Royal Navy, and the US Army and US Navy, and also included a number of civilian specialists and Australian Army personnel.[149] Technicians and operators deployed as teams known as field units, which were attached to operational Australian or US Air Force, Navy or Army units. Field unit personnel were required to report to MacArthur's Chief Signals Officer, at GHQ SWPA. In practice, however, they reported through the operational unit to which they were attached.[150]

Section 22 operators and their equipment continued to serve in USAAF B-24s long after heavy bomber operations had advanced beyond Australia. Several deployed with their USAAF units to the Philippines, where they investigated Japanese radar stations in Formosa (now Taiwan), Hong Kong, French Indochina (now Vietnam) and elsewhere.[151]

By 1945, RAAF Ferret aircraft modified for intelligence-gathering had a range of equipment that could be configured. This included an auto-scanning multiband radar intercept receiver; an auto-scanning microwave intercept receiver; and an advanced auto-scanning, auto-recording radar intercept receiver, incorporating a continuous-drive thermal paper tape on which the received signals were recorded for post mission analysis. The auto-search receivers would be associated with a panoramic adapter in order to demonstrate visually the spectrum of signals around the main frequency under investigation; pulse analyser oscilloscopes to look at characteristics such as the pulse width, shape and repetition frequency of enemy radar transmissions; direction-finding antenna systems, and radar signal tape recorders.[152] By this time, the work of Section 22 was having an impact in the squadron operations rooms, where maps were covered with interlocking circles that indicated the location and effective range of the Japanese radar installations.[153] Signal jammers were also included, where appropriate, on Australian B-24 operations in the closing stage of the war.[154]

With the US interest in Australia waning and the US focus in the war shifting to the north, the RAAF established 201 Flight at Laverton in Victoria on 10 March 1945, in order to continue with the experimentation work undertaken alongside the USAAF. 201 Flight had four B-24 Liberators with which to conduct aerial electronic surveillance of Japanese radar stations. The first two B-24 aircraft arrived in April 1945 and were modified for their new roles. This involved removing the ball turret to fit a radar scanning dome and the installation of an enclosed radar operator's cabin. By the time the modifications were completed in July 1945, some of the unit moved to Darwin, but with the front far away and the war's end evidently approaching, they had little operational tasking and, in the end, did not fly any operational missions. In October 1945, not long after the end of the war, 201 Flight relocated to Laverton, southwest of Melbourne, before disbanding in mid-March 1946.[155]

In contrast to aircraft set up for Ferret missions, a flight of B-24s at Leyburn, Queensland, was set up as 'carpetbaggers', to support clandestine secret agent operations inside enemy-occupied territory. RAAF 200 Flight was raised to support these missions, which were mandated under the auspices of the Allied Intelligence Bureau, charged with managing the special operations functions of the Services Reconnaissance Department (SRD) and Z Special Unit.[156] As a result, their equipment was designed to avoid and jam enemy radars while homing in on drop zones set up by agents on the ground. This configuration was in contrast to that for electronic intelligence missions that required the intercept receivers and analytical sets of the Ferrets.[157]

The activities and functions of Section 22 as well as RAAF 200 and 201 Flights remained little known for decades after the war, but the groundbreaking work on developing airborne radar DF equipment and techniques would prove invaluable. Much of the work undertaken by Section 22 turned out to be the precursor of the capabilities the RAAF would acquire and maintain with its postwar surveillance aircraft platforms, notably the P2 Neptune (1950s to 1970s), P3 Orion (1970s to 2020s) and P8 Poseidon (2020s) aircraft variants. It would also help ensure close collaborative arrangements, particularly on Elint matters, between the RAAF and the postwar national Sigint agency, as well as with counterparts in the United Kingdom and United States.

Looking back on the allied Sigint enterprise in Australia

Compared with the experience of the First World War, where Australia contributed combat forces but provided little of the higher order organisational componentry, in the Second World War Australia's intelligence arrangements were transformed. This transformation built on British and American Sigint developments that had been gathering pace in the interwar years and in the early wartime period. But it was in the period from early 1942 through to 1945 that a dramatic expansion of Sigint-related arrangements for the defence of Australia came into being. Naval, land and air forces worked collaboratively with civilian counterparts, men and women, at home and abroad, to establish institutions that left a legacy for the nation. This legacy included not just a technological transformation but a social one, with women an integral part of that change.

Although there remained points of friction between the various organisations, the work of FRUMEL, Central Bureau, D Special Section and Section 22, and the service intercept stations, among others, played a vital, largely unheralded role in ensuring Australian casualties were far fewer than in the First World War and that the successes were more important than the public was at first allowed to be told. The Coastwatchers were credited with many of the successes derived from Sigint. There were also active measures to suppress the release of information that would show how successful and vital Sigint was to the war effort. For several decades, historians have either been unaware of or constrained from writing about this important foundation of Allied accomplishments.

7
WARTIME SIGINT SUCCESSES, BUREAUCRATIC & OTHER CHALLENGES

Wartime collaboration with allies in Sigint had enhanced Australia's ability to read Japanese intentions and capabilities and adjust war plans accordingly. The benefits of this were great, but also hard to measure. There were a number of successes, but there were also some limitations to Australia's Sigint capability and some challenges.

Australians knew that collaboration with allies was indispensable, but the difficulties are not to be underestimated. As the junior partner to the Americans and the British, Australians bore the brunt of problems. The experience undoubtedly reinforced Australia's determination after the war to establish its own fully-fledged Sigint organisation while maintaining and enhancing alliance relationships. During the war, certain individuals, including the American Major (later Colonel) Abe Sinkov, Australia's Captain (later Lieutenant Colonel) 'Mic' Sandford, Deputy DMI Lieutenant Colonel Little and Wing Commander Roy Booth, among others, made the arrangements and relationships work in challenging circumstances.

Australia's Sigint limitations

As part of an international network, Australia was afforded excellent opportunities to learn and benefit from the cryptologic breakthroughs achieved by the British and Americans in particular. Part of the arrangement from 1942 onwards was a division of labour; US cryptologists were better trained and resourced for code-breaking, whereas the Australians' strengths were traffic analysis and related interpretative work.

As it happens, the skills required for effective traffic analysis were also the most useful for short notice, tactical intelligence insights. Nigel de Grey observed it well, when he wrote that:

> There could be no question of the Army sections being employed on Army tasks – on the contrary their chief function was to deal with air to ground traffic, the larger part of which was naval-air into the bargain. In fact, although other and at times valuable information was produced throughout all the earliest times, the intercept sections acted as early warning units against enemy air attack. In this they were most successful, obtaining their results largely from traffic analysis. The Japanese never achieved the same signals security as the Germans and traffic analysis played a very active part in the S.W.P.A. throughout the war. In this art, the Australian Army held the lead. Two years in the Mediterranean had set them upon the right road, although the tasks were different. The Americans had never practised the art in peacetime, neither had the R.A.A.F. So from a fairly early date the [Army's] Y units took the field [where] they, and the Bureau in the background, were producing information upon which action could be taken. Moreover it was not long before decrypts began to contribute. The systems which offered the best hope of solution were the naval air-to-ground transposition cyphers. When these yielded they helped entry into a [water] transport code [which merchant ships used to deploy and replenish Japanese forces] and lastly the operational code-book was solved and gave a much wider range of information.[1]

While there were successes in deciphering Japanese naval codes, notably JN-25, until early 1943 very little progress had been made against Japanese Army or Army-Air communications systems. With Japanese forces so widely dispersed, it was not always possible to update cryptographic instructions by 'safe hand' means. The Japanese often had to rely on passing instructions using wireless transmission ciphers already solved by the Allies.[2]

Reflecting on their experience, the two Australian and one American technical heads of the Central Bureau recognised the need for forward units to be technically self-supporting. At the same time, they placed great emphasis on the need for continuity and centralised study. 'If at all possible', they argued,

> it is necessary to ensure that all material is studied, rather than concentrate on one or a few major systems, despite the fact that they may be the most significant from the standpoint of intelligence. The latter procedure, though losing little strategic or tactical intelligence, could well cause the loss of more valuable cryptographic information. In large measure, success against the Japanese was the result of matching duplicate messages in different systems and of learning in one system about changes affecting another.[3]

The isolation of Central Bureau

Apart from a cryptanalytic relationship with India, Central Bureau suffered from having little contact with other world centres of Sigint work on the Japanese target. This was exacerbated by communication problems between Britain and the United States. It was also made worse by communication problems between US agencies, as made clear in the following humorous incident recorded about five months after the establishment of the Central Bureau. In a letter to Major Stevens, the British representative in Washington on Japanese military Sigint affairs, Colonel Tiltman of GC&CS complained about the 'unknown quantity of American arrangements and future intentions in ... respect of low-grade, especially air-to-ground ... as regards Melbourne'.[4] (Tiltman, it seems, was not even aware that Central Bureau was by then located in Brisbane.) The senior army cryptanalyst in Delhi, Tiltman went on, was in constant touch with Australia 'but can't get traffic ... because communications ... are hopeless'.[5] Not unreasonably, Tiltman expected the headquarters of American Sigint in Washington to know the arrangements and intentions of the US-controlled Central Bureau in Australia.

Major Stevens replied that the 'War Department ... was extremely vague on this subject'.[6] The Americans who had been sent to Central Bureau in Brisbane as well as the survivors of Corregidor were 'a gift to MacArthur'. The War Department therefore had no control over them, 'and has not yet succeeded in finding out what they do'.[7] In fact, as MacArthur was senior to General Marshall, the chief of staff of the US Army, there was no way the War Department felt it could send out a unit which it controlled without a presidential order.[8] Stevens suggested that the only way to get the information Tiltman wanted was to ask the Australians.[9] Through this channel, the British authorities ultimately did get some of the facts.[10]

From January 1943, Central Bureau was under the direct control of General MacArthur's GHQ by an operations instruction. This document defined the Bureau's responsibilities for cryptanalysis, control and coordination of interception, and traffic analysis. Significantly, there was no mention of what the British understood as the final process in Sigint production – namely the final processed output of the Sigint enterprise, intelligence reports.[11]

This omission was consistent with the American approach, whereby Sigint, up to the point of decryption and translation, came under the signals branch, and all intelligence work on it fell under the remit of the chief staff officer for intelligence (G2) of the general staff. Under the terms of its charter, therefore, Central Bureau was not a centre for the production of finished or highly processed 'special intelligence'. Accustomed as they were, however, to the British approach, the Australians maintained an all-Australian intelligence section for Australian participants.[12]

This arrangement was not much help, however, where Japanese military intelligence was concerned, as they had no access to the product of the US Army's Signal Security Agency, which provided most high-grade Japanese military intelligence. Nor, for several months, did they receive the product of the Japanese Air Intelligence Section at GC&CS. The cryptanalytic conferences held in mid-1943 had given America responsibility for Japanese high-grade military systems, and GC&CS and India responsibility for high-grade army air ciphers. Australia, it was agreed, should continue its work on all low-echelon traffic in the SWPA, which was mainly air force communications traffic, as well as the already broken Military Sea

Transport System (water transport code). For a long time, Central Bureau had to depend for its intelligence on this locally produced material, plus anything Britain's Wireless Experimental Centre (WEC) in Delhi could supply. Washington regarded Central Bureau only as a cryptanalytic organisation and, as such, not entitled to receive intelligence. The latter was seen to be the business of the G2 staff with GHQ.[13]

GC&CS was also hesitant about including Central Bureau in the full exchange of Sigint results it already had with Washington and WEC Delhi. This was partly on security grounds and partly out of deference to the United States. Britain pressed, though, for including Australia in the unrestricted exchange, subject to Australia satisfying both the United States and Britain of the adequacy of security measures in the SWPA. This was finally achieved in January 1945, when US Special Security Officers and RAF Special Liaison Units were attached to all Allied headquarters entitled to receive Ultra material.

Although active Sigint production continued in centres overseas, for about eight months Central Bureau was largely isolated in the matter of Japanese Army intelligence.[14] As Dr Hooper, head of the Japanese Air Intelligence Section in GC&CS, pointed out in his report of December 1944 on Central Bureau, this was a serious problem. Hooper also suggested that Central Bureau was 'largely unaware' that this was so.[15] Central Bureau's position vis-à-vis the main centres of intelligence production certainly affected its own intelligence functions. According to Hooper's report:

> Apart from the very unsatisfactory conditions governing the dissemination of high echelon material in this theatre, the use made by the Bureau of its own product is less impressive than the production itself. There is at present insufficient speed in dealing with the material from the moment it leaves the translator to the moment when it reaches GHQ: a lot of time is spent in checking translations ... but the translation as issued is in no way a fully annotated intelligence product of the kind known to G.C.&C.S. ... Secondly, insufficient attention has so far been given to the needs and interests of other centres, and there has been genuine

ignorance of the value to be derived from a full exchange of signalled intelligence ... since very little has come from those centres (W.E.C. is just now coming to life, but G.C.&C.S. not yet) and no one in the Bureau has had experience of the type of exchange which goes on between G.C.&C.S. and Special Branch.[16]

Senior Central Bureau staff awareness of disadvantages

Hooper's comments make clear the disadvantage Central Bureau was working under. While some might not have been aware of it, the senior staff of the Bureau certainly were. And this was Australia's predicament: having accepted US military and Sigint leadership, it was reluctant to damage the relationship and split the organisation that was already working, by and large, to the common advantage. But at the same time, it was clear that in not doing so, the Australians ran the risk of losing their separate identity in the organisation and of crippling the service of Sigint to their own authorities.[17]

In the meantime, General Headquarters transferred to Hollandia in August 1944, and in November, following the Americans' successful penetration of the Philippines, pushed on to Leyte. 'The war ... is running away from us', commented Lieutenant Colonel Sandford, necessitating in Central Bureau 'a good many changes'.[18] A letter from Sandford to Travis, the head of GC&CS, on 12 February 1945 makes clear that Sandford knew exactly what was going on and had seemingly come to terms with it. According to Sandford,

> since the arrival of American forces in strength in the Philippines, a political gap has arisen simultaneously with the geographical one. General Akin is inclined to make spot decisions as to intercept facilities, according to the immediate dictates of the tactical situation, and these do not always have the happiest results. However, by and large we are quite fortunate

that we have not suffered still more interference with our plans as the result of the reoccupation of so large an area of American territory.'[19]

Sandford was being diplomatic, and was determined to make the best of it. He went on in this letter to recall a discussion months earlier with General Blamey, during which Blamey had expressed the view that it would be detrimental to Australia's interests and 'contrary to our duty' to pursue separation from US forces. Blamey's view had been reinforced in another discussion Sandford had had with him the night before writing the letter, with the added condition, 'provided it would always be possible to serve British commands as a first priority'.[20] This was a reminder that, for Australians like Blamey and Sandford, the closeness of ties with Britain remained paramount and intact.

Sandford saw that the move to Manila would result in 'American influence growing to the point of complete domination'. This, he surmised, 'is inevitable, but in view of my conversations with Blamey, we can see no alternative'.[21] He concluded by observing that,

> if the Australian component is now slowly being swallowed and eventually becomes part of an American machine from the point of view of policy as well as everything else, that one need not be too depressed since they learned a very great deal from us in the early days and seem now to be working along the right lines.[22]

In fact, Sandford's active defence and promotion of Australia's intelligence interests suggest he was not as resigned to the situation as he may have appeared in this letter to Travis.

Notwithstanding the difficulties in inter-service and international Sigint collaboration during the war, the work of the combined Sigint units to which Australia made important contributions reflect an impressive, largely unheralded story of Sigint at war. This includes remarkable breakthroughs on land, in the air and at sea.

Maritime Sigint successes

During the Pacific War, Sigint produced some stand-out maritime successes; notably the Battle of Midway in June 1942, but also the preceding Battle of the Coral Sea. In these battles, Sigint provided a 'priceless advantage' over the Japanese.[23] On 4 December 1941 Japanese naval code JN-25 had been changed. By February 1942 it had been recovered. The intelligence blackout resulting from the code change underscored the perils of reliance on Sigint as the sole source of reliable intelligence during a crisis.

By mid-April 1942, Japanese messages were again being intercepted, decrypted, translated, re-enciphered, and disseminated by Hypo, the US Navy Sigint station in Hawaii – and this time within six hours of their original transmission![24] This was done in close collaboration with work undertaken at FRUMEL.[25]

Allied deception saves Cocos communications

Endeavouring to prevent the United States and Britain from using Australia as a base from which to launch a counteroffensive, Japanese Army and Navy planners took a number of measures in the Indian and Pacific oceans. Recognising that the cable station at Cocos Island was key to links between Australia and the remainder of the British empire across the Indian Ocean, the Japanese Navy shelled the island on 3 March 1942. The attack damaged the cable station and other buildings but did not cut the cable. The next day, Cocos wired London, saying 'THINK POLICY SHOULD BE TO LET ENEMY THINK COCOS OUT OF ACTION'. London replied via radio message to Batavia (now Jakarta) in clear language with a statement to confirm the ruse: 'AS COCOS DESTROYED COMMUNICATION TO YOU NO LONGER POSSIBLE SEE NO NEED YOU REMAIN BATAVIA ... GOOD LUCK DO NOT REPLY TO THIS.' The Japanese periodically sent reconnaissance flights over to check and, finding the cable station still in ruins, assumed it remained out of action. But, in fact, the station stayed operational, continuing to pass traffic between Britain and Australia for the remainder of the war.[26]

The Battle of the Coral Sea

In an effort to cut off Australia on the Pacific side, the Japanese devised an operation to sever communications between the United States and Australia and capture Port Moresby.[27] Official accounts, written before the Sigint revelations were made public in the mid-1970s, noted intelligence was 'garnered from RAAF and US aerial reconnaissance, coastwatcher reports, and the results of American eavesdropping on Japanese radio'.[28] In fact, while aerial reconnaissance and Coastwatcher reports were crucial as well, Sigint provided much of the necessary detail, revealing the Japanese were committing three aircraft carriers to the Moresby operation: *Shoho*, *Shokaku* and *Zuikaku*. With an invasion planned for around 10 May, US carrier task forces were deployed to intercept them.[29]

On 9 April 1942, FRUMEL intercepted a message from the C-in-C Combined Fleet asking for a report on progress of repairs to *Kaga* as he 'requires her services as soon as possible since she is due to take part in the "RZP" campaign'.[30] (RZP being the place designator for Moresby.) The preparatory work in the lead-up to the battle was the first time Sigint units in Hawaii, Melbourne and Corregidor in the Philippines (evacuated to Melbourne by 6 April) worked together in support of shore-based headquarters in Washington, DC, Pearl Harbor and Melbourne.

Some Allied ships were unable to receive the highly classified Sigint reporting. This notably included the RAN warships that had their 'WT Procedure Y' (Sigint) operators transferred to shore in Melbourne. In addition, sometimes the reporting included inaccuracies, leading the US Navy Task Force commander, Rear Admiral Frank Fletcher, to miss opportunities to strike more convincingly. In the end, though, Japan aborted its mission to seize Port Moresby and control access to the Torres Strait. Naval historian Ian Pfennigwerth argues that this was because the Japanese commander was fearful of the prospects of Allied battleships, which had been spotted as part of Fletcher's force.[31]

In the wash-up, FRUMEL intercepted a 13 May message from a 'concealed originator' at Rabaul who gave Japanese losses in Coral Sea as:

> Shoho hit by 7 torpedoes and 13 bombs and sunk. 22 of her aircraft made forced landings resulting in 16 seriously injured and 64 injured. Number of ship's company drowned not yet known.
>
> Shokaku hit by 3 torpedoes and 8 bombs and damaged in engine room, petrol bunkers etc 4 officers and 90 ratings killed and 96 seriously wounded.
>
> 2. Message from Naval Intelligence, Tokyo, states that according to broadcasts, America and Britain declare the sinking of WARSPITE by 5th Carrier Division on 8th as entirely false, and also that the other battleship supposed to have been sighted on the same day was not there.[32]

The latter part of the message indicates that although the authorities in Tokyo were not averse to spreading false information and propaganda, they would not tolerate receiving it from their own forces. On 22 May, FRUMEL intercepted a hastener from Naval Intelligence Tokyo to the C-in-C 4th Fleet for the serial photographs of the Coral Sea action showing the sinking of *Warspite*.[33] It is hard not to feel sympathy for the C-in-C of the 4th Fleet.

The Battle of the Coral Sea seemed inconclusive at the time, with the aircraft carrier USS *Yorktown* damaged and *Lexington* sunk, while Japan suffered the loss of one carrier, the *Shoho*. On balance, however, the Allies carried the day; the thrust towards Port Moresby had been stymied.[34] As Australian naval historian G. Hermon Gill observed, 'so ended the Battle of the Coral Sea, in a tactical victory for the Japanese but a strategic victory for the Allies. The main object of the Japanese operation, the capture of Port Moresby, was denied them.'[35]

Midway and the cryptanalysts at FRUMEL

From an Allied perspective, the Battle of the Coral Sea was a dry run for Midway, which was not only a decisive victory for the Allies in the Pacific but the beginning of the end for Japan, which from then on would be fighting a defensive war. Although the Battle of Midway is regarded as an American victory, in fact, cryptanalysts of FRUMEL in Melbourne made an important and largely unrecognised contribution to that victory.

Key to the victory at Midway was forewarning of the attack, knowing that Midway would be the target, and detailed information of the forces the Japanese would bring to bear. Messages intercepted and decrypted by FRUMEL analysts, in FRUMEL's own records, signed by Jack Newman (RAN's director of communications and recruiter of female signallers who, as we saw, did not 'hit it off' with ace codebreaker Eric Nave) tell the story – it's a gripping one.

On 14 May 1942, FRUMEL reported that the commander-in-chief of the Japanese Combined Fleets, the redoubtable Admiral Yamamoto, informed the 4th Fleet that bombs and ammunitions for the 'forthcoming campaign' would be supplied.[36]

At this point, there was still uncertainty about the target of the attack, as the Japanese were using the place designator AF to disguise it. But it is clear from the following message that FRUMEL believed it would be Midway: they reported on 18 May that 'on the day of the attack it is intended to – in position fifty miles N.W. of "A.F." (Midway Is.) and then fly off aircraft as soon as possible'.[37]

Crucially, the target was confirmed on 21 May, when FRUMEL reported that 'Naval intelligence Tokyo states Midway informed Pearl Harbor that they had only enough water for 2 weeks and asked for immediate supply'.[38] This message removed all doubt that Midway was the target. The Japanese had taken the bait dangled by American FRUPAC analysts, who had telephoned the Midway garrison on a secure line and asked them to radio in clear for water.[39] FRUMEL provided the confirmation of the target.

Over the following days, FRUMEL discovered indications of the timing of the attack in the painstaking analytical process that went beyond the content of the decrypted message to include some inspired guesswork and occasional brilliance. On 22 May, they reported a message about the 'forthcoming campaign', followed by another stating 'since Zuikaku is likely to take part on completion of (? overhaul) the question of pilot replacements is to be considered'. To this, a FRUMEL analyst added the comment: 'this suggests the campaign may start before Zuikaku's overhaul is completed'.[40]

On 23 May, there was another message requesting guns to be loaded 'by 4th June for transport to "AF" (Midway)', as well as confirmation that the IJN, blissfully unaware that the Allies now knew that the attack would be on Midway, were still using the designator 'AF' for it.[41]

Traffic analysis, recognised by de Grey as an Australian forte, was a strength of Australian intelligence that applied equally to the analysts at FRUMEL. In the next few days before the Midway Battle this prowess came to the fore, as FRUMEL was able to glean crucial information from traffic analysis alone. On 24 May, for instance, FRUMEL reported that:

> Traffic associations indicate that 7th Cruiser Division is to proceed from Japan to Saipan area via Chichijima, possibly in company with C-in-C 2nd Fleet in *Kumano,* 2nd Destroyer Squadron, one Carrier Division, 6th and 7th Air Squadrons, and additional 4th Fleet units. 6th Cruiser Division may also join this force between 15th and 20th June.[42]

The next day, 25 May, FRUMEL reported 'unusually large volume of intelligence traffic originated by Jaluit indicates special effort to obtain information prior to attack on Midway. C's-in-C Combined and 2nd Fleets are included in addresses as well as usual addresses.'[43] The last point indicated that the two C's-in-C were then not in Jaluit's communications zone.

Then, in a report on 27 May, FRUMEL stated that a message from Saipan to the commanders-in-chief of the Combined, 4th, 1st Air and 11th had provided an escort program which had the invading force arriving at Midway at 19:00 on 6 June.[44]

All that was missing was an exhaustive description of the attacking force, and this was provided by FRUMEL on 29 May. FRUMEL reported that 42 fighter planes from the 6th Air Group were to be taken to Midway aboard aircraft carriers *Kaga, Akagi, Soryu, Hiryu, Zuikaku* and *Shokaku*. The 7th Cruiser Division (*Kumano, Suzuya, Mikuma, Mogami*), the 2nd Destroyer Squadron, part of the 4th Destroyer Squadron and part of the 5th Cruiser Division were on their way. The 3rd Submarine Squadron, the 1st and 2nd Carrier Divisions, the 18th Cruiser Division (*Tenryu* and *Tatsuta*), three or four battleships, associated destroyers, transports and submarines were all taking part.[45]

On 1 June, drawing again on traffic analysis, FRUMEL reported a considerable volume of traffic being broadcast to Japanese forces in the Midway operation, but there was no traffic from them. FRUMEL concluded, 'the operation has apparently begun and forces at sea are keeping radio silence' and the following day 'the calm before the storm continues'.[46]

After some first sightings and preliminary air attacks on the Japanese on 3 June, the Battle of Midway began the next day. FRUMEL reported 'exceptionally heavy' operational traffic using secret call signs, but noted that 'so far', 1st, 2nd and 5th Carrier Divisions had not shown up and that traffic routing indicated that the Headquarters of the Combined Fleet was in the area of the Jaluit Atoll. By traffic routing again, FRUMEL located the chief of staff of the 1st Air Division on the cruiser *Nagara*, from which they deduced that the carrier *Akagi* had been sunk or put out of action. Later messages from Admiral Yamamoto to the *Akagi*, the 4th Destroyer Division and the Commander of the 1st Air Fleet suggested that *Akagi* was probably badly damaged and being escorted back by destroyers, or that the destroyers were standing by her.[47] In the end, the Japanese were to lose the carriers *Akagi, Kaga, Hiryu* and *Soryu*, as well as the heavy cruiser *Mikuma*, but these losses would only be discovered later.[48]

On 8 June, on the basis of DF fixes on the cruiser *Nagara*, the 1st Air Fleet, the C-in-C of the 2nd Air Fleet, Admiral Yamamoto, the Commander of the 7th Cruiser Division, the *Genyo Maru* tanker and five submarines, FRUMEL reported that 'the Midway Occupation Force appears to be retiring towards Saipan'.[49]

FRUMEL then proceeded to uncover the extent of the losses of the Japanese forces, reporting on 2 July that long messages from the personnel officers to the commanding officers of *Akagi, Kaga, Soryu* and *Hiryu* added *Hiryu* to the list of 'suspected sinkings'. These losses and more were confirmed in a message from the commander, 1st Air Fleet, which was sent on 6 June but not decrypted and read until 28 September. As he said, it was 'a terrific battle'.[50]

Perhaps the most riveting message was decrypted and read on 4 October. Admiral Yamamoto informed IJN headquarters that its losses amounted to one carrier sunk and one badly damaged and one cruiser badly damaged, 35 aircraft not returned, total losses 94. The low-key comment of the FRUMEL cryptanalyst: 'C-in-C, Combined Fleet is apparently feeding false information to his own subordinates, presumably for purposes of morale or propaganda.'[51]

In the meantime, much of the Sigint effort during the war involved more mundane but nonetheless worthwhile achievements, the cumulative effect of which was to wear down Japanese ability to wage war and sustain its campaigns. Not the least of these efforts involved the destruction of Japanese convoys and shipping. One rear admiral, Ralph E. Cook (USN), served as a junior officer in FRUMEL and later recalled, 'thanks to our information, submarines knew when and where to prey on Japanese ships'.[52]

Tragically, aboard many of the Japanese ships sunk were Australian and other Allied prisoners of war, as well as Southeast Asian labourers in transit to work in support of Japan's war effort in mining and other industries. According to estimates, the Allies ended up killing over 20000 such people, as they sank Japanese ships transporting them to work in Japan, helping cover shortfalls generated by the war effort. Although sometimes the Allies knew of the presence of prisoners of war, thanks to Sigint reports, the ships were sunk regardless. In a callous equation, this was justified at the time because interdiction of critical strategic materials was considered more important than the deaths of prisoners of war.[53]

Japan–Singapore convoys via Manila were also subject to Sigint interception, and considerable success was achieved with them, particularly during the second half of 1944 when important convoys between Japan and Singapore sailed via Manila and Miri. Along the way, Sigint played a significant role in slowly strangling Japan's war effort.

Air Sigint successes

Intelligence collaboration between the USAAF and the RAAF in the SWPA from 1942 to 1944 was close. A directorate of intelligence was established, pooling their resources for common objectives and this level of collaboration was maintained, particularly while RAAF resources were seen as necessary by MacArthur and his commander Allied Air Forces SWPA, Lieutenant General George Kenney. As the war went on, though, the US largely excluded the Australians from planning for the Philippines campaign and the combined directorate was disbanded.[54] In terms of Sigint, however, where the USAAF and RAAF worked together in Central Bureau, Sigint support continued for both air forces for the remainder of the war.

Sigint made a major difference to battles at sea and in the air. It was also behind a number of successes for which credit has been directed elsewhere. In the official histories of the 1950s and 1960s, for instance, radar stations are credited with detecting Japanese aircraft on bombing runs to Darwin and remote sites elsewhere. This was the case throughout the SWPA, particularly at the various New Guinea and Darwin bases.[55] In fact, Sigint provided advanced warning of innumerable air raids and a constant flow of tactical air intelligence. This resulted in successful Allied air interceptions and the destruction of enemy aircraft on the ground.

Various indications led to these warnings. Japanese weather reconnaissance over Allied territory was a sure indication that a raid would follow, subject to suitable weather conditions. Sometimes, suspicious bearings on enemy aircraft justified a preliminary warning. As procedures were refined, warnings were issued as soon as a known enemy strike radio frequency was heard 'warming up'.

In *Katakana Man*, Jack Brown noted that the RAAF Number 1 Wireless Unit, working closely with Central Bureau, deployed to Townsville, then Port Moresby and on to Nadzab in New Guinea in September 1943 and, again, on to Biak in the Dutch East Indies in July 1944, before accompanying MacArthur's Headquarters to the Philippines in October 1944.[56] The unit operated on a 24-hour basis. Brown recalled that:

Allied commanders were always receiving information from us on Japanese battle orders of attack, so the Allies could attack quickly. We knew the make-up of convoys, the cargo they were carrying, their routes, sailing days and estimated days to arrival at destination, the serviceability of aircraft, the numbers of aircraft on Japanese airfields, where their supplies were stored, and what their supplies were. We also advised of sneak raids or full Japanese aircraft air raids almost as soon as they left their bases or just before they left.

The effectiveness of the work of the kana operators, direction finders, linguists and code breakers proved to be so successful that the Japanese battle plans were often an open book to the Allies. We operators, who knew so much, had been told in Townsville that if we were ever taken prisoner we had to do away with ourselves.[57]

In early March 1943, Sigint reporting from FRUMEL gave early warning of a Japanese convoy carrying over 6000 troops from Rabaul to Lae. The Japanese troops were deploying with IJN escorts and supported by aircraft. Their deployment from Rabaul was intended to counter the successful Australian and US offensives along the New Guinea coast from Salamaua to Lae. Primed with this information, RAAF and USAAF aircraft attacked the convoy and destroyed them in detail. The story is captured in Michael Veitch's *The Battle of the Bismarck Sea* and is considered one of the best examples of Sigint directly contributing to the outcome of any battle in the New Guinea campaign.[58] In addition, it is arguably the best example of Sigint directly supporting an air strike operation against enemy naval vessels.

Sigint and the death of Admiral Yamamoto

Perhaps the most spectacular example of the successful use of Sigint in the air war was the operation mounted to intercept and kill Japanese Admiral Isoroku Yamamoto, the architect of the 7 December 1941 attack on Pearl Harbor. The daring attack, which took place on 18 April 1943, involved a long-range flight by 16 USAAF P-38 Lightning twin-engine fighters from Guadalcanal to attack the Mitsubishi bomber carrying Yamamoto on a visit to the troops stationed in Bougainville. From a variety of collection sites, Sigint gave the USAAF precise details of the time and place the flight would be transiting. Admiral Yamamoto's plane was shot down and he was killed.[59] The Allies were conscious of the risk of exposing their use of Sigint, as the downing of Yamamoto's plane was hard to explain in any other way. Fortunately for the Allies, the over-confidence of the Japanese Navy in the security of its codes overrode concerns about a possible breach of the JN-25. Japanese Navy officials blamed the army for passing messages about the intended flight using army codes which the Navy felt were less secure, although, in fact, they were better protected than JN-25.[60]

There was also some dispute as to which part of the Allied Sigint network deserved most credit. Some accounts ascribe responsibility to an American monitoring unit at Wahiawa, Hawaii and Dutch Harbor, Alaska for having intercepted the messages concerning the flight on 14 April, but these decryptions were incomplete.[61] Australian sources claim RAAF No. 1 Wireless Intercept Unit near Townsville intercepted messages about Yamamoto's itinerary. In *Katakana Man* Jack Brown claims a 19-year-old Victorian, Keith R. 'Zero' Falconer, had taken the message – Falconer was nicknamed Zero because he never made a mistake in training. What we do know is that there was 'confirmed interception of just two signals that followed the original message. These contained the admiral's itinerary, and provided some important confirming details; both of these reportedly came from FRUMEL.'[62] The Coastwatcher network was used as cover for these successes, coupled with USAAF sorties, which continued over the area for some time afterwards, to give the impression it was just a lucky break.

Land-based Sigint successes

Early progress was made against Imperial Japanese Army (IJA) codes in April 1943, followed by a series of successes. For Central Bureau, 'there was always some success being achieved on high command material'. At times, some systems would remain unbroken for limited periods, while at other times everything being transmitted was readable.[63]

The IJA made it difficult for Allied Siginters, employing a red-coloured army administrative code book (*Rikugun angosho*) for the Army General Staff, and a tan one for the Army Water Transport. The code book was described as a compact, practical and easy-to-carry document, with three columns per page and a total of 10 000 four-digit code entries. Part I included code group categories, including place names, dates, unit designations and other four-digit code groups. Part II duplicated the Part I entries in the Japanese *Kana*. Part III listed the four-digit codes numerically. Part IV listed sender call sign code groups, and Part V included decoding information for call signs. Rounding out the collection was a separate additive book used for enciphering. The IJA frequently tightened its encryption methods, admonishing officers and men to destroy cryptographic materials to prevent them falling into enemy hands, by ordering that losses be reported and, late in the war, by introducing Hebern wired rotor-type cryptographic machines for higher headquarters.[64] It is little wonder then that IJA codes were difficult to break and the early years of the war saw little headway made against them.

Drawing on research in Washington and GC&CS in the United Kingdom, the Central Bureau studied the Japanese Army Water Transport Code to gain initial entry. In April 1943, Central Bureau and Arlington Hall solved the indicator of the Water Transport Code. This then enabled cryptanalysts to strip cipher, reveal code values and, by June 1943, to read some message plain text.[65] Painstaking and meticulous review exposed several flaws in the Japanese code systems, including the transmission of part of the message in plaintext, as well as inconsistencies in enciphering and occasional duplicate messages which allowed matching and duplicate messages in different systems.[66]

Then, in early 1944, the Australian 9th Division (which had captured the 621 Radio Intercept Company in North Africa in July 1942) made the first significant capture of enemy cryptographic material in Sio, then the headquarters of the 20th Japanese Division. A set of code books (*Rikugun angosho* No. 4) was fortuitously found in a rusted box found in a water-filled pit. The enemy was entirely unaware of this capture, with one overheard stoutly certifying he had overseen the complete destruction of the documents (a process supposed to have been verified by the acquittal of the code book covers). The documents were captured in good shape (without their covers), and were valid through to the end of March.[67]

A little while after this discovery, a charred and waterlogged copy of a code book from March to May 1944 was recovered by divers from the safe of a sunken barge, the *Yoshino Maru*. With extreme difficulty, the damaged fragments were carefully read and available for use by May. The badly charred pages were treated with a chemical solution and photographed in the very short time the newly exposed text was readable – this work was 'invented' and undertaken at Central Bureau.[68]

These two captures laid the foundations upon which solutions to future period code books were based. The principal work was done in Washington, though other stations, including Central Bureau, made contributions to the exploitation, by decrypting, translating and disseminating locally intercepted reports.[69]

Sinkov stressed that to appreciate the enormity of the tasks, 'even though we obtained captured material, we were never free from slow reconstruction and analysis work because we studied many systems'. Further, he said,

> solving a code message or a code system is different from solving a cipher. Normally when you solve a cipher, you can read the entire communication without any problem. In solving a code it's a process of little-by-little reconstruction of the book. It means that messages are readable only in part in the early stages and there will be gaps. The extent of the gaps will diminish as you get further into the code book.[70]

In essence, these discoveries and the reconstruction and analytical work gave US and Australian forces a most welcome boost, enabling more confident operational planning. Sigint successes like these helped accelerate the pace of success on the ground, providing the Allies with unique insights and enabling them to repeatedly surprise Japanese forces and thus reduce the risk of own-force casualties.

The importance of personal relationships

As the war progressed, Australians were building the personal relationships and networks for cooperation that would guarantee them the seat at the table that had been missing when the Holden and initial BRUSA agreements were negotiated. As this and earlier chapters attest, one of the key players was Captain (later Lieutenant Colonel) 'Mic' Sandford, who joined the Australian Army Special Wireless Section (ASWS) on Crete as its intelligence officer and head of the intelligence section. Sandford, who had trained at GC&CS, was probably the first Australian officer to undertake training during the war in Sigint procedures with an ally.[71]

Commanders Nave and Newman had participated in a conference on the use of Sigint in Singapore in October 1940. The conference covered the study of call signs, frequencies, operating signals, wireless routines, direction finding bearings and cryptography.[72] In March 1941, Newman again travelled to Southeast Asia, this time meeting Dutch authorities who had established *Kamer 14* and discussing the prospects of later exchanges of cryptographic material with them.[73]

In September 1941, Professor Room and an outstanding Japanese speaker, Lieutenant A.B. 'Jim' Jamieson, attended a cryptanalysis course at FECB.[74] En route to Singapore, Room and Jamieson visited *Kamer 14* as, by then, the Dutch had established good liaison with FECB. The Dutch had made considerable progress on Japanese diplomatic and naval codes independently of the British and Americans, but it is unlikely that any exchange of material took place. The Dutch knowledge was probably due to the defection of a Dutch merchant marine radio operator in Kobe, Japan, who handed his code books to the German consul.[75]

In September 1943, the WEC located outside Delhi hosted a coordination conference between the Allied Sigint centres in the Asia–Pacific. The conference decided on a division of labour between regions based on the supported Allied commands, such as MacArthur's SWPA and Slim's XIV Army in Burma, among others. Central Bureau was to have been represented by its senior US army officer and assistant director, Colonel Joe R. Sherr. Tragically, however, Sherr died in the fiery crash of his C-47 Dakota transport on the way to the conference.[76]

On 13 March 1944, the United States convened the Second Joint Allied Conference at Arlington Hall, Virginia. The United States delegation included Central Bureau's senior cryptanalyst, Colonel Sinkov. Australia, as part of the United Kingdom delegation, was represented by one of Sandford's deputies, Major S.R.I. 'Pappy' Clark. In all, there were 35 attendees at what was perhaps the most secret conference of the war.[77] The conference built on previous high-level meetings to deliver an agreement to streamline and strengthen sharing between the Allies, including a clear division of labour between the various Sigint centres. The results obtained by any one centre were radioed to all the others so that each had all available information.[78]

Clark then visited Ottawa in March 1944, where he discussed the deployment of a Canadian Army Sigint unit to Australia. Clark's visit to Ottawa was at the instigation of GC&CS head, Edward Travis, who knew that Britain would propose a Canadian deployment to Burma.[79] The arrangements were subsequently discussed in London, when Prime Minister Curtin conferred with British authorities. In a memo, the chief of the general staff told the secretary that:

> [As] it was considered that Canada would eventually be interested in the Pacific theatre, it was agreed that the War Office might make a formal request to National Defence HQ Ottawa for the despatch of a Canadian Special Wireless Group with additional intelligence personnel complementary thereto.[80]

The Canadian government agreed in December 1944,[81] and the following February over 300 Canadians with the 1st Canadian Special Wireless Group deployed to Australia arrived in Brisbane.[82] From May, the group then began to operate from Darwin, working with Australian forces through to the end of the war.[83] Such an outcome would seem highly unlikely in the absence of strong personal relationships between two closely aligned but geographically distant partners.

Australian visits to allies were reciprocated. In March 1944, Edward Travis visited Australia, including Central Bureau in Brisbane. A great raconteur, Travis told a group of officers at Nyrambla about a visit by Churchill to Bletchley Park. Churchill, after receiving a briefing on work against high-grade German ciphers, remarked: 'What I don't like is all these people having all these secrets', to which Travis replied: 'Well Sir, if we didn't have all these people, we wouldn't have all these secrets'. Churchill's response, if there was one, was not reported.[84] This exchange captures a central challenge of Sigint: balancing priceless access to intelligence with the need to know and the protection of sources and methods.

Major Geoffrey Ballard, a veteran of the ASWS in the Middle East in 1940–41, became probably Australia's first Sigint liaison officer to represent an agency abroad. As Central Bureau liaison officer to the British Sigint centre at HMS *Anderson*, near Colombo, Ceylon, Ballard's appointment came about because of the convergence of the operations of Admiral Louis Mountbatten's South East Asia Command and MacArthur's SWPA after their victories in Burma and the Philippines respectively. As the war moved towards Japan and the liberation of other occupied territories in Asia, coordination of the Sigint effort became more pressing.[85] At the same time, it remained challenging, particularly due to growing security concerns.

Sigint security measures

Management of the secure handling of Ultra and Purple technology and reporting was critical to the successful prosecution of the war. But from early on, concerns were expressed about poor security in Australia. By mid-March 1943, in stark contrast to the practice of the United Kingdom and United States, Australia's military commanders made Ultra reports available to Australian military officials and not their political masters, who were to be 'kept in ignorance of this type of information for security reasons'.[86] The net result was a growing perception that Asutralia's security was not what it should have been.

The incompetence of the counterespionage and protective security apparatus of state in Australia during the war was a thorn in the side of Australia's British and American intelligence partners. The RAF's Group Captain F.W. Winterbotham, who worked with GC&CS and would later write the groundbreaking book *The Ultra Secret,* played a key role in controlling access to Ultra material.[87]

The distribution list for Ultra reporting was restricted to the Australian DMI, DNI and Director of Intelligence RAAF, as well as MacArthur and his G2, Major General Willoughby. It was also extended to the DMI, New Zealand, and the New Zealand Naval Board. The only organisation on the list was Central Bureau.[88] Ultra material from Japanese diplomatic traffic would not be shared with any Australian civil department until July 1943 nor, officially, with any Australian ministers, even after some civilian officials received it. Even then, the addition of Colonel W.R. Hodgson, the civilian secretary of the Department of External Affairs, to the list in mid-1943 was damage control, following his discovery of the existence of Trendall's bureau (D Special Section) and its work.

This decision to exclude Australia's political leadership from access to Ultra material seems extraordinary, but it appears this was done with the approval of Prime Minister Curtin. Curtin appears to have been briefed on Sigint activity and evidently received information from the British High Commissioner, Sir Ronald Cross. The decision to exclude the Australian cabinet from access to Ultra intelligence was made with Curtin's tacit if not explicit approval.[89]

While Australia and its allies achieved remarkable Sigint successes during the Second World War, not everything went the Allies' way. Late in the war, signs emerged that security information was leaking through Soviet and Chinese Nationalist channels. There was evidence that the Japanese had intercepted and decrypted Chinese communications. The Chinese Defence Attaché, Colonel Wang Chih, wrote detailed reports which the Japanese intercepted, analysed and found very useful.[90]

Evidence of leaks of Ultra information from Canberra

Evidence emerged from Ultra intercepts of additional leaks, indicating that the Japanese also had access to Soviet messages. On 24 November 1944, a Japanese message from Japanese-occupied Harbin, in Manchuria, to Tokyo gave details of MacArthur's plans for operations in the Philippines. The source was listed as the 'Soviet Ambassador' in Australia. Security officials reckoned that this could only be the case if the Japanese had broken Soviet diplomatic codes or if the Soviets (with a diplomatic presence in Harbin) had been sent the details from the Soviet Embassy in Canberra and then, in turn, passed the information directly to the Japanese. Yet for the Soviets to be passing such information, there had to be an individual or network of agents in Australia passing information to the Soviets in Canberra.[91]

On 16 December 1944, the following signal marked 'immediate' was sent from Central Bureau in Brisbane by the visiting British investigating officer, F.W. Winterbotham, who wrote to GC&CS concerning apparent Ultra leaks in Australia, declaring:

> Have investigated matter here and patently obvious all this information originates with Soviet repeat Soviet Ambassador CANBERRA. Cannot trace time lag or accuracy but suppose someone CANBERRA who knew future plans passes information to Russians who send it to Moscow.[92]

The message in Japanese giving details of MacArthur's plans and listing the Soviet Ambassador in Australia as the source was intercepted by the Australians, who recognised that the implications were dire. This matter was not satisfactorily resolved, but the revelations caused great consternation and, eventually, a major upheaval and reorganisation of the postwar Australian intelligence community. The result of the investigation, however, which had begun in Australia and other Commonwealth countries, was some time off.

Writing to the acting Minister for the Army, Senator J.M. Fraser, on 6 January 1945, General Blamey outlined these and other concerns, many of which arose from Sigint analysis. In his letter, Blamey explained that action was 'being taken to restrict the dissemination of information that may be published'.[93] This was a first step, but there was a long way to go. The disinclination of the Australian authorities to deal head-on with the security leaks would have adverse consequences for Australia beyond the end of the war.

The External Affairs Minister and Attorney-General, Dr H.V. 'Doc' Evatt, remained sceptical about the leaks, protective of his staff, and trusting of the Soviet officials he dealt with in Canberra. Frances Bernie, who worked in Evatt's office, made secret copies of sensitive correspondence which she handed on to the spy ring coordinator, Walter Seddon 'Wally' Clayton (known in Russian coded correspondence as KLOD). Clayton passed the documents to the Russian 'Tass' journalist-cum-spy, Feodor Nosov, who, in turn, passed them to the Russian ambassador in Canberra. The ambassador duly reported back via the Russian diplomatic post at Harbin, in Japanese-occupied Manchuria. From there, it is believed, the Russians passed the information to the Japanese with the apparent motive of slowing down the Allied Pacific campaign, buying time for the Soviets to join after the defeat of Nazi Germany. The Soviets then planned on seizing Japanese territory of strategic importance to the Soviet Union, and did so in August 1945, just as the Japanese surrendered.[94]

From the perspective of Australia's allies, both British and Americans 'were aware that security in Australia was not generally understood as it was, for instance, in England'.[95] In the coming months and years, concerns about lax security in Australian government circles would continue to grow. These concerns would persist after the war, reflecting the naivety or outright treachery of those passing on sensitive classified information and the incompetence of those responsible for protective security in the Commonwealth Security Service.[96]

Meanwhile, the focus of D Special Section switched from the war in the Pacific to Russian traffic, with much of the intercept being passed directly to London. With the closure of Central Bureau in late 1945, the senior Army member, Lieutenant Colonel 'Mic' Sandford moved to Melbourne to head D Special Section.[97] Over the next year, D Special Section would become identified with Central Bureau. The DMI used the term 'Central Bureau' to designate their Sigint activities in the immediate postwar period.[98]

Unbeknownst to Australians, however, another source of information, derived from intercepted and decrypted Soviet diplomatic reporting, suggested the scale of the leaks was greater than had been appreciated by many in government. Following the Nazi invasion of Russia, the Soviet Union became a partner of the Allies in their fight against Nazi Germany and, eventually, Imperial Japan. But that did not stop the Soviet Union and its Western allies from spying on each other. Professor F.H. Hinsley's official history, *British Intelligence in the Second World War,* records that virtually all work on Russian codes and ciphers stopped from 22 June 1941, the day Germany attacked Russia.[99] However, from 1942 and into early in 1943, fearing the Soviet Union might consider another deal with Nazi Germany akin to the Molotov–Ribbentrop Pact of 1939, the US Army's Signal Intelligence Service examined and exploited Soviet diplomatic communications.[100]

Venona and the Australian connection

We now know that 'Venona' was the code word eventually settled upon to represent a collection of over 2000 partly decrypted Soviet secret messages, mostly concerning clandestine activities of the Soviet security and military intelligence arms known as the NKVD and the GRU.[101] The Venona program, as it came to be known, would reveal Soviet atomic bomb espionage and a range of espionage matters in the United States as well as in Australia, although much of the latter would be revealed only after the end of the war.[102] The British would join the Venona program in 1948, and further exploitation of the high-grade Soviet intelligence and diplomatic service communications material uncovered would continue for decades afterwards, ending only in 1980.[103]

Venona was made possible by the German onslaught on the Soviet Union in 1941. Prime Minister Baldwin's revelation in Westminster back in 1927 had taught the Soviet Union to rely on one-time pads for their encrypted messaging. At this low point of 1941, however, they produced a 'second' batch of these one-time pads to cover a critical shortfall in cipher keying material. The second copies of one-time pads ceased to be used for live communications in 1948 – a development some attribute to the Cambridge spy ring, which included Kim Philby, Guy Burgess and Donald MacLean. American military historian Matthew Aid observes, however, that William Weisband, a Russian linguist working at the US Army Security Agency (ASA), 'told the KGB everything he knew about the USA's Russian code-breaking efforts at Arlington Hall'. Aid notes that 'for reasons of security', Weisband was not put on trial for espionage – and was considered 'the traitor that got away'. The project was so sensitive that the circle of indoctrinated personnel was far smaller than was the case with wartime Ultra Sigint reporting.[104]

Considered the 'holy grail' of counterintelligence, Venona decrypts revealed clues to the identities of thousands of Soviet spies across the globe, including Australia.[105] Intercepts from 1943 to 1948 revealed over 200 messages that were decrypted and translated by US Army Sigint personnel in Arlington Hall, covering traffic to and from the Canberra KGB residency. Much of this was intercepted and decrypted in near-real time. Eager to protect this sensitive and valuable source, however, investigators sought corroborating evidence before any action was taken.[106]

After the revelation of the key role of Sigint in wartime, it may seem surprising that there was not more curiosity and speculation about the role of Sigint in the Cold War. Because of the secrecy surrounding Sigint, there is a tendency for its role to be overlooked and, as we have seen, for history to be misread as a result. In Australia, the role of Sigint was, for some time, of global importance, but, until recently, these secrets were tightly held.

In the meantime, while the full import of these leaks had not yet sunk in, there was a need to establish Australia's intelligence organisation on a postwar footing. Australia had learned much from its wartime experience. But finding the right balance between a more independent, national Sigint organisation and the continuing need for collaboration with the United States and Britain, remained a challenge.

A major preoccupation was that the organisation and status of Australia's Sigint would ensure it had greater equality with its allies. A report of 3 August 1945 by Australia's Joint Planning Committee recommended that the existing constitution of the Joint Intelligence Committee as a sub-committee of the Joint Planning Committee should be upgraded to a Joint Intelligence Committee, responsible directly to the Chiefs of Staff Committee. Furthermore, 'in view of the need to combine military and political intelligence, it should include a member of the Department of External Affairs'.[107] This issue would be considered further after the war, as we will see in the next chapter.

Australia's communications security continues to be abysmal

Meanwhile, and despite the remarkable Sigint successes at sea, on land and in the air, there was continuing concern about cavalier attitudes towards the enemy's ability to monitor and act on Allied communications. There was, as we have seen, carelessness on all sides regarding communications and particularly signals security, with the potential for dire consequences. As the danger is invisible, the need for those handling classified information to resist a false sense of security cannot be overstated. There can be few more horrifying examples of the consequences of breaches of security than Operation Lagarto, described as having 'no redeeming feature,' and Operation Cobra, which followed it in East Timor.[108]

Lagarto was an Australian special operation. It consisted of a Portuguese–Australian party, led by Lieutenant M. de J. Pires, inserted into Timor on 1 July 1943 by its handlers in the Services Reconnaissance Department (SRD), which was part of the Allied Intelligence Bureau (AIB). The AIB was the cover name for an international and inter-agency organisation established in Australia with input from the British, American, Australian and Dutch forces for special intelligence collection and reconnaissance, as well as sabotage and psychological warfare.[109] The purpose of Operation Lagarto was to evacuate refugees, establish and operate a secret network, maintain the morale of the Portuguese Timorese and devise operations to cover enemy movements. On 3 August, Lagarto succeeded in evacuating a group of 87 Portuguese and Timorese refugees, but this, unfortunately, was to be the only success to the operation's credit. At the same time as the refugees were picked up, Australian signaller Sergeant A.J. Ellwood disembarked to join Lagarto. On 10 August, the Japanese captured and tortured Timorese into giving away the whereabouts of the party. Chiefs and local inhabitants of the places through which it had passed were captured and tortured until the party itself surrendered, on 29 September.[110]

One of the first security breaches may have been unavoidable: About to be captured by the Japanese, Ellwood tried to dispose of his diary, cipher, signal plans and private papers, but there were no matches, so he could not burn them. Instead, he scooped a hole in the sand and buried them.[111] When the party surrendered, the Japanese searched the area and dug up Ellwood's cipher and signal plan.[112] Ellwood eventually succumbed to brutal torture and interrogation and the Japanese were able to open communications with Australia on his wireless link.

What is truly shocking, however, and a cautionary tale on the dangers of ignoring or overriding security controls, was the role of Lagarto's handlers in SRD, Melbourne. Signals from both Ellwood and Pires on 26 and 27 September had told SRD categorically that they were about to be captured. When the Japanese took control of their communications after capturing them, no explanation was offered, or indeed sought by SRD, as to how Lagarto had avoided capture. Having failed to 'join the dots' at this point, or unwilling to believe the worst, SRD went on to send signals which disclosed details of almost all subsequent operations in Timor, throwing away the lives of operatives on missions which were doomed before they left Australia.

On 24 December, for example, Operation Cobra's prospects were ruined when SRD signalled 'Cashman, Liversidge, Heathcote Shand are now with us ... as soon as you are able to report conditions you should be seeing each other'.[113] The purpose of Cobra was to have been a long-term reconnaissance of the east end of the island, to locate an entry point for a longer-term party and establish a local intelligence network. Cobra arrived in Timor on 29 January 1944. Locals of the Lagarto party, which was of course in Japanese hands, met the Cobra party, leading them into an enemy ambush and capture. Cobra's first signal was on 8 February 1944, 12 days after arriving and five days after it had been agreed communications would be established. 'Normal' signalling continued after this until 12 August 1945.

One message decoded by the Bureau from the Japanese representative at Dili revealed that the enemy was reading the Australian guerrilla code in Timor.[114] In February 1944, a Japanese signal that had been intercepted and decoded said that Lieutenant Cashman, the leader of Operation Cobra, had been captured. The special authenticator word 'slender' had been allotted to Cobra for the purpose of challenge. 'Slender' was to be included in a normal context in signals from the base to question the freedom of the party. If the party's reply did not contain this word, it was to be considered compromised.[115]

On about 26 February, Cobra was challenged by a signal including the phrase 'Slender girl sends greetings'. This signal was acknowledged on 28 February 1944, but 'slender' was not repeated in the response, indicating the party was compromised.[116] Nevertheless, SRD signalled back on the same day: 'Our 6. Please ack greetings in last sentence from this particular girl. It will relieve her feelings and ours. We do not know her name but she is not rpt not the fat rpt fat one.'

No reply had been received to the last SRD signal when on 2 March SRD sent another signal giving war news ending with the sentence: 'Jap chances escape encompassing troops very slender rpt slender rpt slender'. On 6 March Cobra responded with a message ending: 'Our chances of getting natives very slender rpt slender'. Meanwhile, all of these messages had been shown to the Cobra leader by his interrogators. He had managed to pass off the first and second challenges. However, when the third challenge arrived with 'slender' repeated twice, the Japanese realised it was a challenge and brutally tortured and starved the party leader, forcing him to confess the meaning of the challenge which the Japanese used in the reply of 6 March. On 7 March, on receipt of the 6 March reply to the challenge, SRD signalled:

> Your 6 big relief. Col. Maj. C. and all here sick at heart due to intercept Jap cipher naming you personally and apparently claiming your capture Jan 29. You must have moved just in time. Good work and congrats.[117]

In a mirror image of SRD's complacency about the security of its communications and persistent reluctance to believe the Cobra operation had been compromised, despite the evidence, the Japanese attitude towards this message similarly was disbelief. They believed their cipher could not possibly be cracked, and even if it could, they could not believe that Australia would be so inept as to send such information to the field.

The last signals received over the Cobra and Lagarto links on 12 August 1945 were malicious taunts: 'For ACB from Nippon. Thanks for your information this long while ... Nippon army.' And from Lagarto, 'Nippon for LMS. Thanks your assistance this long while. Hope to see you again. Until then wish you good health. Nippon Army'.[118]

Although SRD HQ suspected it had been captured, Cobra was maintained as if free until 12 August 1945. Subsequently, two of its five members were recovered as prisoners of war, but all the other brave and capable participants died in captivity.

This security debacle provided ample justification for a communications security monitoring function to be included as an integral part of Australia's postwar defence and security arrangements. The natural repository for that function was in the national Sigint organisation, once established, where the motto 'reveal their secrets and protect our own' would resonate for generations and where the legacy of Cobra and Lagarto would provide a sober reminder of the importance of vigilance.

Australian Sigint as war's end approaches

Notwithstanding the tragedy in East Timor, as the end of the war approached, a sense of confidence and optimism took hold. In addition, there was considerable reflection on the highs and lows of the extraordinary wartime experience of secret collaboration, and the remarkable breakthroughs that had made such a difference to the course of the war. Much thought was given as to what to do with this extraordinary capability once the war ended.

The experience of two nationalities working side by side, as one British Sigint expert observed, had:

> Presented problems of its own, such as the fact that the Australians were in American eyes the junior partner in the firm, although initially they had greater practical experience, and that Washington never perhaps fully appreciated the value of Australian contribution in the low echelon field, since they were inclined to be ébloui [dazzled] by their own success in high echelon codes and cyphers,

these rubs at no time reflect upon the type of organisation set up ... it exemplifies the system of a functionally divided inter-service centre controlling the whole of the Signal Intelligence work in its area, from the disposition of its field units to the dissemination of the final results, and through the normal service channels administering its own personnel. It further possessed its own communications within the theatre, which extended roughly from Brisbane to Okinawa and from Guam to Borneo.[119]

Meanwhile, FRUMEL, always a smaller and less mobile organisation, had become an entrenched feature of the defence establishment in Melbourne. Based at Albert Park, southeast of Melbourne's central business district, since 1944, its facilities would be enormously useful once the remaining Australian elements of FRUMEL and Central Bureau sought to consolidate after the war. The closure of several special wireless units in May 1945 heralded the scaling down and consolidation of capabilities, knowledge and skills after the war – consolidation that would centre on the residual elements of FRUMEL and Central Bureau.

In the meantime, by early 1945 the Royal Navy's British Pacific Fleet (BPF), under Admiral Sir Bruce Fraser, had arrived in Australian waters with the intention of supporting US naval operations in the Pacific. Their intelligence organisation was initially set up at Sydney but material was supplied directly from the US Navy's Joint Intelligence Centre Pacific Ocean Area (JICPOA) to BPF ships which were forward deployed. Sigint supplied traffic analysis and tactical intelligence of enemy aircraft movements likely to affect the fleet's operations. But it appears the focus was on engagement with the US Navy, rather than with Australian elements such as FRUMEL.[120]

By the end of the Second World War, many postwar arrangements had fallen into place. Agreements had been signed, the division of labour on targets agreed, technical cooperation and collaboration were becoming common. Sharing arrangements were in effect and many of the fundamental rules were being settled, not least security and handling arrangements which Australia still at times struggled to implement. But while Australia had neither initiated nor led many of the wartime successes, it was now at the table. It was seen to have made a major contribution to the wartime effort, albeit not always in its own right or as the independent partner it was to become.

Looking back on Sigint successes and failures

The Sigint arrangements during the war established trusting, collaborative engagement between Australia and the United States that matched the engagement between Australia and the United Kingdom before the Pacific War. Australians had made direct contributions to the Allied victory. This, and the work with their American allies, left a lasting impression in Australia and a willingness on the part of policymakers, politicians and military commanders, to work closely with the United States in the future.

The three armed services had grown and developed exponentially. New technology had been introduced that changed the conduct of warfare, notably the invention and use of atomic weapons. In terms of Sigint, the spectrum of functions and capabilities had been refined and greatly expanded. The three armed services gained experience in a wide range of functions, notably with radar and sophisticated traffic analysis, that had barely been contemplated before the outbreak of war.

From 1942 onwards, the division of responsibilities between the Australians and the Americans worked quite well. However, as the war came to an end, and American forces departed, Australia was left with a national Sigint enterprise that was highly developed in part, particularly traffic analysis, but lagging in other areas, notably cryptanalysis and the use of machinery in the aid of deciphering. This meant that in order to maintain a national Sigint entity after the war, Australia would need to fill the gap left by the Americans.

With the United States reluctant to maintain military engagement with Australia in the years immediately after the war, Australia's new Prime Minister, Ben Chifley, looked once again to Great Britain for assistance. But exactly what form that would take was not immediately apparent. Reflecting on the wartime challenges Australia faced, expert on defence and security Desmond Ball considered that

> because of the propensity of other intelligence agencies to be somewhat less than completely forthcoming in these relationships, Australia must have an indigenous capability for operating its own national technical means in areas where intelligence is critical to the defence of Australia or where US and/or British collection and assessment may be deficient from Australia's perspective.[121]

Ball declared that 'Australia's assessment capability must be entirely independent'. For him, the experience of the Second World War suggested that 'a more independent, dedicated effort in electronic intelligence can only be of benefit to Australian security planning and operations'.[122] The postwar architecture was much more substantial than that which had preceded the war, but dependence on great and powerful friends would continue, including requirements for technical, logistical and personnel support.

Despite the outstanding contribution made by Sigint units to the war effort, the burden of secrecy meant that they received none of the normal rewards and commendations for their extraordinary achievements. Australian Sigint service personnel were even sometimes publicly denigrated by their combat unit peers for not having done the hard work of others. Their formal service records often omitted details of their operational service. Most bore the insults in silence. One RAAF Sigint operator, Jack Brown, had the following recollection after Anzac Day in 1948:

> Following the march we went to the Returned Services League clubrooms in Adelaide and one of the guys grabbed my tie and bit it off and then said, 'Now you have been operational'. What an insult! It was fifty years before I went into an Anzac Day march again.[123]

Great work had been done by thousands of men and women and with little, if any, recognition. But now, they and their successes can be recorded and remembered.

There was also a closeness amongst Siginters that, not surprisingly in a mixed environment, often saw romance bloom.

The professional bonds which had developed since the outbreak of the war in the Pacific were suspended at the end of the war, as US personnel were returned home promptly. Little did they realise that with the war having ended, the ties of trust, compatibility and interoperability which characterised the international collaborative Sigint enterprise, would prove to be lasting, although there was some turbulence in the early postwar years.

In the meantime, thousands of Australian men and women had made a largely unheralded contribution to the defence of Australia and to the Second World War victory. Many may not appreciate the significance of the work undertaken by women in Central Bureau, FRUMEL and elsewhere. Their experience demonstrated that many of the myths and misconceptions about the role of women in society were due for a dramatic overhaul. That would take some time after the war, partly because women involved in the war effort were not able to speak in public about their work or declare: 'we also served'.

8
POSTWAR SIGINT TO VIETNAM

While the wartime and the postwar intelligence arrangements might seem like a natural progression, there was a period of uncertainty after the end of hostilities in August 1945. Demobilisation quickly followed Japan's surrender. For many, in fact, it had started in 1943, when the Australian armed forces' commitment reached its peak and the federal government recognised the need to redeploy national resources to primary industry and other non-military needs. That demobilisation continued through 1946 and was largely wound up by 1947. In the meantime, FRUMEL, Central Bureau and the associated Sigint-related entities were dwindling.

The events and experiences of the previous decade had left an imprint. As former head of MI6 Alex Younger might say, we have had a glimpse of this imprint on the nation's soul, or, as novelist John le Carré might have said, the nation's subconscious. These experiences in the nation's history, and particularly the history of Australian Sigint – Australia's junior partner status with the United Kingdom and United States; the well-founded concerns about Australia's lax security; having to manage with inadequate resources; resourcefulness and unexpectedly high performance in challenging circumstances; the remarkable impact of a few gifted people – all continue to exert a powerful influence.

Demobilisation and salvaging what mattered most

As postwar demobilisation gathered pace, each service consolidated its own Sigint unit capabilities. These efforts were not helped by the rapid demobilisation and dismantling of structures. As suggested in the previous chapter, while Sandford was a skilled diplomat in his dealings with senior figures in Allied intelligence, at the same time, working closely with Assistant DMI Little, he was quick to seize opportunities to defend Australia's interests. He and Little were instrumental in salvaging the remaining Sigint organisation in Australia, sometimes in opposition to those intent on demobilisation. The secrecy surrounding the Australian Sigint capability meant that they were among the few who were willing and able to defend it. The essence of Nave's SIB, like the man himself, survived by being taken under the wing of the Australian Army. SIB was dissolved, but the diplomatic sub-unit survived as 'D' Special Section, with the alternative name of Special Intelligence Section. Little decided that the diplomatic group should continue for the benefit of the Commonwealth government, 'but think it best to keep it under Army, away from Central Bureau as if under Central Bureau it would again be under GHQ which might act similarly to USN'.[1] This decision was strongly supported by Sandford, who 'loaned' men and women from Central Bureau to SIB, then to 'D' Special Section, during periods of high demand. Later, Little and Sandford would work together to preserve as much of what remained of SIB as possible during the immediate aftermath of the war so that it could serve as the basis of an Australian national Sigint capability.

In July 1946, cabinet gave 'in-principle' approval to a Signal Intelligence Centre 'as part of a British Commonwealth Signal Intelligence Organisation'. The in-principle approval was accompanied by the now familiar reluctance to spend on intelligence. The treasurer's approval had to be obtained before authorisation was given to set up the organisation.[2]

Notes on the agendum tell us that the British Commonwealth Signal Intelligence Organisation would be based on centres in the United Kingdom, Canada and Australia, with the New Zealand contribution integrated in the Australian centre.[3] Following a visit to Australia in December 1946 by Sir Edward Travis, now director of GCHQ, and Major General Strong, director of the UK's Joint Intelligence Bureau, the Joint Intelligence Committee produced detailed recommendations, including the establishment of a Joint Intelligence Bureau and a Signal Intelligence Centre.[4] The recommendations for the latter included the three intercept and DF stations at Harman (Canberra), Cabarlah (southeast Queensland)

and Pearce (near Perth in Western Australia) with plans for a fourth, probably near Darwin in the Northern Territory.[5] Importantly, under the heading 'Status of the Melbourne Centre' the recommendation was that

> it will contain all the elements necessary for the production of signal intelligence and will be a self contained unit so far as the main tasks on which it is engaged are concerned. It could be expanded rapidly in an emergency, to the extent that work normally undertaken by another Centre could be transferred to it if conditions demanded.[6]

The last part of this recommendation may be an allusion to the fact that FECB did not find a haven in Melbourne after the fall of Singapore, which, as noted, Nave and others saw as a missed opportunity.

Delays in obtaining approval to these detailed proposals, mainly caused by objections from the Department of External Affairs, prompted Frederick Shedden to write, in an apologetic personal letter to Sir Edward Travis, 'I have never before experienced, in one subject, which appeared so straightforward, so much time and work for the small progress made to date'.[7]

For the Army, in 1947 the remainder of the Australian Special Wireless Group became 101 Wireless Regiment (later 7th Signal Regiment), the Army's primary Sigint unit. Established at Cabarlah, Queensland, the regiment conducted strategic Sigint collection against 'certain Asian targets' under the technical control of the DSB. It would be the principal supporting unit for the deployment of Army Sigint teams on operations in Malaya, Singapore, Borneo, Vietnam and others.[8]

For the RAAF, this involved establishing No. 3 Telecommunication Unit (3TU), a RAAF unit formed from what remained of the wartime special wireless units. Initially formed on 15 October 1946 as a separate unit within the command of Headquarters of the RAAF's Western Area, 3TU became operational on 6 September 1947. Located at RAAF Station Pearce, just outside Perth, the unit performed specialist communications tasks in strategic and tactical support to the Australian armed forces both in Australia and overseas. Started with less than 20 officers and other ranks, including signals operators,[9] 3TU was functionally controlled by the Department of Defence (Air Force Office), but DSB was in operational control. The Operations Flight included sections covering set room (for the radio receivers and associated equipment), traffic analysis, reporting, non-Morse search, and development and training sections. The achievements of the unit and the high standards of operational capability were a reflection of the cooperation and goodwill between Air Force Office, DSB/DSD and the station. ASD records that 3TU 'made a significant contribution to the Australian [Sigint] effort, both directly and as a specialist training provider', providing communications support 24 hours a day from 1948 for over four decades, until the closure of the unit following the relocation of DSD to Canberra.

For the RAN, it ended the war with an HF receiving and direction finding site established just outside Darwin, within the naval base HMAS *Coonawarra*. In the years that followed, the encroachment of suburban Darwin, particularly in the 1960s, meant that this site was no longer tenable, and in the early 1970s the decision was made to relocate the station to a greenfield site north of Darwin at Shoal Bay. Shoal Bay Receiving Station (SBRS) was partially constructed when Cyclone Tracy struck Darwin on Christmas Day 1974. The existing station at HMAS *Coonawarra* was completely destroyed by the cyclone and the fit-out of SBRS was hastened, with the station becoming operational by April of the following year. As with 3TU, operational control was vested in DSD, however the management, administration and command of the station remained with Navy. The role of SBRS expanded over the course of its existence, with satellite collection gradually replacing HF reception as its primary purpose. Command of SBRS transferred from Navy to ASD in 2002.

Postwar national and international Sigint arrangements

Australia's junior partner status with the United Kingdom and the United States had tremendous advantages, but there were also costs. As we have seen, the combined effects of being a junior partner and the secrecy of Sigint generally led to Australia's contribution being underestimated or overlooked. The Battle of Midway was one example. In that conflict, Australia was not one of the main players, but Australian Sigint nonetheless had an important – and to this day, largely unrecognised – role. Similarly, the contribution of Australian Sigint from FRUMEL to the success of the USN submarine campaign in the Pacific is little understood or recognised.

Even where Australian forces were the only victors in a battle, Sigint still has tended to be overlooked. Its secrecy and technical complexity has meant that few outsiders, even within the armed forces, let alone the general public, understand fully what it achieved, or the consequences of ignoring it. For those who did understand, years of protecting secrets often instilled a visceral reluctance to mention them, much less take pride in recounting them, even when it was safe to do so.

The Five Eyes – the United Kingdom, United States, Australia, Canada and New Zealand – was and remains the most powerful intelligence alliance that has ever existed. Australia owes its membership of the Five Eyes to the familial ties with Britain and to the trust built with Britain and the United States during the two world wars. The sense of community and camaraderie among Five Eyes partners working on the same issue, often on a first-name basis and knowing what the others are doing on a given day, is unique.

Australia's membership of the Five Eyes is not a secret, and yet despite numerous revelations over the years, many of its practices remain undisclosed. Membership of the Five Eyes arrangements gives Australia access to a treasure trove of intelligence it would otherwise be denied. For successive postwar generations, it has been seen as an offer too good to refuse. But the price of membership of this most exclusive of clubs has been a degree of independence and national self-confidence. The national interest demands that Australia cultivate these important relationships with much larger Five Eyes partners, namely the United States and United Kingdom, while asserting and maintaining its independence. Some would argue that Australians have not done enough of the latter, but the advantages of Five Eyes membership are undeniable and widely accepted across the political spectrum in Australia.

The Five Eyes collaboration was formally brought into being with the BRUSA Agreement of 17 May 1943. After the war, Brigadier Bertrand Combes was commissioned to assist Defence Secretary Shedden and produce a 'Report on Joint Intelligence Organisation', which included recommendations on Sigint. This was the first in the series of landmark reports which shaped Australia's postwar intelligence architecture in the second half of the twentieth century and first decades of the twenty-first century. The arrangement included the establishment of a Joint Intelligence Committee (JIC) and the appointment of a Controller Joint Intelligence (CJI) to oversee, on behalf of the JIC, the Joint Intelligence Bureau (JIB – the precursor of the modern-day Defence Intelligence Organisation) and a yet-to-be-formed Signals Intelligence Centre. The first CJI would be Brigadier Frederick Chilton and the first director of the JIB was Lieutenant Colonel Allan Fleming, a former journalist and wartime intelligence officer.[10]

On Sigint, Combes stated categorically that 'the main point is that Australia must form part of the Empire Scheme: she cannot afford either to stand alone or be out of it altogether'.[11] The Chifley Cabinet eventually approved an agendum on 19 July 1946 which included the proposal from a British Commonwealth conference that the British Commonwealth Signals Intelligence effort be based on centres in the UK, Canada and Australia, with the Australian centre to be established in the Melbourne area.[12]

A few days later, on 23 July 1946, government in-principle approval was given for a new peacetime Signals Intelligence Centre. This was initially drawn from the remaining wartime Australian Sigint elements, mainly from Central Bureau and FRUMEL. Since describing the national Sigint agency as the Signals Intelligence Centre revealed the purpose of the agency, it was given a cover name, the Defence Signals Bureau (DSB), and established formally in Melbourne on 1 April 1947. The cover name would become its official title. A Cabinet committee finally approved the proposals on 12 November 1947.[13]

The DSB then became the repository of the remnants of the centralised wartime Sigint entities. The DSB moniker was to be used 'where it is necessary to conceal the nature of its functions'.[14] The transformation from its establishment in 1947 into the Australian Signals Directorate (ASD) of today deserves to be better understood. Equally deserving is the role of Sigint forces in the three armed services. If Australia's national Sigint agency was, and still is, too little known and acknowledged, the frontline work of signals and intelligence forces in the three services is even less so.

Mindful of the security challenges in Australia, the United Kingdom offered to help set up the new bureau and even provide its first director.[15] In fact, GCHQ's Travis proposed that a British officer be chosen as the first Director DSB. The reasoning was that Britain would share its Sigint reporting only if it had complete confidence in DSB's director. Besides, initially DSB was not a purely Australian Sigint centre, as it included a significant number of seconded British staff and New Zealanders, many with cryptological skills that Central Bureau and FRUMEL had previously relied on the Americans to provide. The British director assigned to the task was Lieutenant Commander J.E. 'Teddy' Poulden, an experienced Sigint officer and regarded as 'one of the ablest signals officers of his generation'.[16] Poulden had been responsible for Sigint support for Mountbatten's wartime South East Asia Command centred on the WEC in India. His strong technical background promised to compensate for the shortfall in these skills following the departure of the Americans.

Poulden's appointment came as a surprise in Melbourne and a reminder that Australia was still a junior partner. Australia, after all, had senior Sigint managers with wartime experience who could have been appointed. Nave, the brilliant codebreaker, and the popular and diplomatic Sandford, were examples. The choice of Poulden, however, guaranteed continuing British support and helped allay the perennial concerns about Australian security. The leading Australian military historian David Horner records that Charles Spry, then the DMI and JIC member, and later head of the Australian Security Intelligence Organisation (ASIO), was very happy to have a British officer in the position. Spry considered Australia would then have the best possible access to British intelligence and that, in due course, an Australian would be appointed to succeed him.[17] Notwithstanding the appointment of Poulden, concerns about Australian security continued to dog DSB.

Despite being established at the beginning of April, it was not until October that DSB commenced operations, with the collection of Sigint drawing on collection sites established during the war as part of the armed services special wireless collection capabilities.

Lax security, Soviet decrypts and British intervention

As we saw in the previous chapter, during the war the United States engaged a compartmented counterintelligence program called Venona. It was a relentless attack on Soviet diplomatic ciphers, with Britain initially providing the cryptanalytic expertise, supporting the United States with the processing capability.[18]

In 1947, the British decrypted some KGB messages containing classified British military estimates from the Soviet Embassy in Canberra. It appeared that an Australian was passing the information to the embassy. Sir Percy Sillitoe, Director General of MI5, was despatched, along with the head of MI5's protective security division, Roger Hollis (later head of MI5 from 1956 to 1965), to discuss the matter with Prime Minister Chifley. Sillitoe was hamstrung, however, by instructions to conceal the source of the information. When Chifley asked for proof, Sillitoe could only offer an unconvincing cover story about a possible mole.[19]

Sillitoe presented the case to Chifley and government officials, including Chilton and Shedden, but avoided revealing the Sigint source of their concerns. The Australian government rejected the accusations. Whitehall, therefore, found another solution: Britain would help reform Australian security and vouch for it in Washington. Sillitoe and Hollis convinced Australian officials and politicians that Australia could solve its security problem only by creating a local version of MI5: ASIO. Then British authorities launched a diplomatic campaign, aimed at persuading their American counterparts to accept the proposed solution.[20]

The British took security extremely seriously, but their attitude to Australia's lapse was much more forgiving than that of the Americans, who distrusted the Chifley Labor government. The Americans regarded Labor as too friendly by far towards the Soviet Union. They were suspicious of the Minister for External Affairs, Dr H.V. Evatt and of his Departmental Secretary John Burton, in particular.

Britain's Prime Minister Clement Attlee was obliged to intercede on Australia's behalf, pointing out in a letter to President Truman that

> The intermingling of American and British knowledge in all those fields is so great that to be certain of denying American classified information to the Australians, we should have to deny them the greater part of our own reports. We should thus be placed in a disagreeable dilemma of having to choose between cutting off relations with the United States in defence questions and cutting off relations with Australia.[21]

This appeal by Attlee was not enough to sway Truman, but he accepted Attlee's proposal that Sir Frederick Shedden, who remained the Secretary of Australia's Department of Defence, visit Washington to plead Australia's case. Shedden, however, failed to win over the Americans.[22] As Shedden wrote to Sir Edward Travis, 'Personally, I feel the lack of information about signal intelligence, now that we have run into trouble.'[23] This awkward circumstance persisted through 1948 and well into the following year, when ASIO was eventually established by Chifley. Teething problems, initially associated with continuing concerns about Australia's security, meant that while DSB could collect, there was little sharing of Sigint – at least not until after the December 1949 election and the defeat of the Chifley government which brought about a sea change in international Sigint information exchange.[24]

The victorious Menzies government was able to distance itself sufficiently in the eyes of the Americans from perceived leftists in the Chifley government. This was critical, because through the Venona program, which was to play a key role in uncovering a global network of Soviet spies, the source of the leaks was known to be two Soviet sympathisers in Australia's Department of External Affairs associated with Labor leader 'Doc' Evatt. With Menzies in power, the United States allowed a limited resumption of cryptological exchange with Australia.[25]

Canada's Sigint – parallels and differences

In the meantime, Australia's fellow New World federal bicameral, Westminster-style parliamentary democracy and wartime ally, Canada, ended up in a similar predicament as a member of what came to be known as the Five Eyes Sigint community. But Canada took a different route, avoiding the security doubts faced by their Australian counterparts. This was despite the revelation of Soviet spies in the Canadian capital Ottawa when the cipher clerk in the Soviet Embassy, Igor Gouzenko, defected in 1946. With so much in common, the comparison of this 'strategic cousin' of Australia is worth noting, particularly for the light it sheds on Australia's predicament.[26]

Australia and New Zealand initially were prepared to participate in the Sigint arrangements under the auspices of the UK, reflecting the dynamics agreed at the Commonwealth Signals Intelligence Conference in February and March 1946. In contrast, though, Canada signed a bilateral Sigint sharing agreement with the United States in 1949 which came to be known as CANUSA.[27] This agreement involved dissemination not only of the Sigint reporting but also of the technology that supported this collection. The agreement covered traffic analysis and information on communication practices of target countries.[28] These arrangements reflected the greater degree of independence from Britain of Canada's wartime Sigint architecture, and its closer partnership with the US Signals Security Agency, albeit as a junior partner. This included the wartime Sigint body which operated under the cover name of 'Examination Unit of the National Research Council', incorporating four intercept bases. These were at Rockcliffe in Ottawa (which had been monitoring Vichy French diplomatic traffic), Amherst, in the east coast province of Nova Scotia; and on the west coast in the province of British Columbia, in Riske Creek and West Point Barracks, in the capital Victoria (monitoring Japanese traffic).[29] Canada's distinctive arrangements with the United States also reflected the desire of both parties that the United Kingdom was not to be considered a third party but that both would keep each other informed of arrangements each made with the UK Sigint authorities or those of other Commonwealth countries, namely Australia and New Zealand.[30]

Canadian intelligence historian Wesley Wark has described the CANUSA as 'effectively a declaration of SIGINT sovereignty by Canada and an expression of a desire to be treated as an equal in a tripartite pact with the US and UK'.[31] As we have seen, Canada had gone through its own espionage scandals, with the defection in 1946 of Gouzenko. But after some initial stumbles, in the end, with the help of a royal commission, this matter was managed in a way that brought credit to Canada.[32] The precedent of a royal commission on espionage would resonate in Australia a few years later, when Vladimir and Evdokia Petrov defected.

In addition, Canada's wartime Sigint enterprise had been more independently managed than had Australia's. As Wark explains, 'Canada had created a small civilian SIGINT organization during the war, known as the Examination Unit, which focused on diplomatic communications but also did some work on military and clandestine signals'. This was complemented by traffic analysis run by different branches of the Canadian military. After the war, these disparate Canadian elements were amalgamated into something euphemistically called the Joint Discrimination Unit.[33] In turn, that became the equally cryptic Communications Branch of the National Research Council or CBNRC. The CBNRC would later become the Communications Security Establishment (CSE) in 1975.[34]

What is particularly noteworthy is the contrast between the Canadian and Australian experiences. At no stage did Canada subordinate its modest national Sigint instrumentalities to either UK or US control. Canada also had the advantage of being so close physically to the action in Washington, as well as to the United Kingdom. At war's end, it maintained a balanced set of skills to cover the spectrum of cryptologic tasks required for a modern, but still quite small, national Sigint capability.

Australia's contribution, on the other hand, particularly to the peak Sigint bodies in Australia, FRUMEL and Central Bureau, was as a subordinate and partial contributor to a US-led enterprise under General MacArthur and Admiral Nimitz, where many of the technologically more demanding functions were undertaken by US counterparts. Their departure at war's end left Australia with incomplete skills and capabilities to establish an effective independent national Sigint agency, in sharp contrast to its Canadian cousin.

For the Canadians, mindful of their relatively mature Sigint capabilities, the BRUSA arrangement was a bilateral one which 'served as a terrific spur' for them. It demonstrated that the wartime arrangements would continue after 1945 and that there were prospects for Canada to 're-fashion its own wartime links'. Canada had participated in the 1946 London conference but with a view to carefully managing its affairs independently of UK interests.[35]

For Canada's independent Sigint effort to survive, it needed 'the right US agreement' that met its requirements. It would take three more years of negotiations to finally reach agreement. By 1949, the US recognised the 'very definite benefits' of encouraging Canada to maintain and expand its Sigint enterprise, notably its northern Sigint intercept stations, in a manner complementary to US requirements.[36] Wark has observed that 'an intangible in all this, operating on both sides of the path of CANUSA was the Second World War experience of close cooperation, including intelligence matters, between Canada and the US. This cooperation had created a legacy of trust, so fundamental to any intelligence alliance.'[37]

The parallels with Australia were clear, although their different circumstances led them to arrangements that were hard for Australia to insist upon for DSB, particularly in the first few years.

New national intelligence community arrangements

By 1949, the Cold War was making itself felt in the Canadian capital Ottawa as much as in the offices of Australia's government in Canberra (with much of the national defence and security apparatus of state still in Melbourne). From the residual intelligence architecture of the Second World War came the early indications of what would be known as Australia's National Intelligence Community. That name would not be given to the community of Australian intelligence agencies until 2017, but the outline of it can be seen in the chronological chart (see figure 4).

The chart demonstrates the linkage of DSB and its successor organisations with its wartime antecedents. It also shows the parallel development in other parts of Australia's national security and intelligence architecture. The DIO precursor, the Joint Intelligence Bureau, was the principal recipient of DSB Sigint reports. It absorbed the functions performed by the Combined Operations Intelligence Centre as well as certain analytical functions performed by the Australian component of the Allied Intelligence Bureau. The secret intelligence collection and reporting functions of the AIB would continue once the Australian Secret Intelligence Service (ASIS) was established in 1951. The mapping functions would continue but in a disparate manner, mainly in the service cartographic and hydrographic entities.

The RAAF maintained an aerial photography capability built around three units which could do much more than just aerial photography. First, a long-range photographic reconnaissance capability was established to support its long-range strike capability. Initially this was with RAAF 87 Squadron, equipped with Mosquito aircraft in the 1950s; then 6 Squadron, with camera equipped Canberra aircraft in the 1960s; and subsequently F-111 bombers especially equipped with cameras and called RF-111s, introduced in the 1970s. The RAAF converted four such aircraft and they stayed in service until 2010. Second, the RAAF operated a Target Intelligence Centre (TIC) which was raised at RAAF Fairbairn in Canberra as part of the Directorate of Air Force Intelligence. This was established in the 1950s and maintained through to the 1990s. Its role was imagery-derived target intelligence to support targeting for the strike capability. It changed names and command and control arrangements with its transfer from Air Force to the newly formed Australian Theatre Joint Intelligence Centre (ASTJIC) in the late 1990s, before being folded into the Defence Imagery and Geospatial Organisation (DIGO) as a result of the early 2000s review of Sigint arrangements by Commodore Denis Mole. In July 2010 No. 460 Squadron was reformed within AGO, under RAAF command and control arrangements, to continue this targeting role. Third, the Central Photographic Establishment (CPE) was formed in 1949, which is where all RAAF reconnaissance and hand-held imagery was stored and archived until it closed in the mid to late 1990s.

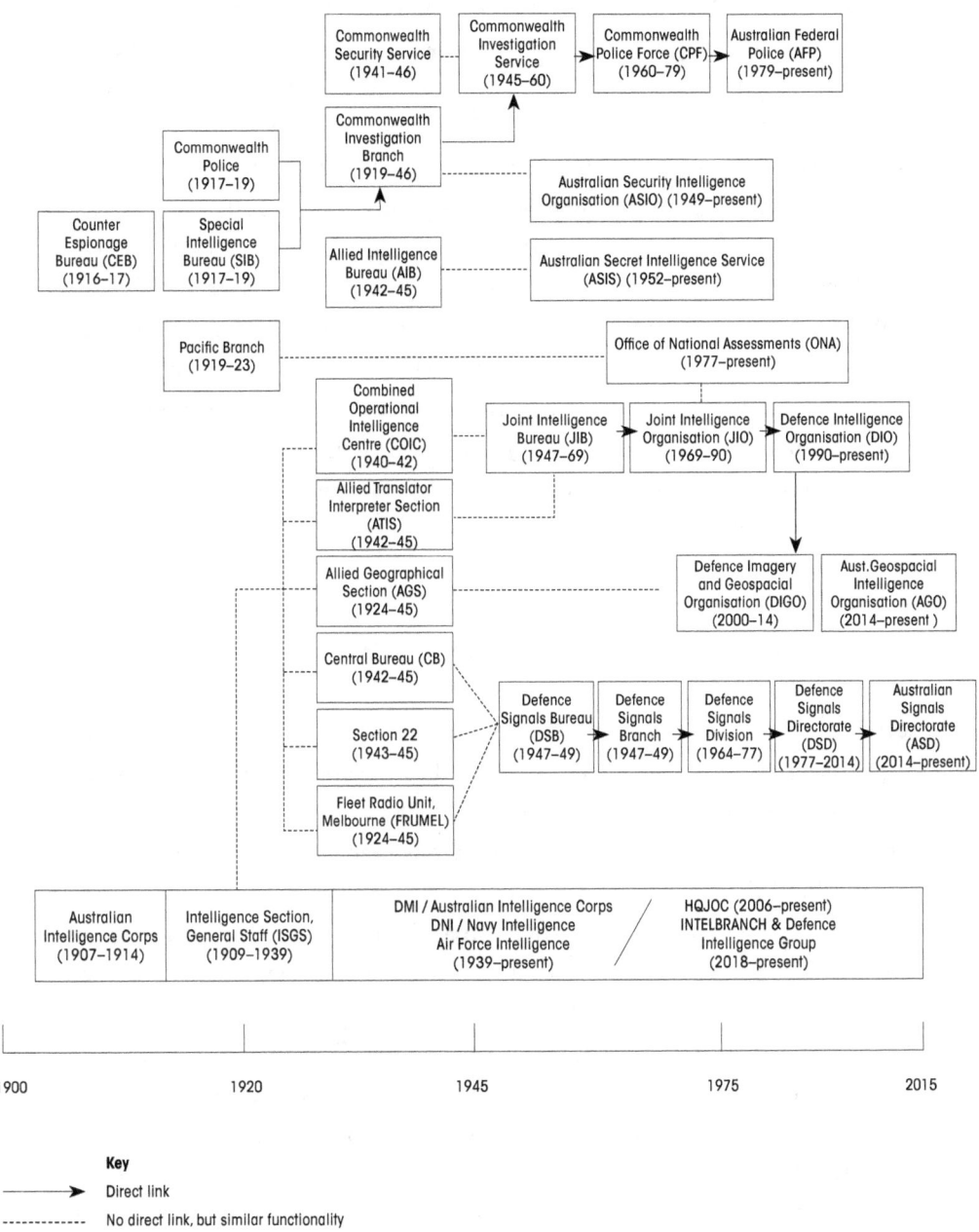

Figure 4. Chart of the history of the Australian Intelligence Community

Eventually DIO incorporated an Imagery Exploitation Centre as well. The DIO imagery component was separated from DIO and merged with the aerial photographic, cartographic and hydrographic elements at the start of the new millennium. This composite group was the Defence Imagery and Geospatial Organisation (DIGO), which was later renamed the Australian Geospatial-Intelligence Organisation (AGO).

The establishment of an Australian Federal Police force (AFP), security intelligence organisation (ASIO), Secret Intelligence Service (ASIS) and Office of National (intelligence) Assessments (ONA), later to become the Office of National Intelligence (ONI), took place gradually, but the roots are visible in figure 4. While these agencies would develop extensive connections with international counterparts, it is in the Sigint domain that the trusted arrangements, built on the legacy of the ties of the Second World War, are the oldest and strongest.

The Korean War

Shortly after Menzies came to power, war broke out on the Korean peninsula. The peninsula was partitioned at the end of the Second World War, after 35 years of Japanese occupation. The power struggle between the Soviet-backed Communist North Korea and its 'eternal leader', Kim Il Sung, and the US-backed anti-Communist South led by Syngman Rhee, drew Australia in as an ally of the United States when the North Koreans invaded the South on 25 June 1950. The invasion was eventually repulsed by the forces of the United States, South Korea and 15 other countries, working together under the United Nations' flag. This included Australian air, land and naval forces stationed in Japan after the war who were just about to return to Australia but managed to be redirected before that repatriation plan could be implemented.

On 26 October, Chinese forces retaliated massively, throwing back UN troops.[38] This included Australia's 3rd Battalion, Royal Australian Regiment (3 RAR), which had advanced north, and approached close to the Yalu River on the Chinese border. The war seesawed through 1950 and 1951, before settling into a static defensive war lasting from 1951 to

1953 along what is now known as the demilitarized zone or DMZ. And of course, although the July 1953 armistice holds to this day, the Korean peninsula remains deeply divided with no final peace settlement in sight.[39]

For Sigint practitioners, the Korean War was nothing like the heady experience of successful interception of high-grade enemy communications during the Second World War with Ultra and other high-level Sigint. Nonetheless, there were valuable lessons for Australian Sigint. Still the junior partner, Australia's main motive for joining the war was to cement ties with the United States, although support for the United Nations and Korea's strategic importance were asserted as important reasons at the time as well.[40] Australia's contribution was small in comparison with its own in the Second World War and with those of allies.[41] However, Australians like Captain A. Findlay, an Australian Signal Corps member who led a wireless detachment from Japan to establish a site in Pusan, were among the first to give active communications support to South Korea.[42] This support would be focused on own-force communications, but elementary Sigint also came into play.

By the time of the Korean War, all three Australian services included forces which could have provided what could be called electronic warfare support, but the requisite equipment and teams were not sent to Korea. Australian naval Sigint, which had reached a high level of sophistication by the end of the Second World War, was a pale reflection of its former self and the RAN had disbanded the WRANS, which had provided the majority of FRUMEL's intelligence personnel.[43]

The same could be said for the other services. No Sigint or electronic warfare (EW) land unit deployed from Australia to support the troops in Korea. In the previous war, that capability had been the preserve of higher echelons. Similarly, while the RAAF also deployed to Korea and contributed notably with Mustang fighters and later Meteor jet fighter aircraft, no Sigint or Elint elements were deployed as part of the RAAF forces sent to Korea, even though the RAAF had been actively involved in the development of such capabilities in the Second World War.[44]

At the beginning of the war in Korea, intelligence on the side of the United Nations Command was worse than patchy. One explanation was that until war broke out, Korea was not a priority for the United States.

At a press conference in January 1950, Secretary of State Dean Acheson described an American sphere of influence in the Pacific that did not include Korea.[45] Not surprisingly, therefore, there were no high-priority intelligence requirements on Korea as the demand for information surged.[46]

Operating along the border between the demarcated zones in the days leading up to the outbreak of war was a UN-mandated team involving two Australians: Squadron Leader Ronald Rankin and Major Stuart Peach. They travelled to the 38th parallel that separated the partitioned peninsula in June 1950 and reported back to the United Nations on what they saw. Their report assisted the United Nations Security Council to decide to come to the aid of the Republic of Korea.[47] Beyond this eyewitness account, however, there were few, if any, technical means of collection to flesh out the intelligence picture of the conflict which erupted on 25 June that year.

There is little material available describing the source of Sigint used by UN commands, but US commanders passed 'appropriately disguised' Sigint to units under their command. As Ian Pfennigwerth notes, 'the provenance of this information has not been identified'. Reports suggest 'GCHQ provided a great deal of intelligence, particularly on Chinese intentions and activities', although attributable references for these claims are not available.[48]

Australia's signals contribution consisted of troops like Captain Findlay, trained to manage Australia's 'own-force' military communications. Their job was not to eavesdrop on the enemy and there were no cryptanalysts among them. However, like naval Signalman John Varcoe, the sailor immortalised in the statue in Sydney's Martin Place, the signals forces in Korea were resourceful. In addition to their official own-force communications role, they were not above intercepting enemy force transmissions when they could, echoing their signals forebears in Mesopotamia in 1917.

Meanwhile, the entire US intelligence community was appalled by the performance of the Sigint system in Korea. The Armed Forces Security Agency (AFSA), the newly formed American national cryptologic organisation, had failed to predict the outbreak of the war.[49] Figure 5 provides an outline of the connections between US, UK, Australian and other UN intelligence collection and reporting agencies in the Korean

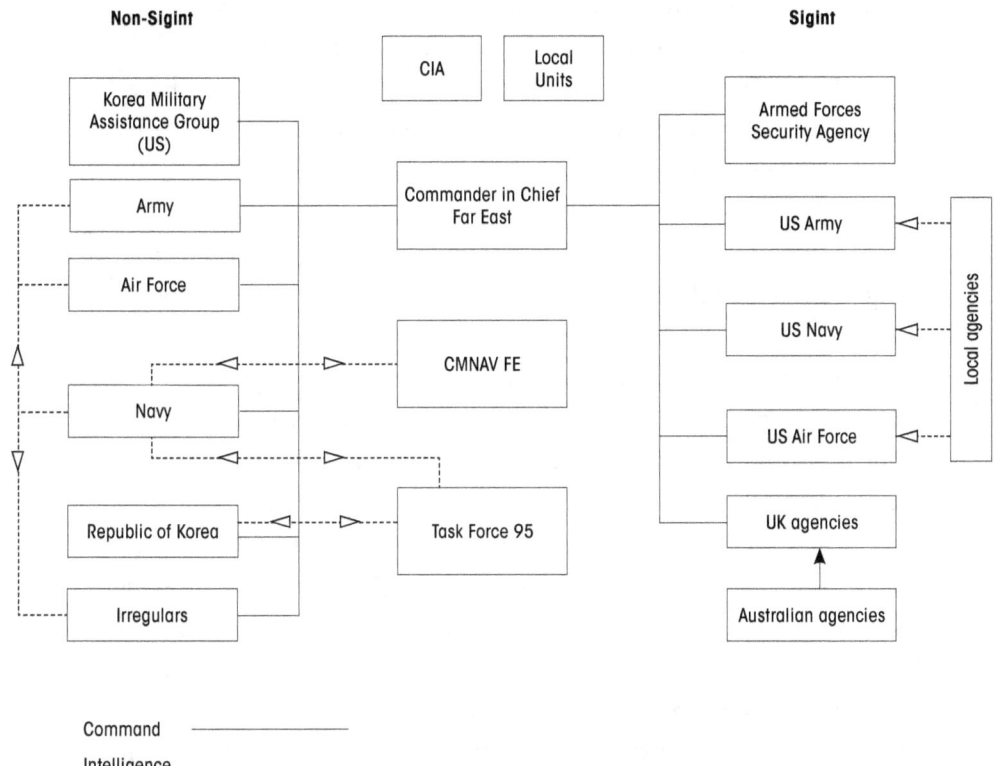

Figure 5. UN intelligence organisation, Korea
SOURCE Ian Pfennigwerth, *Missing Pieces: The Intelligence Jigsaw and RAN Operations 1939–71* (Canberra: Department of Defence, 2008), p. 159.

War. A contributing factor was the United States government's reluctance to spend on defence after the Second World War, and no exception was made for Sigint.[50] AFSA was given neither the budget nor the legal authority to do its job.[51]

Writing in June 1952, General A. James Van Fleet, Commander of the US Eighth Army, the main ground force formation during the war, expressed the frustration of those military commanders who remembered the Second World War:

> It has become apparent ... that during the between-wars interim we have lost, through neglect, disinterest and possible jealousy, much of the effectiveness in intelligence work that we acquired so painfully in World War II. Today our intelligence operations in Korea have not yet approached the standards that we reached in the final year of the last war.[52]

The shortcomings of AFSA led to demands for the creation of a more integrated National Security Agency (NSA) which emerged from AFSA in 1952.

After the initial intelligence failure, there were some successes – by July 1950 codebreakers in Washington succeeded in penetrating North Korean communications,[53] and AFSA supplied crucial intelligence for the Battle of Pusan Perimeter in August and September.[54] However, its forecasting in October and November of the likely Chinese military intervention was ignored or misunderstood.[55] An influential proponent of the view that China would not invade was MacArthur's intelligence chief, Major General Willoughby.[56]

Then, in mid-1951, all access to enemy communications was lost when the Soviet-trained North Koreans changed all their codes and ciphers, adopting instead the unbreakable one-time pad cipher systems. Thus, for the last two years of the war, the United States and its allies had to rely on low-level voice intercept and traffic analysis for insights on enemy capabilities and intentions.[57]

In these challenging conditions, Australian signallers made themselves useful. Captain Bruce 'Buck' Rogers, Royal Australian Signals Corps Officer for the 1st Battalion, Royal Australian Regiment (1 RAR) in 1952, explained in a letter:

> Intercept is normal practice for both sides ... and Charlie sends an astonishing amount of valuable information by radio in clear. The intercept is good enough to have it [translated by Chinese Nationalists] and passed to the patrol commander on the valley floor in time for him to act on it. Such things as 'do not fire until the

enemy is within 30 feet' and 'let the patrol through the outposts and then attack it from behind', and so on.[58]

Apart from learning at first hand the value of high-level Sigint from its absence, Australia was a participant in a historic turning point with global ramifications: the United States changed course completely from post-Second World War disarmament to rearmament in order to stop Soviet expansion, tripling US military outlays and doubling the number of troops in Europe to bolster the North Atlantic Treaty Organization (NATO). The war also strengthened the Sino–Soviet alliance and enmity between the United States and China, and entrenched the Cold War, in which Australian Sigint would be of global importance for a time.[59]

In the meantime, British and American naval ties that proved valuable in Korea clearly benefitted from the legacy of close collaboration during the preceding war. In 1948 an agreement had been reached between the US Navy's Office of Naval Intelligence and the Admiralty's DNI to establish and maintain an active intelligence exchange between their fleets in the eastern Atlantic and Mediterranean.[60]

Australia's land force contribution in the Korean War had been closely aligned with that of its British counterparts. The 28th Commonwealth Brigade, to which Australia was the principal force contributor, was an integral formation as part of the 1st Commonwealth Division, commanded by a British major general. Similarly, maritime forces had worked closely with their British and American counterparts, capitalising on the Second World War experience.

Korea, ANZUS and intelligence ties

Beyond the actual force contributions in support of the UN-mandated defence of South Korea, the Korean War was also important for Australia as an opportunity to strengthen ties forged with the United States during the Second World War. Australia was willing to commit forces promptly in 1950, in part, because it hoped to cement support in Washington for greater security ties with the antipodean Anglosphere countries of Australia and New Zealand. The abrupt departure of the US forces at war's end led to Australia warmly inviting back British counterparts; but with America ascendant, seizing opportunities to strengthen ties with the United States made eminent sense. Historians have long appreciated how these dynamics contributed to the United States' willingness to form a military alliance with Australia and New Zealand in September 1951.

The ANZUS Treaty had become negotiable owing both to the reassessment of American strategic security interests following the outbreak of the Korean War and the military cooperation Australia and New Zealand offered under UN auspices, and conducted under US leadership. As Australia's official historian of the Korean War Robert O'Neill observed on Australia's involvement in the war, 'the major dividend of participation was diplomatic rather than military or strategic'.[61]

Reflecting American wariness of overcommitment, the ANZUS Treaty, however, lacked a clause guaranteeing mutual obligations for protection in the face of an armed threat by an adversary aimed at one of the parties. Indeed, ANZUS also lacked mechanisms for the coordination of mutual defence. There was little substance beyond the 800-word document signed by US Secretary of State Dean Acheson in San Francisco on that first day of September in 1951.

This absence of mutual defence clauses as well as robust mutual defence mechanisms, procedures, headquarters and command and control arrangements stood in marked contrast to the provisions that applied to NATO. Article five of that treaty deemed an attack against one ally to be considered an attack against all of them. Absent such a clause, Australia had what Allan Gyngell described as its 'fear of abandonment'.[62] While it may not have been apparent to casual observers at the time, the Australian government's approach to the United States reflected a nervousness about British capacity limitations and resolve to come to Australia's defence, and that unease was not completely settled by the signing of the ANZUS Treaty. Intelligence, and particularly Sigint, would serve to calm some of these fears as part of a deeper, and more expansive network of 'ties that bind' than the authors of the ANZUS Treaty may have envisaged.[63]

Decades later, Desmond Ball would write about the 'strategic essence', whereby the Sigint connection, as codified in the 1947 UKUSA agreement, 'remains the most important international agreement to which Australia is a party'. Ball argued that 'this logic underlay the agreements on the North West Cape communications station in 1963, and the satellite intelligence facilities at Pine Gap (1966) and Nurrungar (1969) in central Australia'.[64]

Tripartite Conference and closer collaboration

Circumstances surrounding the Tripartite Conference of 1953 illustrate that the close collaboration, the strategic essence of the US–Australia ties, was an iterative process and amounted to more than the signing of the ANZUS Treaty. But the importance of these ties predates the agreements over North West Cape, Pine Gap and Nurrungar. Indeed, it was the events of the 1953 Tripartite Conference which would see Australia firmly established as an enduring close and trusted UKUSA security partner. This was an event that followed Australia's active participation alongside US and UK forces in the Korean War and built upon that connection, as well as the preceding history of close wartime collaboration.

Before the Korean War, Australia's relationship with its Sigint partners remained in flux for some time after the defeat of the Chifley government, and this uncertainty continued even after Australia's commitment of forces alongside the United States in the Korean War. The principles of Sigint collaboration between the United Kingdom and the United States were reviewed at the BRUSA Conference in 1952, and the approach to Australia was carefully considered. Full exchange was not resumed with Australia until it was reinstated as a full partner in time for the Melbourne Tripartite Conference of September 1953, concluded after three years of Australians fighting alongside US and UK forces in the Korean War.[65]

The 1953 Tripartite Conference in Melbourne included representatives from DSB, GCHQ and the newly minted NSA.[66] The conference established a basis for collaboration between DSB and NSA.[67] The main recommendations of the conference were in Appendix J, which set out the

'Principles of UKUSA Collaboration with Commonwealth Countries other than the United Kingdom' (formerly known as BRUSA). Fine print was required, and in Annexure J1, the specific arrangements affecting Australia and New Zealand were outlined.[68] The first paragraph shows that five years after the establishment of the DSB, Australia was still far from having an independent national Sigint organisation:

> It is noted that Defence Signals Branch Melbourne (D.S.B.) is, in contrast to Communications Branch Ottawa, not a purely national centre. It is and will continue to be a joint U.K.-Australian-New Zealand organization, manned by an integrated staff. It is a civilian organization under the Australian Department of Defence and undertakes Comint tasks as agreed between the Comint governing authorities of Australia and New Zealand on the one hand and the [London Signals Intelligence Board] on the other. On technical matters only, control is exercised by Government Communications Headquarters on behalf of L.S.I.B.[69]

As a mark of trust and collaboration, a Special US Liaison Officer Melbourne (SUSLOM) was assigned to DSB in 1954. This was followed the next year by an Australian liaison officer at NSA, designated the Australian Sigint Liaison Officer Washington (AUSLOW).[70]

By that time, Teddy Poulden had been replaced as director by an Australian, Ralph Thompson. Having served in the Middle East during the Second World War with No. 4 Special Wireless Section, Thompson had joined the Australian Mint after the war and was appointed to head DSB from 1950 onwards. He would remain in the chair for 28 years, overseeing the development of what has been described as 'a highly professional operation with technical innovations that led the world'.[71]

While much remained to be done, the Tripartite Conference of 1953 proved to be a watershed moment, turning Sigint arrangements that had been transactional in wartime into something more permanent and trusting. It put Australia on level pegging with Canada in terms of the relationships with the first party signatories of the UKUSA agreement.

The Petrov affair

The defection in 1954 of Vladimir Petrov, Third Secretary of the Soviet Embassy in Canberra and a KGB colonel, along with his wife Evdokia (a Soviet cipher operator and senior to Petrov in KGB rank), more than justified Australia's reinstatement. The most significant Soviet defection to the West at the time, it was a timely intelligence victory for Australia.

Petrov provided documents to ASIO which confirmed what Venona had already revealed: a Soviet spy network was operating in Australia. The Petrov Papers also suggested that members of the ALP had passed information to the Soviet Union. Menzies announced a royal commission on espionage on parliament's last sitting day before the 1954 general election. The commissioning of a royal commission made sense, but the timing would see it described as a political stunt to improve Menzies's re-election prospects.

The value to Western intelligence of Petrov's defection went beyond the information he gave the royal commission. On the basis of the Petrov interrogations, ASIO prepared detailed reports on aspects of Soviet intelligence which were shared with overseas agencies. Both Petrovs provided very useful information on Soviet codes and ciphers. Above all, because of Petrov's background in KGB headquarters, he contributed, without knowing it, to the effort to break encrypted KGB cables under the Venona program.[72]

In due course, the national Sigint agency did become Australian, and no longer a joint UK–Australian–New Zealand organisation. As we shall see, however, more than 20 years later, Justice Robert Marsden Hope, leading Australia's first royal commission into intelligence and security, would still be calling for greater independence for Australian intelligence, including Sigint.

Evolving relations with the Brits

When the Petrov affair made headlines, Australian forces remained in Korea monitoring the DMZ, but the Korean War armistice had provided a breathing space for a reallocation of Australian military forces. This included the deployment of an infantry battalion group to the British Commonwealth Far East Strategic Reserve – a predominantly British force, with Australian and New Zealand attachments, based largely in the Malayan Peninsula.

With common language, procedures and equipment, in addition to other cultural ties, relationships between the Commonwealth Division's forces on the battlefields of Korea became very close and collaborative. That collaboration with Britain was closest in the Sigint domain. John Ferris has observed that Britain helped to 'guide Australia from receivership' in the Sigint domain, and in so doing,

> Britain integrated Australia, British and New Zealand Sigint into an independent Commonwealth capability in Asia. This approach helped all parties. By 1954, NSA has reopened an intelligence exchange with Australia, but with restrictions on what could be exchanged. With British assistance, DSD became large and competent. Australian Siginters worked alongside British and New Zealand counterparts at Hong Kong and Singapore. DSD directed these organisations, in liaison with GCHQ. A GCHQ officer became deputy director of DSD, an influential position for a generation.[73]

Amongst the closest ties with British counterparts were those of members of the RAAF. Between 1949 and 1985, RAAF telegraphists and signals operators enjoyed postings to Hong Kong where they worked alongside British counterparts as integrated staff with the RAF Sigint station, 367 Signals Unit, and 121 Signals Squadron detachment (Australian Army). This was initially located in the New Territories and then at Little Sai Wan on Hong Kong Island.[74]

Meanwhile, back in Melbourne, British influence would continue, but with a significant change at the top. The highly experienced and capable Ralph Thompson, an Australian, was now in charge. He appreciated the benefits for Australia's Sigint enterprise of continued connections with the United Kingdom. Ferris notes that 'for Britain, this cooperation bolstered an important area, which for Australia was a key border. Australian

Siginters and soldiers supported British and American waters in Southeast Asia during the 1960s, for what all thought were common purposes.'[75]

Australian support featured particularly during the Malayan Emergency. This campaign, pitting armed members of the Communist Party of Malaya against British and locally-recruited security forces, saw greater Australian involvement as the commitment to the war in Korea was coming to an end. Commencing in 1948, the Emergency continued through to 1960 and during this period, RAN operations incorporated naval Sigint personnel for the first time since their withdrawal from RAN cruisers in 1939.[76] These specialists, who monitored possible 'Communist Terrorist' (CT) transmissions, were sent to Singapore 'as part of Australia's commitment under the UKUSA Agreement'.[77]

As a consequence of the wartime collapse of Britain's plans for the defence of its possessions in the Far East, Australia insisted on being more involved in decisions affecting British Commonwealth defence matters in the Pacific. This had a direct effect on DSB which acquired more responsibilities under the UKUSA Agreement. These responsibilities stretched their intercept and analysis capabilities, leaving DSB with an acute shortage of trained personnel, notably Mandarin linguists. However, these issues appear to have been overcome by April 1955, when both the JIB and DSB were working towards a combined strength of over 1100 staff.[78]

The Suez Crisis, greater autonomy but continued collaboration

Notwithstanding the closeness that developed working alongside UK forces, Australia, like the United States, was kept in the dark and surprised by British plans during the Suez Crisis of 1956. Britain's Prime Minister, Anthony Eden, together with the leaders of France and Israel, conspired to bring down the Egyptian government of Gamal Abdel Nasser and to reverse the nationalisation of the Suez Canal. But before the scheme was hatched, Australia's Prime Minister Menzies led an international delegation to Egypt. Menzies hoped to reason with Nasser, but his position was compromised by Anglo–French–Israeli secretly coordinated military actions, on which he was not briefed. The situation was further complicated by President Eisenhower, who was determined to avoid resuming the wartime closeness with Britain that had strengthened the latter's position. Thus, once the invasion of Egypt began, Eisenhower responded by pressuring the British and French, and threatening their financial ruin. In effect, Eisenhower, with Eden's blundering help, split the alliance temporarily, at least at the political level. This reduced Britain's already weakened global influence.[79] The end result was a turning point for the United Kingdom and its residual imperial aspirations.

Ferris notes, however, that GCHQ and NSA continued technical cooperation during the Suez Crisis as the political clash 'took place above their heads'.[80] Nonetheless, afterwards, the UKUSA was 'no longer an alliance between equals'. The NSA would from then on be the driving force.[81] For Australia, the eclipse of Britain's power following the Suez Crisis meant that DSB had to find its own way, no longer so dependent on Britain. That self-reliance would become stronger after Britain's 1967 decision to withdraw from 'east of Suez'.

In the meantime, as with the operational cooperation between GCHQ and NSA, British secretiveness over the Suez Crisis did not prevent expanding collaboration on Sigint closer to home, where Australia's collection and reporting priorities lay. From the mid-1950s, over 200 Australian military personnel were posted to British intercept stations in Hong Kong and Singapore.[82]

Against this background of shifts in Australia's key relationships, in the 1950s and 1960s there were conflicts within and between Indonesia, Malaya and the remaining British possessions in Borneo. As has been noted, there were changes in the priority given to Australia's intelligence targets. In July 1959, for instance, China, Indonesia, and Malaya-Singapore were the highest priority. After Indonesia's leader Sukarno had gained control of Dutch New Guinea in 1962, he confronted a federated Malaysia. As a result, the intelligence priority had switched to China, Indonesia and Southeast Asia, and a reassessment in 1963 confirmed that Australian intelligence priorities were all in Southeast Asia.[83]

Finding its Far East security interests threatened, Britain moved to defeat Sukarno. In 1965, Australia was drawn into the so-called *Konfrontasi* (or Confrontation) and Australians were deeply involved in the security forces' Sigint activities. Australian activities also expanded, with Sigint collection stations at Darwin and Perth providing better coverage of Indonesian targets. To a great extent, the intelligence used by the British Commander-in-Chief Far East (CINCFE) and his staff to direct the operations during the Confrontation came from Australia or was produced in stations at least partly operated by Australians.[84]

Another Australian source of Sigint was 693 Signals Troop, which deployed to provide communications support in place of a UK signals troop.[85] This 693 Signal Troop was a descendant of the Australian Special Wireless Group raised during the Second World War. The Army's primary Sigint collector was 101 Wireless Regiment. In 1959 it established a

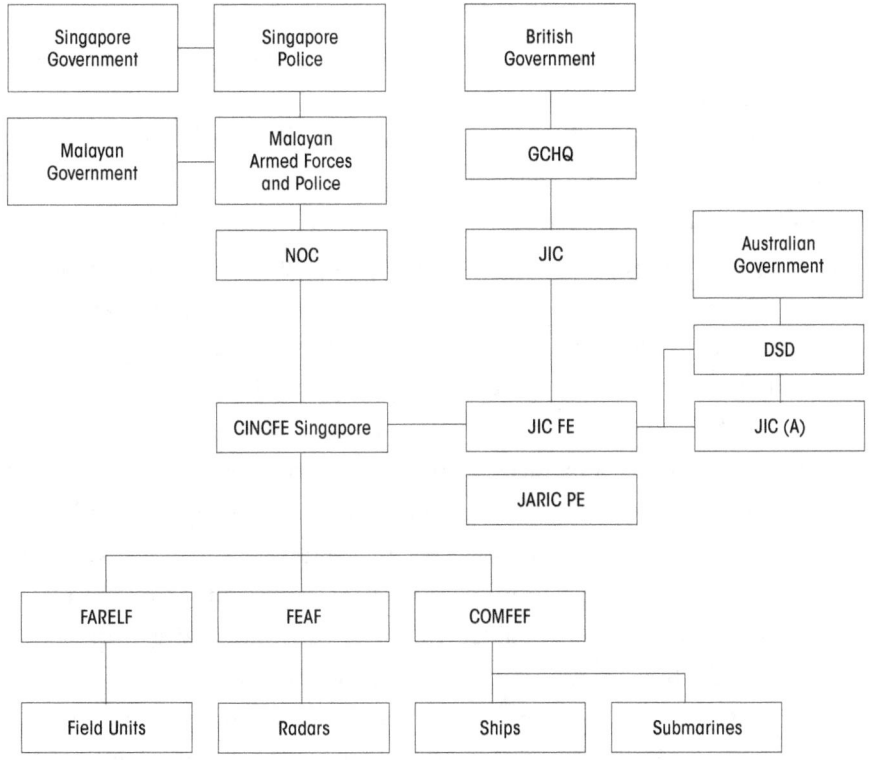

Figure 6. Far East intelligence organisation, 1964
SOURCE Ian Pfennigwerth, *Missing Pieces: The Intelligence Jigsaw and RAN Operations 1939–71* (Canberra: Department of Defence, 2008), p. 204.

subordinate unit in Singapore, known as 201 Signals Squadron, which conducted similar collection. It was from this sub-unit that 693 Signal Troop was raised for deployment to Borneo.[86]

In Borneo, as in Korea, Sigint gathering could be basic, such as tapping telephone lines on the Indonesian side of the border. Security forces could also form a clear picture of the casualties inflicted through intercepted radio traffic.[87] Information from Sigint was tightly held. At headquarters of the British Director of Borneo Operations (DOBOPS), briefings up to Secret level were 'the equivalent of reading a good newspaper account'. The select few who were 'in the club' (that is, cleared above Secret) were privy to everything, including the highly secret Commonwealth Operations in Kalimantan, known as Claret.[88]

Sigint made an important contribution during Confrontation, compensating for deficiencies on the ground.[89] It provided leads on where to look and how to interpret information from observers. This also applied to naval operations along the coastline.

As we now know, communications regarding Indonesian military movements were often intercepted, and these assisted the ordering of ground force retaliation and harassment shoots by naval vessels. Occasionally, there were engagements with Indonesian gun emplacements at the coastal border with Sabah. In such cases, the intelligence triggered an immediate and direct response in accordance with the rules of engagement.[90]

Indonesia targeted not just territory in Malaysian Borneo but also in Peninsular Malaysia and Singapore. Singapore had a number of tempting sabotage and infiltration targets, but Indonesian communications were easily intercepted from sites in Singapore and East Malaysia. In addition, Indonesian signals security was lax, making the cryptanalysts' work relatively easy. Knowledge of impending attacks derived from Sigint, with other intelligence reports, enabled forces to assemble and go where they would have the best chance of intercepting the raiders.[91]

Operations during Confrontation succeeded because intelligence gathered from wireless radio intercepts and interrogations was tactically applied. Individual patrols benefitted directly from electronic interception-derived information.[92] Not everyone, however, was convinced of the value of Sigint. According to Major General Walter Walker, Commander British

Forces Borneo, much of the Sigint was stale by the time it reached him, and he found the material provided by the SAS and his unit commanders fresher and more useful.[93] On the other hand, the British Director of Borneo Operations commended the Sigint effort as a significant force multiplier that helped British and Commonwealth forces gain the initiative.[94] Even former Indonesian officers acknowledged the security forces' Sigint effort was far better than the Indonesian equivalent and, because of their technical superiority, the security forces were able to gather high-grade intelligence.[95]

In 1964, the 101 Wireless Regiment was renamed the 7th Signal Regiment and 201 Signal Squadron became 121 Signal Squadron, the latter continuing operations in Singapore until 1974. The strategic collection sites at Cabarlah, Singapore and Hong Kong gave the Army's Sigint personnel in-depth exposure to the types of communications, especially of Asian communist insurgent groups, that they would face in South Vietnam.[96]

Australian Sigint contributions in the Vietnam War

As *Konfrontasi* subsided, the war in Vietnam was intensifying. Australia's involvement in the Vietnam War has been much written about from a number of angles, by official historians and others. Australia's participation this time was at the behest of the United States, under the rubric of stopping the spread of communism in Asia. Australia initially sent the Australian Army Training Team to South Vietnam in July 1962 – a seasoned group of Australian soldiers assigned to work closely with South Vietnamese combat units as advisors and trainers.[97] This was followed by the deployment of the 1st Battalion, Royal Australian Regiment (1 RAR) in June 1965 to Bien Hoa province near Saigon, where it worked as part of the 173rd US Airborne Brigade.[98] In response to American requests for further support from friendly countries in the region, Australia's contribution included RAN and RAAF contingents. The naval forces operated along the Vietnamese coastline and RAAF contingents operated from Phan Rang Air Base, with helicopters deployed in support of the land campaign. In terms of the land forces, Australia replaced 1 RAR with the (larger, brigade-sized)

1st Australian Task Force (1 ATF) in May 1966. One dimension, however, which has received little coverage is the use of Sigint.

It is worth noting that this was the first time since 1945 that Australia had sent to war a formation with a full complement of arms and services, including two specialist intelligence units: the 1st Divisional Intelligence Unit and 547 Signal Troop. In previous campaigns, such specialised units had been retained at higher echelons of command and control (division and corps headquarters, and higher), and not assigned to specific brigades. But this case was different: 1 ATF was operating with a higher degree of autonomy than brigades had deployed in corps and army group campaigns in the Second World War and the Korean War. Allocating such specialised support at this level made sense politically, but also reflected technological innovation that made it easier to deploy such elements to an unprecedentedly low part of the spectrum of combat forces. It was also the first time since the Second World War that these intelligence units were supporting Australian, rather than Allied forces.[99]

When Australian forces first deployed to Vietnam with the Australian Army Training Team Vietnam in 1962, Sigint was still little known or understood. DSD hid its functions behind its anodyne title, and the service's Sigint and electronic warfare functions remained veiled, even to most military practitioners. Still, while US collection efforts in Vietnam were far more extensive, Australia did have its own intelligence collection arrangements, with support from DSD, including the British stations in Hong Kong and the Australian station in Singapore. There was a general understanding that North Vietnam's major operational moves required the tacit and actual support of the People's Republic of China. Consequently, intelligence on Chinese activities would have been a useful contribution to the Allied pool.[100]

Notwithstanding DSD's capabilities, few in the Australian military, let alone the general public, had much, if any, exposure to Sigint support for operations. Indeed, wartime Sigint successes would not be revealed to the public for another decade, when F.W. Winterbotham published *The Ultra Secret* in 1974.

On land, Australian combat forces had teamed up in 1965 with the US Army's 173 Airborne Regiment. These forces deployed without Australian-sourced Sigint support. The following year, Australia expanded its land forces commitment to a brigade-sized task force. The deployment of troops and equipment, largely by sea, was supported by RAAF Neptune aircraft providing a continuous airborne anti-submarine warfare screen with RAN escort destroyers for the voyage of HMAS *Sydney* from Townsville to Vung Tau in South Vietnam.[101]

Meanwhile, part of the reason for the limited understanding of Sigint, particularly among land force commanders, was that Australia's wartime commanders with Sigint experience had long since retired. In Korea a decade earlier, Sigint had been managed at the highest levels, with few, if any, Australian commanders having the chance to work closely with Sigint capabilities. Australia had not deployed any such units to Korea during the war and US assets in that war were held at corps and army group level, several echelons above Australia's most senior commanders in Korea.

The prowess of 547 Signal Troop

Despite lack of experience with Sigint, planners and commanders had the foresight to include 547 Signal Troop, along with the 1st Divisional Intelligence Unit, in the plans for the deployment of the 1st Australian Task Force (1 ATF) to Phuoc Tuy province in 1966. It would be some time, however, before commanders were ready to make effective use of it. They would learn by trial and error.

Raised specifically for Vietnam from the 7th Signal Regiment,[102] 547 Signal Troop had 45 members for most of its time in Vietnam. As the conflict deepened, they specialised in search and interception (that is, scouring radio frequencies and then recording, deciphering, analysing, translating and reporting on transmissions identified), airborne radio DF, the operation of experimental high-frequency DF equipment, and analysis.[103]

Unlike its allied counterparts, the troop included men with over 20 years' experience of Sigint, including tours of duty overseas. They were considered by other nations to be highly professional and were held in very high regard.[104]

Before long, the troop's results were noticed favourably by Allied forces. Lieutenant General Bruce Palmer, Commanding General of the US regional military command known as 2 Field Force Vietnam (II FFV), insisted that a liaison officer visit the troop every morning by helicopter for the latest results. On several occasions, the troop received commendations for the standard of information it delivered. Palmer's subordinate divisional commander, Brigadier General O'Conner, Commander of the 9th US Infantry Division, had also been impressed by the troop's work. Research elements of his division and of the 2nd Cavalry Division worked as outstations of the troop at times.[105]

One of the most important contributions of Sigint to the Australian land force campaign in Vietnam took place shortly after 1 ATF's arrival, as it was getting established in and around a small hill in central Phuoc Tuy province known as Nui Dat. It is now evident that the Viet Cong hoped to score an early victory against the newly arrived Australians. A large force was mustered but the battle that followed at Long Tan, just 5 kilometres to the east of Nui Dat, disrupted the Viet Cong plan.[106]

The Battle of Long Tan took place on 18 August 1966, but nearly three weeks before the battle, 547 Signal Troop picked up indicators of substantial offensive action targeting the Australians. Radio schedules and corroborating intelligence indicated that they were in a rest-and-retraining cycle. Around 29 July, 275 VC Regiment started sending more traffic and longer messages.[107] Captain Trevor Richards, troop commander and senior DSD officer, recounted:

> Now we have found over the last couple of weeks ... that we can now read this system. The VC are still using it and we are finding now that the 274th Regiment is passing operational orders over this network ... It is possible after receiving a series of these messages to start pinning down roughly where the various battalions are.[108]

547 Signal Troop had, in fact, located a much larger formation than expected, from the North Vietnamese Army, which was using the VC Division and Regiments for its communications, but there was no way of knowing this at the time. The troop reported the signs of increased

enemy activity. Due to the secrecy surrounding Sigint, although a limited and sanitised version of the available information was passed to battalion commanders, they were not told its source and discounted it, especially when their patrols failed to turn up signs of enemy activity. This included, at first at least, D Company of the 6th Battalion, Royal Australian Regiment (6 RAR), patrolling near the Long Tan rubber plantation.

At Long Tan, D Company 6 RAR encountered a much larger force. Vastly outnumbered, they fought heroically against the 275 VC Regiment. After the battle, when the Sigint reporting had proved to be correct, Australian and Allied staff officers began to pay more attention to it.[109]

547 Troop had other, fortunately unalloyed, successes. A year after their arrival in Vietnam, the head of NSA commended them for their 'professional approach, exceptional expertise and outstanding performance of a difficult task'.[110]

Air, land and sea direction finding

An important technological development in Sigint in Vietnam was the Weapons Research Establishment's accurate airborne radio direction finder (ARDF). This technology incorporated DF techniques used in earlier wars, now available in smaller and transportable equipment that could be set up inside a small aircraft. 547's ARDF operators, all volunteers, flew on Cessna aircraft operated by 161 Reconnaissance Flight, at least once a day at low altitudes along straight paths within machine gun range over enemy territory.[111]

In one instance, the ARDF located the headquarters of 274 VC Regiment 38 kilometres northwest of the Australian tactical area of responsibility in neighbouring Bien Hoa province, where an infantry battalion of the Royal Thai Army Volunteer Force was located. The Thai position was lightly defended, but once the message was passed up the chain to Bien Hoa and across, the Thai position was reinforced in time to inflict significant casualties and blunt the attack.[112]

According to Peter Murray, the Officer Commanding 547 Signal Troop in 1968, the ARDF was probably the single most important intelligence gatherer apart from the 'Mark One Eyeball' of infantry patrols. Cessna aircraft with the equipment on board would be flown up to six times a day to eavesdrop on enemy radio traffic. The aim was to pinpoint enemy headquarters and unit locations for taskforce intelligence and operations staff involved in planning. Murray recalled, 'the [Special Air Service] went out and took a radio station which had 14 people. They came out with the radio set and their cipher books. The cipher books opened up a lot of information to us.'[113]

On another occasion, a Viet Cong ambush was discovered in the area over which an Australian reconnaissance aircraft was flying. A US company of infantry was moving towards it. At Troop Headquarters, two Australians worked frantically to identify the troops on their way to be ambushed. The company was identified as part of the US Army's 199th Light Infantry Brigade. The company was redeployed, and a counter-ambush staged, in which 30 Viet Cong and one American were killed. The company commander sent a message of thanks to the Australians, with a comment from the brigade commander 'instead of a US body count we were able to do a VC count'.[114]

In addition to the ARDF, and also worthy of mention, was the 'single station locator'. This was another piece of equipment developed in conjunction with the Weapons Research Establishment which utilised a four-antenna circular array derived from the Second World War German *Wullenweber* aerial system. It was designed to pinpoint the source location of a radio wave by measuring the phased time and angle difference between two incoming radio waves striking the antenna in the circle and calculating the angle of deflection from the ionosphere. This was similar in concept to the artillery locating battery which used a string of sensors to locate enemy guns, measuring the time difference between arrivals of sound waves from the rounds fired. The technique of putting it all together in a container and placing it within the Nui Dat compound was considered very successful.[115]

The achievements of 547 Signal Troop in Vietnam ensured that such units were maintained as a regular part of the ADF's suite of little-known specialist capabilities that were included in operational planning contingencies in the years that followed.

While these land campaigns were being fought, at sea, the RAN operated recently acquired US-built destroyers on the gun-line along the Vietnamese coast from 1967 to 1971.[116] When operating north of the so-called DMZ, RAN vessels were supplied with information on coastal defences from electronic intelligence detections. Historian Ian Pfennigwerth notes that:

> North Vietnamese army coastal artillery batteries were believed to be radar-directed, and the characteristics of these emitters were recorded and analysed by USN airborne collection assets. The information was passed to ships equipped with EW receivers, which could thus be alerted when a coastal defence site illuminated them. The high number of EW detections of NVA artillery sites caused 7th Fleet to order a 'crash' program of fitting radar noise jammers to counter the threat in 1967. RAN ships had well-trained and effective EW teams as part of their ships' companies, but it appears that no electronic intelligence operators were embarked for special collection tasks.[117]

Limitations on Sigint in Vietnam

Despite the numerous tactical successes of Australia's forces in Vietnam and the important role of Sigint in the outcome of those battles, there were instances of Sigint-sourced information being insufficiently understood and used. The most notable in Australia's experience in Vietnam concerned the Battle of Long Tan in August 1966. After that encounter, however, the use and effectiveness of Sigint became more widely appreciated by commanders and other staff who were brought into the tight circle of people in the know and briefed on it.

The Sigint enterprise had been well enough resourced, it seemed, to help provide the combat forces with more situational awareness, down to a lower tactical level than had been the case in previous conflicts. In the Second World War, however, the idea of a specialist Sigint unit being assigned to a brigade-sized task force such as the 1st Australian Task Force in Vietnam was unheard of. Yet despite this more comprehensive and timely Sigint support, the overall prospects of success were largely unaffected.

The failure to win the war exposed some deep flaws. One was an over-reliance on Sigint at the higher echelons of command and direction of the war, which made US intelligence vulnerable to communications deception. In addition, restrictions on the distribution of Sigint to operational commanders often prevented it from being used effectively as tactical intelligence.[118]

The contrast with the experience gained in the Second World War is instructive. During the earlier conflict, high-level code-breaking of Enigma, Lorenz and Purple machines had prised open the secret plans of the senior commanders of the Axis forces, enabling high-level efforts to counter their plans and formulate effective strategies. In the Vietnam War, however, the inability to read high-level codes left Sigint units and agencies focused on the tactical level communications, where lower-grade codes and communications networks tended to be easier to decipher and interpret for immediate tactical use. The problem this time was not seeing the wood for the trees. An over-reliance on tactical intelligence leads left some inundated with urgent tactical messages which obscured the higher level trends and led to a misreading of them. The challenges of intercepting Soviet and Soviet-aligned and encrypted communications, with their emphasis on one-time pads, would persist after the war.

Many veterans felt they had done what they could, but there were some deeper, more fundamental problems, beyond the issue of difficult-to-break Vietnamese encrypted communications. This had to do with the way the war itself was conducted – a factor that is beyond the scope of this study. These complicating factors meant that their good work and best intentions ultimately made little difference to the course of the war. Nonetheless, it is fair to say that timely and accurate Sigint support to the Australian forces deployed there provided them with a higher level of situational awareness that helped to reduce the risk of even greater own-force casualties.

A foreign policy challenge: DSD in the eye of the storm

As Australia's commitment in the Vietnam War ended, it faced a foreign and defence policy challenge. Australia had long relied on the concept of forward defence, alongside 'great and powerful friends', the United Kingdom and United States. Britain's withdrawal from 'east of Suez' around 1970 was, in hindsight, the natural conclusion of Britain's post-1956 Suez Crisis decline. This led to a slashing of its Sigint presence in Singapore and Hong Kong. Similarly, the US withdrawal from Vietnam and President Nixon's so-called Guam Doctrine, enunciated during a visit to Guam in 1969, saw the United States, interest in Southeast Asian affairs decline. Foreign help for DSD seemed to have vanished. This was compounded by the election of a new government under Labor leader Gough Whitlam, who 'broke decades of conservative domination' of Australian politics, and exposed a 'generational gap' in Australian attitudes towards intelligence and its allies.[119] As John Ferris has observed,

> DSD confronted a perfect storm, causing common UK-Australian purpose and effort to collapse. The matter was compounded by Whitlam publicly announcing that DSD conducted Sigint. This, the first such statement by any leader of the Five Eyes, challenged all of their cover stories for Sigint. It raised hackles at the highest levels in Britain, the United States and Singapore – where the Sigint station was abandoned. These Sigint issues had no role in the collapse of Whitlam's government, but intelligence and mistrust over it did.[120]

Thereafter, Australia adopted a more independent role in the UKUSA. As Ferris points out, 'its Siginters took responsibility for the Indonesian target, which was a national priority and cooperated with GCHQ at Hong Kong. Australian leaders found DSD useful to its strategic policy and "central to Australia's status as a self-reliant regional power".'[121] Its status and importance was still not immediately apparent to most, however.

Consolidation of postwar Sigint

Australian Sigint had gone through a dramatic transformation from the end of the Second World War to the close of Australia's military commitment in Vietnam. The precarious state of Sigint as the Second World War ended was followed by security leaks which threatened to exclude Australia from the close-knit intelligence relationship with the United States and United Kingdom. That danger passed once the Australian government established ASIO and committed to strengthening domestic security arrangements. The move made sense for national security as well as reassuring Australia's traditional security partners.

Australia's ad hoc security and intelligence arrangements were replaced by institutional ones, with Sigint a permanent feature of the government bureaucracy and armed forces. This period witnessed the transformation of Australia as technically and legally independent but still with close ties to the United Kingdom and not yet a treaty ally of the United States. The intervening years saw Australia involved in the Korean War, securing the ANZUS Treaty in 1951 and consolidating postwar Sigint links, notably with the 1953 Tripartite Conference, with both the United States and the United Kingdom as part of the Five Eyes intelligence-sharing arrangements.

The United Kingdom featured prominently in the growth of Australia's Sigint capabilities in the late 1940s and 1950s. Closely connected culturally as well as technologically and procedurally with the United Kingdom, Australia was committed to supporting British security endeavours in the 'Far East', notably in Korea, during the Korean War, in Hong Kong, and in British-controlled Southeast Asia. The Suez Crisis in 1956 and the declaration of independence of Britain's territories now known as Malaysia, Singapore and Brunei was a turning point. Along the way, as Indonesia's Sukarno challenged the legitimacy of the federation of Malaysia, he launched *Konfrontasi*, to which Australia responded in support of the British and Malaysians, between 1963 and 1966. Then, the war in Vietnam, from 1962 to 1972 realigned Australia's defence capabilities, including Sigint, towards the United States.

Australia's national Sigint capability during this period increased as it deepened its ties with the United States. As we will see, this was most tangibly demonstrated by the establishment of intelligence facilities at Pine Gap and Nurrungar, in the mid-1960s.

9
REFORM, COMPUTERS & MILITARY SIGINT SINCE THE 1970S

Between 1945 and 1972, Australia's intelligence arrangements underwent a major reorientation. In the decades that followed, a transformation took place, in large part due to the reforms of the mid-1970s and 1980s led by the royal commissioner, Justice Robert Marsden Hope. In addition, technological transformation, with the rise of computers and the evolution of use of the electromagnetic spectrum from analogue to digital, had meant that the Sigint enterprise also had to adjust and adapt.

As figure 7 illustrates, a proliferation of devices were operating across the radio frequency spectrum, presenting new challenges and opportunities for Sigint agencies around the world.

It was around this time, in 1982, that the term 'cyberspace' was first used. Appearing in a science fiction novel, the term was used to describe the creation of a computer network in a world of beings with artificial intelligence.[1] Within two decades it would be part of the common language.

What this meant, in part, was that whereas the task of intercepting communications on the radio frequency spectrum had been relatively straightforward in the mid-twentieth century, by the end of the century and thereafter, the challenges were growing exponentially. No longer was the Sigint task just a matter of monitoring analogue HF and VHF communications, for instance. Although the radio frequency spectrum itself had not increased, technology that capitalised on the differing strengths and features of various parts of the spectrum was being used by more and more machines, countries and industries. This greater use of the spectrum was facilitated by the digital revolution, where analogue use of the electromagnetic spectrum was replaced by digital, with a plethora

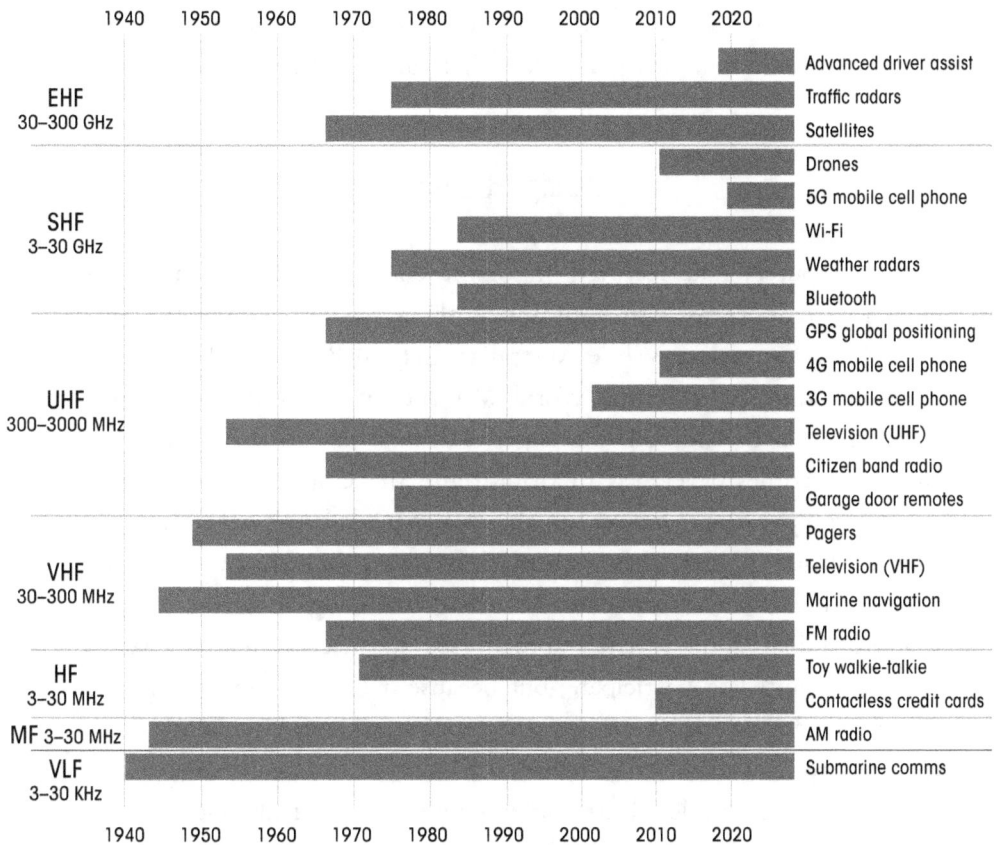

Figure 7. Use of the radio frequency (RF) spectrum over time
SOURCE Derived from the ASD display at the Museum of Australia in Canberra.

of permutations of combinations of transmitting and receiving what, in essence, amounts to ones and zeroes.

More than 40 years after its first use in science fiction, cyberspace had become critical to the way society works. The demand for change gathered pace with the proliferation of satellite communications technology, the advent of the World Wide Web and cyber – all unimaginable a century earlier. Along the way, society had gone from being web-enabled to web-dependent and web vulnerable. This triggered enormous demand for cyber security support – a function that the national Sigint agency was ideally placed to perform. Computers had long featured in the technology-oriented DSD, and would be an integral part of this transformation.

Computers

Basic computers had featured in Sigint endeavours in Vietnam and elsewhere in Southeast Asia, building on the technological breakthroughs of a generation earlier. While Australia's Central Bureau had used IBM computers during the Second World War, they were mostly controlled by US personnel. In addition, the larger and more powerful Hollerith and Bombe machines developed in Britain and the United States provided critical computational power to the Allied code-breaking efforts, but were not used in Australia. Such machines were retained in the United States and United Kingdom. After the war, technological advances meant that Australia's Sigint partners abroad used advanced computer systems. Australia wanted to participate, but because it had not invested in this equipment during the war, it was still dependent on its UKUSA partners for access to such advanced systems.

NSA's first parallel electronic computer with a drum memory, the Atlas 1, was delivered in December 1950. NSA introduced its successor, Atlas 2, in 1953 – the first core memory computer. By 1958, NSA received its first computer that relied on transistors. In 1962, a computer system with a fully automated tape library known as Harvest was installed – a computer strongly influenced by the design of the IBM System 360 machines.[2] In describing developments in GCHQ, intelligence expert Richard Aldrich notes that it was a step behind its US counterpart, the NSA, but sought to keep up. By 1977, GCHQ took delivery of its first US-built Cray supercomputer. A supercomputer

> breaks down a problem into many tasks that can be done at the same time. By using different parts of its brain in parallel, it can undertake vast calculations unbelievably quickly. The main applications are code-breaking, designing nuclear weapons and weather forecasting. At the time the world of super-computers was led by Seymour Cray.[3]

From then on, Cray supercomputers became increasingly powerful. The Cray X-MP was produced in 1982 and would be used for such things as digital effects in movies. The Cray 2 followed in 1985. It was nicknamed 'Bubbles' as it was the first to use liquid immersion cooling. The Cray 2 was a 'four processor vector architecture with a 256 million 64-bit memory'; at the time, this was the largest central memory on any computer. It had a 4.1 nanosecond clock speed and reached a peak speed of 1.9 gigaflops (units of measure for calculating the speed of a computer equal to one billion floating-point operations per second). The system's circuitry ran so hot that the liquid would boil. Seymour Cray put it on display, making the boiling liquid visible through glass panels.[4]

Australia's national Sigint agency was that much smaller and that much further away, but it was not long before Australia also had a supercomputer. DSD's acquisition of the Cray X-MP 2.2 'Marsik' in 1986 was the beginning of a new era for Australian Sigint.[5] DSD was also an early user of 'big data' and cloud technology.[6]

The first Cray Y-MP was built for the US National Aeronautical and Space Administration (NASA) in 1988. It made a cameo appearance in the 1992 movie *Sneakers*, featuring Ben Kingsley and Robert Redford and was the world's first computer to sustain over 1 gigaflops. It had eight parallel vector processors (employing multiple threads in parallel and generating multiple results concurrently), denser circuits and a large central memory. It could reach a peak speed of 2.67 gigaflops.[7]

The Cray C90 appeared in 1991, with the initial production model sold to the European Centre for Medium-Range Weather Forecasts, although the Ford Motor Company was the system's first commercial customer. This parallel vector system had 16 new CPUs capable of 1 gigaflop each and 2 gigabytes of central memory. It could perform five times faster than Cray's previous best. One of its main features was a dual-vector pipeline, which allowed each of the CPUs to deliver two vector results per functional unit every 'clock period.'

In the following years, Cray supercomputers became increasingly powerful. The 2018 HPE Cray EX, was six billion times faster than the original Cray 1. Known as the 'Shasta' computer, this model was built for 'exascale era workloads' (calculations involving numbers with at least

18 zeros), supporting diverse processor technologies, interconnected with other supercomputers to generate an extraordinary level of computing power. In 2019 Hewlett Packard Enterprise (HPE) bought the company.[8]

While the specific computer models used in Australia are not widely known, and this catalogue of Cray models does not serve to confirm the configuration of computers used in Australia, Cray computers have featured widely in the UKUSA intelligence arrangements.

New Zealand Sigint

While links with the United States had become the predominant ones for Australia's Sigint enterprise, particularly with the United Kingdom's withdrawal from 'East of Suez' by the early 1970s, Australia's Sigint links with New Zealand were also important, albeit more modest, and tended to receive little attention. Leveraging on its mid-twentieth-century achievements, the New Zealand government had maintained a Sigint capability since the Second World War. The Government Communications Security Bureau (GCSB) website reports that:

> There was a long recognised need to ensure that the Government was protected from 'bugging' (technical security, or TECSEC) and that its sensitive messages could not be read by third parties (communications security, or COMSEC). Until the establishment of the GCSB, these services were provided by bodies such as the New Zealand Defence Force and the New Zealand Security Intelligence Service (NZSIS). In 1977, Prime Minister Robert Muldoon approved the formation of GCSB, but its functions and activities were kept secret.[9]

For a small organisation like GCSB, collaboration with partners has long been seen as essential. It is not possible for an organisation the size of GCSB to collect intelligence on all threats to New Zealand's national security interests. However, through New Zealand's long-standing partnerships, it is able to 'draw on greater support, technology and intelligence than would otherwise be available'.[10]

The government of Prime Minister David Lange came to office in 1984 with a strong platform against nuclear weapons. The next year his government refused to allow the guided missile destroyer USS *Buchanan* into New Zealand ports. Even though the ship was not armed with nuclear weapons, it was capable of carrying them and the US Navy maintained a 'neither confirm nor deny' policy. The United States had counted on the compromise of an evidently older-style warship (highly unlikely to be equipped with nuclear weapons or propulsion) as being sufficient, but Lange insisted on saying 'no'.[11] The United States responded by suspending ANZUS Treaty obligations in 1985 and Lange then introduced legislation confirming the position in 1987. Notwithstanding this alliance break, reporting indicates that New Zealand continued its Sigint cooperation with the NSA without interruption.[12]

While Sigint ties survived, other intelligence links suffered. This complicated the situation for Australia, requiring the duplication of many tasks as it was required to manage separately its bilateral intelligence relationship with New Zealand. Ensuring the exclusion of US-sourced material from the sharing arrangements which Australia maintained with New Zealand proved an onerous but necessary task for Australia's intelligence agencies. Cooperation with the United States was reported to have been restored in 2009 when then Secretary of State Hillary Clinton visited New Zealand.[13]

Throughout this period and later, GCSB has maintained two communications interception stations: a high frequency radio interception and DF station at Tangimoana, near Palmerston North, and a satellite communications interception station at Waihopai, near Blenheim in the South Island. Much like its Australian counterparts, but on a smaller scale, GCSB employs staff in a wide range of disciplines, including foreign language experts, communications and cryptography specialists, engineers, technicians, legal, policy and support staff.[14] In recent years, GCSB has also hosted New Zealand's National Cyber Security Centre, echoing practice in Australia and elsewhere of linking cyber security with the national Sigint agency.[15]

While Sigint collaboration with Australia dates back to the days of the Second World War, in recent decades the necessity and usefulness of such collaboration has been reinforced by the shared experiences of operations in places like Afghanistan. This applies even more to operations in the region, notably in Solomon Islands, Bougainville and East Timor, discussed below.[16] But before discussing the surge in operational tempo in the 1990s, we now turn to the dramatic change in intelligence arrangements as a result of landmark formal reviews of the 1970s and 1980s.

Hope: reform in the 1970s

When Gough Whitlam became prime minster in November 1972, DSD was 25 years old. Like the other intelligence organisations in Australia, DSD had been established under executive authority, as opposed to formalised legislation debated in parliament. In addition, it had never been subject to a formal review. In 1974, Justice Robert Marsden Hope was commissioned by Whitlam to undertake a royal commission on intelligence and security.[17] While ASIO had been subject to the Royal Commission on Espionage following the defection of the Petrovs in the 1950s, that inquiry had focused on the 'nest of spies' rather than on ASIO or the intelligence community itself.[18]

Hope was highly complimentary about what was by then called the Defence Signals Division. By contrast, he was critical of ASIO, prompting recommendations for major reform. Hope's work established robust accountability mechanisms for the Australian intelligence community, including the Inspector General of Intelligence and Security (IGIS), the Parliamentary Joint Committee on Intelligence and Security, the Administrative Appeals Tribunal and others. These were designed to prevent and, where necessary, manage the fallout of mistakes in the intelligence community, including Sigint.

DSD was part of the Royal Commission into Intelligence and Security in 1977, and again of a second royal commission led by Hope nearly a decade later. Hope laid the foundations for today's Australian intelligence community, including the Australian Signals Directorate.

While holding DSD in high regard, Hope was concerned about its lack of independence, both within the Australian government, particularly vis-à-vis the powerful Defence Department, and within the UKUSA arrangements. Hope recommended that DSD should have greater independence from defence system line control and should be encouraged to take its place, as an intelligence agency, with its own statutory authority, alongside other agencies of the intelligence community.[19] Hope also favoured a more independent approach for DSD, with the national interest in mind:

> Australia's intelligence interests do not, and cannot, coincide with those of any other country. Therefore, although we can and should benefit from exchange of information and views with friends and allies, we need our own intelligence collection and assessment capabilities. We also need constantly to reassess the benefits to Australia from intelligence relationships with other countries against the costs.[20]

There was scope, he considered, for DSD to be more independent and ambitious in its operations. Hope recognised that participation in UKUSA was overwhelmingly in Australia's interests and therefore Australia should continue to participate. But he also pointed out that the arrangement might not last forever and Australia might not always receive all the Sigint its partners possessed. Therefore, he urged, consideration should also be given to the establishment of some alternative resources which, while enhancing DSD's general Sigint capability, would ensure Australia's access to Sigint of critical importance to its security and well-being.[21]

Hope's warnings about UKUSA seem not to have been questioned, either at the time or in subsequent inquiries. They were apparently accepted on their merits without prompting any particular action. It may have appeared that UKUSA would last forever and that there were no alternatives that could compete with it. Hope's recommendation that DSD enhance its general Sigint capability was certainly followed, however, as we shall see.

For Hope, DSD was far from being the 'problem' agency, either in his first report in the mid-1970s, when ASIO's operations and management came in for scathing criticism, or in the second, in the mid-1980s, when

ASIS was under scrutiny over the Sheraton Hotel training exercise debacle. The incident was widely publicised, including the hotel management's call for police help in response to ASIS's bungled activity which had been launched without permission of the hotel management or advice to the local police authorities.[22] In contrast to ASIS, Hope found that DSD was a very capably managed agency and was believed to be so by staff. There are echoes here of the high morale and camaraderie of Central Bureau. Moreover, Hope observed, DSD's partner agencies, and, in particular, NSA and GCHQ, 'clearly hold DSD in high regard'.[23]

Pundits gave the long-serving director, Ralph Thompson, much of the credit for not only keeping DSD's profile low, but also for the professionalism of the DSD team. The royal commission recorded the following observation on the exceptional quality of the management of DSD at that time, which had turned a potential weakness into a strength:

> The twin constraints of specialisation and secrecy combine to develop in DSD a degree of exclusiveness and inward looking that could prove dangerous for morale. So far as I can tell, that has not been the effect on DSD's morale. On the contrary, the management has contrived to engender a strong feeling of *esprit de corps* based upon these same constraints.[24]

Hope found, however, that DSD had tended to keep to itself and to approach other parts of government only to the degree felt to be absolutely essential, and to reduce demands for funds and staff to levels it felt would be unexceptionable to Canberra. These attitudes resulted in budgetary deprivations and low staff classifications.[25]

This observation by Hope is reminiscent of the managing with inadequate resources that we have encountered before in the history of Australian intelligence. Sigint, the most secret and least understood area, tended to suffer particularly from budget constraints. Between the wars, the Australian Navy had attempted to develop some Sigint capability in the face of the government's reluctance to spend. Decades later, Australia's financial contribution to the joint facilities at Pine Gap is not publicly known, but is reputed to be minimal.[26]

In 1977, following Hope's recommendation that DSD be given greater national recognition, Prime Minister Malcolm Fraser acknowledged the agency's existence publicly for the first time in an address to parliament. He explained that DSD was

> An organisation concerned with radio, radar and other electronic emissions from the standpoint both of the information and the intelligence they can provide and the security of our own Government communications and electronic emissions. It is an agency which serves wider national requirements in response to national priorities ... In close conjunction with the defence force DSD provides a capability which is just as much an integral and essential part of a modern defence posture as a capability in air or ground defence or maritime surveillance. That capability is a sophisticated one for which long periods of training and development are required ... [and] the preservation of secrecy as to the agency's operations is vital.[27]

Despite this public outing, DSD was not given full statutory authority, thanks to resistance from the redoubtable Defence Secretary, Sir Arthur Tange. It was renamed the Defence Signals Directorate (DSD), but reported to a deputy secretary of the Defence Department rather than to the secretary or the minister.[28] That particular change would have to wait for nearly 40 years.

Hope seems to have foreseen the increasing public interest in the activities of intelligence agencies, questions about whether these activities are in the national interest and legal, with growing demands for transparency and accountability. In 1977, he proposed rules to ensure that DSD kept strictly to its foreign intelligence mission, adding that 'it is essential that those responsible for the management and operation of DSD understand that orders are not always orders and that unlawful orders are not orders at all'.[29]

Hope focused sharply on the lawfulness of DSD, noting that while it had been lawfully established for a lawful purpose, 'There is room for doubt whether DSD has been properly part of the defence system, because part of its work is not for defence purposes, or for purposes incidental to defence purposes'.[30]

DSD's report card from the 1977 report was overall positive, but Hope made some strong recommendations for improvements, some of which, such as the move to Canberra, were only implemented years later.

More Hope in the 1980s

While initially commissioned by Whitlam, Hope reported on his royal commission to Whitlam's successor as prime minister, Malcolm Fraser, who was in office from November 1975 to March 1983. When Fraser's successor, Bob Hawke, won office, it seemed appropriate to consider how effectively and thoroughly Hope's reforms had been implemented in the intervening years and to consider whether further initiatives were warranted.

In the report of the Royal Commission on Australia's Security and Intelligence Agencies of 1983–85, DSD again received high marks. By this time, Thompson had been replaced. Tim James was director from the mid-1980s until 1991, when he was replaced by Martin Brady, who would serve as director from 1991 until 1999. Brady would oversee DSD's successful move from Victoria Barracks in Melbourne to the Russell precinct in Canberra.[31]

Hope reported that DSD had been impressive 'by any Australian standards' in 1974 and, a decade later 'is impressive now'. He went on to say that fears that the loss of senior officers who had been with DSD from the beginning would adversely affect the organisation had proved groundless, as those officers had been replaced by others just as good.[32]

As in the first royal commission, Hope was concerned about DSD's independence and status:

> DSD is an outrider organisation of the Defence Department, responsible not directly to the Minister, but to the Department. This situation has advantages as well as disadvantages, but it produces a number of problems if only by reason of the fact that DSD is required to service the whole government, as well as the Defence organization.[33]

Hope sought to ensure that DSD would be able to perform its whole of government role in the future. He nevertheless foresaw difficulties, including one which did not eventuate – that is, a staffing crisis due to too little movement at its upper levels for too long.[34] He also anticipated problems associated with DSD's lack of autonomy in the Defence Department,[35] and a sense of isolation stemming from lack of contact with the wider family of government.[36]

The move to Canberra

As a remedy to its isolation, Hope recommended that the agency move from Melbourne to Canberra. The move of DSD's headquarters from Melbourne to Canberra began in 1992 and was completed in 1993. DSD was the last member of the intelligence community to make this move.[37] As was the case with ASIO, which also moved to Canberra during this period, DSD's relocation to Canberra led to some separations and a surge in recruiting. New recruits commenced work in the new purpose-built facilities alongside DSD's sister defence intelligence agency, DIO, in the Russell precinct, where Defence headquarters was also located. Allan Gyngell and Michael Wesley indicated that as a result of the move, DSD had 'become much more effectively integrated into the rest of the intelligence community as well as the Defence Department'.[38]

The move from Melbourne would also trigger readjustments in the way Defence Force personnel in the Sigint enterprise were managed. Special wireless units had operated from Western Australia during the Second World War because of the value of monitoring enemy high frequency radio waves from there. After the war, the RAAF's 3TU operated from there until the unit was disbanded on 1 December 1991, with its personnel integrated into DSD Canberra.

A decade earlier, Joint Telecommunications Unit Melbourne (JTUM) had been set up at 'K' Block, within Melbourne's historic Victoria Barracks on St Kilda Road, with more than 250 signals operators and specialist communications personnel from all three services. The commanding officer was a lieutenant colonel, or equivalent rank, and rotated between the RAN,

Army and RAAF.[39] The communications personnel were the majority of the workforce (93 perc-ent), supported by a training section of eight and an administration section of nine. The station operated 24 hours a day in shifts of eight hours. In June 1988, the overall personnel figures for JTUM had fallen to a record low of 64 per cent of the establishment. This was due to the services' inability to recruit into the ADF in general. Despite being understaffed, the unit remained operational until it was disbanded after ten years' service, in February 1991, when DSD moved from Melbourne to Canberra and its functions merged with those performed by the staff from 3TU.[40] Similarly, the majority of naval personnel employed at Shoal Bay Receiving Station in Darwin were relocated to Canberra and integrated into the larger ASD workforce, leaving a small crew to manage the site.

The military influence in DSD was tempered by recruitment of more civilians who made their careers there. One such was Ron Bonighton, who joined DSD after graduating from Melbourne University in 1967. He then worked in intelligence production, management and in intelligence liaison in Australia and overseas. In 1988 he was appointed head of DSD's Planning and Coordination Branch. Later that year, he was posted to Washington, DC, as Australian Liaison Officer to the NSA. On return, he was promoted to Deputy Director DSD. After leaving DSD to work in Defence for two years, Bonighton returned to replace Martin Brady as Director in November 1999, remaining in that position until 2002. Commenting on his experience, Bonighton said:

> When I started at DSD, back in the late 1960s, we had a few people still with us who served in Central Bureau. Of course they have all long since retired, but the current generation of Siginters remembers them with great respect. Theirs was a tradition which we have been proud to inherit.[41]

Thereafter, he was promoted and appointed Deputy Secretary Intelligence and Security in the Department of Defence, with oversight of DSD, DIO and the newly created Defence Imagery and Geospatial Organisation (DIGO), later renamed the Australian Geospatial-Intelligence Organisation (AGO).

Bonighton would be replaced by Stephen Merchant, who had previously served in other parts of Defence intelligence. Merchant was Director DSD from April 2002 to September 2006, succeeding Bonighton as Deputy Secretary Intelligence and Security from 2006 to 2011.[42] Later, with Michael L'Estrange, he led the 2017 Independent Intelligence Review which would recommend the establishment of ASD as a statutory authority, as Hope had recommended years earlier.[43] In the meantime, while the intelligence organisational arrangements were consolidated and expanded, other arrangements had contributed to strengthening the ties between the US and Australian intelligence communities, with far-reaching consequences.

Joint US-Australian facilities and operations

Australia's presence alongside US forces in the Vietnam War was symbolic of the deepening relationship with the United States and by the mid-1960s, this relationship had matured noticeably. Australian ties with the United Kingdom remained strong, particularly while the UK maintained a substantial presence in Australia's near north, Britain's Far East. But from 1956 onwards, the relationship with the United States grew steadily, with Australia featuring in US Cold War calculations, and intelligence ties expanding in scope. The most obvious manifestation of this expansion was the construction in the 1960s of joint facilities at Pine Gap in the Northern Territory, Nurrungar at Woomera in South Australia, and the North West Cape in Western Australia.

The era of satellite communications was in its infancy when its use for monitoring Sigint became evident. US intelligence agencies invested in a global network of satellites and monitoring stations that could operate continuously, regardless of time zone. The importance of Australia's geographic location also became clearer than ever.

The 'Agreement with the Government of the United States of America Relating to the Establishment of the Joint Defence Space Research Facility' at Pine Gap in the Northern Territory was signed in Canberra on

12 September 1966.⁴⁴ The Joint Space Research Facility was commissioned in 1967, becoming known officially as the Joint Defence Facility Pine Gap in 1988. Nurrungar was commissioned in 1969 and decommissioned in 1999, with its functions absorbed by Pine Gap.⁴⁵

The Naval Communication Station Harold E. Holt, on the North West Cape of Western Australia, six kilometres north of Exmouth, was originally leased in 1963 and commissioned as a United States facility in 1967. The station was equipped with very low frequency antennae to communicate with submarines in the Indian Ocean.⁴⁶

As mentioned earlier, Desmond Ball had argued it was these arrangements, building on the UKUSA Agreement of 1947–48, rather than the ANZUS Treaty and related defence cooperation agreements, which comprised 'the ties that bind'. Reflecting on developments at the end of the millennium, Ball further noted that these arrangements, coupled with privileged access to the highest level of US defence technology, helped Australia develop its own technical capabilities, making the US relationship 'now, somewhat paradoxically, indispensable to Australia's self-reliance'.⁴⁷

In addition to its importance for intelligence collection for operations, Pine Gap played a pivotal role in managing the strategic relationship between the United States and the Soviet Union. Australia's involvement in US space and Sigint programs was still seen by policymakers in the context of the strategic balance between the superpowers and the verification that facilities such as Pine Gap enabled.⁴⁸

This verification capability was enhanced by a space-based infrared system (SBIRS) which also led to the replacement of the facilities at Nurrungar by SBIRS satellite communications terminals at Pine Gap on 1 October 1999. The SBIRS satellites were reported to be able to detect faint infrared luminosity at the launch of missiles and then track them through their flight.⁴⁹

The original terms of the agreement on the facilities left them principally in the control of US government officials. But with the election of a more independently-minded government under ALP leader Gough Whitlam in late 1972, some adjustments were to be expected.⁵⁰ Yet, contrary to public perceptions that the Whitlam government wanted to close down the facilities, it became clear during his term as prime minister that he was prepared to defend them.[51] Former Labor Party leader and Defence Minister Kim Beazley noted, however:

> The Whitlam government experienced something of a shock when, during the Middle East war of October 1973, North West Cape was used without prior Australian knowledge. Its concerns were resolved by an agreement that didn't hinder the use of the facilities but assured Australia of forewarning. Whitlam had cast some doubt on the continued operation of the various joint facilities when agreements fell due for renewal. He didn't operate on that doubt, but some in the Labor Party harboured concerns that the facilities may have played a role in his eventual dismissal.[52]

Conspiracy theories live on in certain circles.[53] What is clear, though, is that the United States had been concerned about undue exposure of some of the classified aspects of Pine Gap and had expressed its concern through ASIO channels. But, as discussed in John Blaxland's *The Protest Years*, incomplete insights and prejudices and a predisposition to conspiracy, do not amount to a strong case.[54] The longstanding bipartisan support for the joint facilities under a succession of political leaders is a more compelling one.

In the meantime, North West Cape became a joint facility in 1974 under the Whitlam government, and an Australian facility in 1993 under the Hawke government. In July 2008, a treaty gave the United States access to and use of the Australian facility for 25 years. In November 2012, the US Defense Secretary Leon Panetta and Australia's Defence Minister Stephen Smith signed a memorandum of understanding on the establishment of a jointly operated C-band radar space surveillance installation.[55] By this time, the official approach to Pine Gap–related matters had become a little more relaxed.

Much of the demystification of Pine Gap is due to the forensic scholarship of the late intelligence and strategic studies scholar, Professor Desmond Ball. He first brought Pine Gap to notice in 1980 with his groundbreaking work *A Suitable Piece of Real Estate: American Installations in Australia.*[56] While insiders criticised this publication, practitioners came to see it as a useful and publicly available point of reference. Even though some of the details were inaccurate, there was enough information to give the Australian public a helpful picture of how the intelligence arrangements had developed since the signing of the ANZUS Treaty in 1951.

When Bob Hawke became Prime Minister in March 1983, he was 'well aware that threats to remove the joint facilities could have severe electoral consequences'. Kim Beazley observed that ALP leaders

> formed the conviction that the facilities were critical for global stability, vital to the Western alliance, important for the achievement of arms control agreements and, as the decade went by, increasingly of direct relevance to Australia's defence. They were also a ticket to the top table.[57]

ALP leaders recognised the facilities were not only important for the United States, but allowed Australia flexibility in foreign policy initiatives that didn't bring the facilities into contention. Beazley noted Bob Hawke's ministers 'developed a mantra along those lines that informed their debating points as criticisms emerged in public campaigns during their time in office'.[58]

Ball followed up his 1980 book with a range of related publications.[59] Other writers have also examined information available to the public about the facilities at Pine Gap, including David Rosenberg's *Inside Pine Gap*[60] and Tom Gilling's *Project Rainfall*.[61] In his book, Gilling claimed the 'top-secret project known in US intelligence circles as RAINFALL' was a 'CIA listening station at Pine Gap officially called the Joint Defence Space Research Facility, but it had nothing to do with research and was joint in name only: Australians were hired as cooks and janitors, but the spies were all American'.[62] This misrepresents the situation at Pine Gap.

A robust debate has continued over the years between commentators and members of the public who are critical of the joint facilities, on the one hand, and successive Australian governments and some security pundits, on the other. Particularly for those who felt they had been kept in the dark, or who already had concerns about the joint facilities, there is a narrative which taps into the imprint left by Australian Sigint history, including resentment about being still the junior partner, fears of continued dependency and a lack of national self-confidence. Australian governments, while often constrained for security reasons from entering into the specifics of the debate, have sought to respond.

In June 1984, Bob Hawke stated that the joint facilities contributed directly to Australia's national security and that some functions performed there contributed to deterrence, and the monitoring and verification of arms control agreements.

Hawke again addressed parliament in November 1988 on new arrangements at Pine Gap, including the appointment of Australian Defence officials to senior management positions. The deputy chief of the facility would be an Australian. Some years later, Stephen Smith, as Defence Minister, declared that these changes 'confirmed that the joint facilities served Australia's national interest and the depth and substance of our bilateral, strategic, alliance relationship with the United States'.[63]

A central part of the infrastructure behind the facilities has been the US Defense Satellite Communications System (DSCS). Australia has hosted a number of satellite ground stations as part of the DSCS, including the facilities originally at Nurrungar in South Australia and Pine Gap in the Northern Territory, as well as the satellite terminal facilities at Watsonia in Melbourne (where John Blaxland was posted as a young Signals troop commander in the late 1980s and first learnt about Sigint). Ball claimed that Australia had 'insufficient control over the DSCS deployments and operations', and that the Australian government had 'been remiss in failing to inform the public about the extent of Australia's role in the DSCS system and the implications of this involvement'. He argued Australia should take advantage of the DSCS facilities to support its own defence communications requirements.[64]

Technological changes in the 1980s led the United States to alter the character of operations. Until then, the information collected was largely historical. Now, Pine Gap operations could produce information on battlefield situations in real time. Beazley observed that ministers 'couldn't discuss that, but could negotiate a situation in which assertions of full knowledge and concurrence continued to be real'. 'In exchange for certainty and continuation', the renewal agreements for Pine Gap and Nurrungar included the incorporation of Australian personnel on every one of the four shifts and in charge of two of them. The Australian deputy in both facilities was in a position of command in the absence of the American commander. Furthermore, as some of the functions of the facilities served US nuclear war planning, Australia received regular briefings on those functions from the Pentagon. As the Defence Minister, Beazley sought regular written reports on the facilities' activities.[65]

With greater direct involvement, the government sought clarity on the shared arrangements and the issue of 'full knowledge and concurrence.' According to Stephen Smith, 'Full knowledge equates to Australia having a full and detailed understanding of any capability or activity with a presence on Australian territory or making use of Australian assets'. 'Concurrence', he argues, 'means Australia approves the presence of a capability or function in Australia in support of its mutually agreed goals', although 'concurrence does not mean that Australia approves every activity or tasking undertaken'.[66]

As the Cold War ended, the Iraqi invasion of Kuwait in August 1990 triggered a multinational US-led and UN-endorsed coalition to oust Saddam Hussein's forces from the territory in February 1991. During this period, Ball noted 'there was an unprecedented concentration of intelligence collection activity in the Gulf region'. Billions of dollars were spent monitoring political developments and military deployments in the lead-up to the war. Once war commenced, technical intelligence was crucial for both the precision air campaign and preparations for the ground campaign.[67]

Australia's greatly enhanced power and reach as a result of its close alignment with the military of the United States and access to American technological and intelligence capabilities were clear to Australian defence planners. But this was almost the last hurrah of the Cold War; Iraqi forces, after all, used mostly Soviet-sourced equipment and procedures. The absence thereafter of a clear and present threat led to calls for a 'peace dividend' and questioning of the necessity for investment in such high-technology capabilities. Technological breakthroughs in the following years meant that resourcing of the Sigint enterprise would remain largely quarantined. This was based on the view that for a restrained ADF posture to work, decision-makers required advanced intelligence capabilities to ensure that anticipated long lead times for a possible threat to Australia were closely monitored.

Military Sigint components of the three services

While DSD played an increasing role as the nation's Sigint agency, the three services – the Navy, Army and Air Force – maintained their own Sigint and EW units to support forces on military operations, notably the ADF elements deployed for *Konfrontasi* and the Vietnam War. These units inherited operational capabilities and functions from their Second World War predecessors, but because of the secrecy of the work, even the practitioners knew little more than what they learned or experienced themselves while serving in these units.

Army

The legacy of the Army Special Wireless Sections lived on in 101 Wireless Regiment, later renamed 7th Signal Regiment. The operational experience of the respective sub-units, 121 Signal Squadron and 547 Signal Troop, in particular (in Borneo, Malaya, Hong Kong, Singapore and Vietnam), was touched on in chapter 8, along with the role of the RAAF's 3TU in Western Australia.

Air Force

In addition to 3TU, the RAAF had maintained and expanded wartime Elint capabilities developed with Section 22 and No. 201 Flight. Initially, Lincoln bombers were fitted with some of the Elint monitoring equipment, but in 1951, the RAAF acquired Lockheed P2V Neptune maritime surveillance aircraft from the United States. Operated by No. 11 Squadron, the Neptunes were long-range aircraft equipped with a suite of radios and antennae, including an AN/APS-20 search radar (for airborne early warning, anti-submarine, maritime surveillance and weather). They also carried the AN/APX-6 coder or transmitter/responder for the identification of friend or foe aircraft (IFF transponder). In addition, the Neptunes had the ability to drop and monitor sonobuoys (buoys equipped to detect underwater sounds and transmit the findings by radio back to the overflying aircraft).[68] The Neptunes worked alongside UK forces in and around Malaysia and with US forces in the Pacific.

Starting with the acquisition of the P2V Neptunes in 1950, the RAAF maintained that capability until the acquisition of the P-3B Orion aircraft in the late 1960s and later P-3C Orion aircraft in the 1980s. The P-3C came with digital computing systems enabling the display, rapid analysis and transmission of data which enhanced their ability to detect and report on submarines and other vessels on maritime surveillance fights. Kit included Doppler radars (for weather pattern detection and monitoring), magnetic anomaly detectors (used to detect minute variations in the earth's magnetic field such as those from a submarine) and electronic countermeasures radars as well as directional acoustic frequency analysis and recording (DIFAR) sonobuoy indicator sets (for direction finding of targets from the aircraft). The Barra passive directional sonobuoy, developed by the Defence Science and Technology Organisation, was also used, with a range of about 200 kilometres.[69]

Project Peacemate involved the remaining specialised AP-3C and C-130 Hercules aircraft assigned to 42 Wing to better reflect their realignment as a specialist Electronic Warfare Intelligence, Surveillance and Reconnaissance (EWISR) unit. The two remaining Orions (as well as two RAAF C-130H Hercules) were upgraded with advanced electronic warfare systems by L3 in the late 1990s and early 2000s under Project Peacemate. For many years, the EW functionality of the AP-3C (EWs) had not been publicly acknowledged by Defence. In the wake of the retirement of the rest of the AP-3C fleet, however, retaining the two airframes under separate arrangements seemed hard to justify. They were scheduled to be replaced by four Gulfstream G550-based MC-55A Peregrines in 2023.[70]

Today, the RAAF has several platforms capable of conducting the intelligence, surveillance and reconnaissance (ISR) role. Dedicated ISR aircraft include the two AP-3C (EW) aircraft of No. 10 Squadron, the 12 P-8A Poseidon aircraft of No. 11 Squadron (replacing the P-3C suite of maritime surveillance aircraft), and the six E-7A Wedgetail Airborne Early Warning and Control (AEW&C) aircraft of 2 Squadron. The Air Force's air combat fleet of EA-18G Growler (electronic attack), F/A-18F Super Hornets and F-35A Lightning II fighters all have a secondary but substantial ISR role. In the near future, Air Force's ISR capability is due to be significantly expanded with four MC-55A Peregrine electronic warfare aircraft to replace the two AP-3C (EW) aircraft and up to seven MQ-4C Triton uninhabited aerial vehicles with significant electronic warfare capability to complement the P-8A fleet. These functions are increasingly connected, providing the ADF with sophisticated and potent capabilities for ISR operations.[71]

In addition to the specific Army and Air Force assets, the Joint Electronic Warfare Operational Support Unit (JEWOSU) provides EW support to the Australian Navy, Army and Air Force and advice to the Defence Materiel Organisation and Defence Intelligence Group. Based at Defence Park in Edinburgh, South Australia, JEWOSU is considered a critical unit that is part of the RAAF's Information Warfare Wing, contributing significantly to the mission effectiveness and survivability of ADF personnel deployed on operations in aircraft, ships and submarines around the world.[72]

Another RAAF capability, to be known as Distributed Ground Station – Australia, comes under Project AIR 3503 and is due soon. It will be housed with 83 Squadron and involve centralised ISR processing, exploitation, and dissemination. This is a first-of-type capability for the RAAF, developed with a company employing veterans with years of service and knowledge of ADF ICT environments. Distributed Ground Station – Australia will be responsible for the analysis of data collected from the various Air Force ISR platforms. It is also to have access to national intelligence material using advanced computer systems, to rapidly fuse collected information and provide decision-makers with enhanced situational awareness.[73]

Navy

One of the most important and yet least known of the services specialised communications units belongs to the Navy. Inheritor of the legacy of FRUMEL, the Royal Australian Navy Tactical Electronic Warfare Support Squadron (RANTEWSS) is based at HMAS *Albatross* in the southern New South Wales coastal town of Nowra. RANTEWSS was established by the 'inimitable' polyglot, the Mauritian Creole, Russian, Spanish, German and French speaking Commander James Armstrong, a brilliant English boffin who shocked his colleagues when he announced one day that his godfather was Donald Maclean, the notorious Soviet spy.[74] Armstrong had served in the Royal Navy for 35 years, earning an OBE, before transferring to the RAN in 1979, earning an OAM for his work in 1987. Arriving in Australia when the movie *Moonraker* was released, to his staff he was a Bond-like figure. After he finally left the service, the new building housing the multi-million-dollar facility for RANTEWSS was named the James Armstrong Building. Whilst Armstrong was reportedly acutely embarrassed by this, no one doubted it was entirely appropriate to name the new facility after him.[75]

RANTEWSS provides both operational radar and communications electronic support to maritime commanders. Its functions include the provision of electronic intelligence (Elint) in support of anti-ship missile defence (ASMD). Separately, electronic protection (EP) for the Navy's vessels is generally achieved through the integration of active and passive countermeasures into capabilities, systems or processes. Examples include frequency agility in a radio, spectrum management processes, emission control procedures, and wartime reserve modes.[76]

Electronic eavesdropping is also technically feasible from submarines and Australian submarine operations commenced once the fleet of six UK designed and built Oberon class submarines was purchased in the late 1960s began operating. In a rare exposé, journalist Geoffrey Barker spoke with former submariners who allowed a glimpse of their work. Barker notes that the patrols may have started at the request of the United States, particularly because the larger US nuclear-powered submarines were less suited to close-in intelligence-collection patrols in relatively shallow coastal waters. British Oberon class submarines (O-boats) were suitable, he notes, but busy conducting electronic surveillance, acoustic signature recording and underwater looks in Arctic waters. Australian O-boats, therefore, were ideally suited to target Cam Ranh Bay and the South China Sea.[77]

Designed to accommodate a crew of 62, including five officers, two of these submarines were designated 'mystery boats' and reportedly equipped with upward-looking cameras, hydrophones and other sensors. They sometimes deployed with more than 70 on board – including trainees, specialist civilians operating intelligence-collection equipment and specialist linguists, fluent in Russian and other languages, able to provide early warning of detection. Their other task was to record the acoustic signatures of Soviet surface ships and submarines. The O-boat would lie submerged and silent, passive sonar hydrophones switched on, recording the sounds of passing ships and submarines.[78]

Initially, however, the program apparently lacked strong political or even Navy support, and 'a lot was done by blokes on an ad hoc basis'. Rear Admiral Peter Clarke was a RN submariner who transferred to the RAN. As Barker recalls, the

> Defence Science and Technology Organisation and the Defence Signals Directorate worked on bits and pieces and so did some navy boffins. It was good stuff, done on a wing and a prayer. They did outstanding work.[79]

Barker cites an O-boat commander as saying that the secret patrols admitted Australia to one of the biggest big games in the Cold War and demonstrated the capacity of the Australian submarine arm at a time of high international tension.[80]

As Defence Minister, Kim Beazley in 1986 was eager to convince the government of the need to replace the Oberon submarines. He took with him submariners to explain their functions and utility to the Prime Minister, Bob Hawke. What became clear was that the Oberons had been effective at monitoring Soviet and Chinese Communist fleet activity in the South China and East China seas. While their specifications remained secret, their capability persuaded the Hawke cabinet. Six Swedish-designed submarines, designated the Collins class, were built in Australia with a range of sophisticated and sensitive sensors. Subsequent governments likewise have agreed on the capability's usefulness, despite the seesawing decision-making over a replacement for the Collins class submarines.[81]

Onshore, the RAN participated in a system developed by the US Navy known as the Ocean Surveillance Information System (OSIS). This was described as 'a synergistic blend of intelligence and cryptologic personnel that provided tailored, fused, all source intelligence to operating units'.[82] Installed in Australia near the turn of the millennium, the various upgrades have led to an automated system that 'receives, processes and disseminates timely, all-source surveillance information on fixed and mobile targets of interest afloat and ashore'. The OSIS Baseline Upgrade (OBU) was designed as an integrated element of naval command and control mechanisms which, operated from a sensitive compartmented information facility (SCIF), enabled the management and dissemination of data at multiple security levels. The next version, known as the OSIS Baseline Upgrade (OBU) Evolutionary Development (OED), sited contractor maintenance support technicians alongside to assist in system upgrades and maintenance. This was a system that provided situational awareness compatibility with the USN Navy, Royal Navy, Japanese Maritime Self-Defense Force and Republic of Korea Navy.[83]

Servicemen and women posted to RANTEWSS, 7th Signal Regiment, and the RAAF's EW-related units and squadrons were also posted to work in DSD/ASD. Gradually, the original leadership, with wartime experience, was replaced by civilian leaders, like the above-mentioned Ron Bonighton. DSB, then DSD and later ASD, became civilian-led and thus closer to the model of Britain's GCHQ than the United States' NSA, which remained part of the US Department of Defense with a uniformed military commander. The UK influence remained, but the three armed services of the ADF maintain their own intelligence collection and reporting capabilities for good reason. And these worked best when closely aligned and connected with their counterpart national agencies which are integral parts of the Australian intelligence community. Nowadays, that coordination is conducted under the auspices of Headquarters Joint Operations Command at Bungendore, a few kilometres east of Canberra.

Joint operations in the Gulf War: a niche role for Australia

In 1990–91, Australia had a small adjunct role to play in the US-led intelligence collection efforts in support of the international coalition to throw back the Iraqi invasion of Kuwait. In this conflict, reports indicated 'spy satellite intercepts of Iraqi military communications gave US generals a capability their predecessors could only dream about – the ability to track just about every important military action Iraq undertook'.[84] In addition, Sigint was credited with tracking Iraqi Scud missiles. Aware of advanced US Sigint capabilities, the Iraqis had sought to hardwire cable communications, but this also hampered their own operations.[85] Key to this capability, according to Ball, was a series of geostationary Sigint satellites called Advanced Orion, which intercepted microwave signals from the far corners of the globe. These were later supplemented by Magnum Orion satellites (with a parabolic reflector span reportedly of about 100 metres).[86]

Australia made a niche contribution to operations in the Gulf, with a specialist intelligence support team working as integrated officers at US headquarters. While Sigint support was integral to the mission, the Australian team that deployed consisted of imagery analysts rather than Sigint specialists, and included two RAAF officers, flight lieutenants Margaret Larkin and Chris O'Brien. There was also one army warrant officer, Gary Shepherd, and two army officers on the team, Major Gary Hogan and Captain David Gillian. Hogan and Gillian would go on to become brigadiers. Another, Army Reserve Captain Greg Moriarty, would go on to become Australia's ambassador to Tehran and Indonesia and subsequently Secretary of the Department of Defence.[87] Another, Pat Scanlan, was embedded on US JSTARS aircraft. In addition, Australia's command and control arrangements for offshore deployments were tested and refined. These included arrangements for collection and sharing of intelligence. Eminent Australian military historian Professor David

Horner observed that 'the Gulf crisis was the first major test of the new command arrangements', and suggested that it would be 'an ideal model for future ADF activity'.[88]

The arrival of the internet

The Gulf War predated the era of the widespread use of the internet, but the concept of the World Wide Web was developed in 1989. This drew on the pioneering computer-to-computer communications network developed in the late 1960s by the US Defense Advanced Research Projects Agency (DARPA), known as ARPANET. The development and application of digital communications technology, premised on the use of a binary mechanism counting ones and zeroes, or on-off mechanisms, gradually enabled the acceleration of innovations in computer-to-computer exchanges of packets of data (known as packet switching).

By the mid-1970s the transmission control protocols or internet protocols (TCPs/IPs) and file transfer protocols (FTPs) were established, setting a common standard for digital communications. The development and production of optical fibre networks in the 1980s and 1990s facilitated the rapid uptake of internet services in the mid-1990s.

In the meantime, analogue mobile cell phones were proliferating and then replaced by digital ones that could connect with and communicate across the terrestrial landline telephone cable network and, in time, via satellite.

The combination of the internet and the World Wide Web, along with rapidly advancing mobile cell phone technology and the switch from analogue to digital, transformed communications globally. The trend accelerated from the mid-1990s onwards. This enabled not just government, but major corporations and small business enterprises to develop their own communications networks and to operate remotely and in a global network. This generated an explosion in demand and supply for increasingly mobile and compact communications technology.

For the traditional Sigint agencies to monitor such communications required substantial research and significant reinvestment and reskilling. This would be a gradual process over decades, but one that would lead to an organisational, technological and operational transformation for DSD.

Beyond Sigint – DSD and the information revolution

An information revolution was underway, with society becoming highly dependent on electronic information and its technology. As this digital era began, the Director DSD, Martin Brady, a DSD 'outsider', recognised digitisation, now commonly referred to as computerisation, was a sea change which demanded an innovative response.

Brady commissioned a major study of the implications of digitisation for information warfare. Ian Dudgeon, a senior official in the Department of Foreign Affairs and Trade, was seconded to DSD in 1995 as Special Adviser to undertake this study. His report, published in February 1997, identified issues of much wider national and international importance. The report's main recommendation was the establishment of a coordinating structure involving government and the private sector to implement policy for protection of the national information infrastructure, now generally known as the internet. That recommendation was endorsed by the Secretaries' Committee on National Security in August 1997. An interdepartmental committee, chaired by the Attorney-General's Department, proposed a national government and industry framework, which included an expanded role for DSD in protecting Australia's information infrastructure. That framework, with some amendments, still exists.[89]

In addition, Dudgeon's report identified that the growing web dependence also increased the vulnerability of the national information infrastructure, which affected the national security community, state and federal governments as well as the private sector. This infrastructure included information storage, processing and transportation, the people who managed and serviced that network, and the information itself. Industries at the core of this infrastructure included banking, telecommunications, transport, energy, water supply, information and emergency services, the stock exchange and air traffic management.

Some aspects of information required iron-clad protection. First, information had to be available, which meant it had to be stored, processed and forwarded when needed. Second, the integrity of the information and supporting infrastructure had to be assured. Third, the confidentiality of the information had to be maintained. Fourth, a way to verify the user was required. Fifth, ways to prevent denial of responsibility, or 'non-repudiation', were needed.[90]

Threats to this national information infrastructure could be accidental or deliberate. They could involve software and hardware destruction. They could also simply be a denial of service. There were vulnerabilities throughout the national information infrastructure from within Australia, and from overseas via connectivity to the global information infrastructure, and the need for a response was clear. The report predicted correctly that computer hacking was 'expected to increase approximately in proportion to the increase in computer literacy and Internet connectivity'.[91]

The response required of DSD, which Martin and his successors directed, went beyond traditional Sigint. It also meant that Sigint practitioners, long accustomed to secure compartmented facilities, away from prying eyes, had to leave their comfort zone. Fortuitously, with the move from Melbourne to Canberra came some turnover of DSD staff, as many older staff, in particular, chose to retire rather than relocate. Those who made the move to Canberra were deeply committed to DSD and determined to rise to the new challenges. Mike Burgess and Rachael Noble, DSD officials who went on to head the organisation, were part of that transformation. It was time for DSD to take on a broader role in national security. This meant a paradigm shift from being a secretive organisation focused on support of higher level government decision-makers to being more open and accessible, providing a cyber service to the community.

In the meantime, the above-mentioned vulnerabilities were a cause of widespread concern. Security pundits and intelligence practitioners contemplated the potential for a cyberwar. Discussion covered the pace of change of the information revolution, the 'cybersphere', the impact on military operations, the foundations of information assurance, encryption challenges and hacking, as well as the vulnerabilities of an increasingly web enabled and web dependent society.[92] Yet, despite Dudgeon's key recommendation, there was still no structure for the coordination and implementation of a national policy to protect and assure the continued operation of critical elements of the national information infrastructure.

One early initiative was the creation of Computer Emergency Response Teams (CERTs) to respond to communications and computer systems breakdowns and outages. They became more important as government agencies, industries, schools and community groups turned to computers. Education and awareness programs were an integral part of computer security, or cyber security, awareness.

The concept of information operations (IO) was another attempt to keep pace as computerisation and the internet transformed social interactions as well as business. At a symposium in November 1997, the then Special Adviser to DSD advocated a national approach to the challenges and opportunities arising from the transformation of telecommunications, information services, banking and finance, transport and distribution, energy and government services.[93] The need for a national policy on the protection of critical information infrastructure became pressing. This included physical security requirements for computer systems, as well as the 'cyber security' for computer systems gateways and networks. The need for effective liaison and coordination between federal and state government agencies and industry, as well as foreign counterpart agencies, was becoming evident.[94] The impetus for a national cyber security mechanism was gaining momentum, leading DSD/ASD to invest in cyber security operations and to the establishment of the Australian Cyber Security Centre.

Meanwhile political events were drawing attention to the use, or misuse, of Sigint, not least in East Timor.

East Timor

Sometime before the advent of mobile phones and the internet, and two years after Australia's participation in the Vietnam War had ended, the future of East Timor became a major issue for Australia. The Cold War was still at its height when Portugal experienced a leftist revolution and quickly divested itself of its colonies, including East Timor. An independence group emerged with sympathies towards Communist China. Against this background and with stable relations with Indonesia a top priority, the Whitlam government considered that East Timor was not viable politically or economically as an independent state. Giving evidence on East Timor, Whitlam told a Senate committee that this had also been the view of the Menzies government in the 1960s. He also told the committee that the only way to stop Indonesia invading East Timor would have been to wage war in the midst of his government's confrontation with the Senate over its refusal to pass the 1975 budget.[95] In other words, such a scenario was never seriously contemplated.

On 16 October 1975, during the invasion of Portuguese Timor by Indonesia, five Australian journalists were killed in Balibo. There were allegations of an Australian government cover-up of these deaths, suggesting intelligence and diplomatic reporting would have tipped off Australian authorities as to what was going to happen there.[96]

There were also specific allegations against DSD. Justice Hope addressed these allegations in the 1984 report of the Royal Commission into Australia's Security and Intelligence Agencies. Mr Ken Fry, at the time the member for Fraser in the House of Representatives, said in evidence before Hope on 7 May 1984 that he had

> been given information that the intelligence community knew, by intercepting intelligence message traffic between Timor and Jakarta, that those journalists (the five Australian journalists who were killed in East Timor on 16 October 1975) were going to be treated as combatants, and they knew that the Indonesians were about to invade the area where the journalists were, and they did not do a damn thing to protect them.[97]

On investigating this, Hope found that DSD had no intelligence that the journalists were going to be treated as combatants and did not know where the journalists were at the relevant time. 'There is no evidence of any improper behaviour on the part of DSD in respect of any of these matters', he declared.[98] Reports of evidence of Australian foreknowledge of a likely Indonesian attack on Balibo, of Australian intelligence bearing 'silent witness',[99] have bubbled up since then, suggesting Australian officials were aware of Indonesia's broad intentions for East Timor and were eager to prioritise good relations with Indonesia.[100] Australia's efforts to maintain such good relations are longstanding. Nonetheless, despite claims a warning from Australia might have saved the lives of the investigating media team,[101] repeated investigations have fallen short of presenting hard evidence that Sigint gave actionable foreknowledge that might have saved the journalists' lives.

By 1998, the situation of East Timor was quite different. The Cold War was over, Vietnam was a member of ASEAN, and the Asian financial crisis had struck with full force on Indonesia's economy. Above all, Indonesia had failed to win over the East Timorese, whose leaders had, in the meantime, been highly effective advocates for self-determination.

By late 1998, most analysts in the Australian intelligence community had concluded that there were several pro-integration militia groups with access to Indonesian military weapons which they would use against East Timorese who were pro-independence. With much of the information coming from 'sensitive sources', Desmond Ball noted Australia's difficulty in engaging with Indonesian counterparts 'at a time when cordial communications were considered imperative'.[102]

On 27 January 1999, Indonesia's mercurial President B.J. Habibie announced that the status of East Timor would be determined by an act of self-determination. Australia's long-held preference was for East Timor to remain part of Indonesia. Having written to Habibie in late 1998 advocating initiatives towards limited self-government at least, Australia's Prime Minister, John Howard, was obliged to support Habibie and a self-determination ballot.

As the security situation deteriorated in East Timor and a United Nations mandate was established for the UN Assistance Mission to East Timor (UNAMET), the then secretary of the Department of Defence, Paul Barratt, pointed to the fragility of sensitive information that can be divulged:

> The leaking of sensitive information about our capacity to monitor
> interactions between elements of the Indonesian military (TNI)
> and pro-Jakarta militias in East Timor denied us a valuable source of
> information on the eve of our despatching unarmed police to serve
> as monitors during the 1999 plebiscite, because the Indonesians
> plugged the gap in their security.[103]

This experience demonstrated the truism that the secret of success in the espionage business lies in keeping one's successes secret.

Media reports suggested 'two P-3C Orion Sigint aircraft began regular collection flights around Timor' and new computer equipment arrived from the United States for improved 'intelligence processing and dissemination'.[104] Ball claimed that 'at different times during the year, the US agreed to the realignment of one of its geostationary SIGINT satellites controlled from Pine Gap in central Australia, to provide coverage of signals from the VHF up to the super-high frequency (SHF) band (i.e., from walkie-talkies to satphones)'.[105]

Several conversations redolent of collusion were reportedly held between Indonesian officials and militia leaders.[106] Exasperated by the reluctance to publicly divulge compelling evidence, Ball declared

> the need to protect sources and methods is never absolute, and
> the injunction against actions which might compromise SIGINT
> operations is really not so compelling. Secrecy may be critical where
> cryptanalytical activities are involved, but the great volume of DSD
> intercepts during the Timor crisis involved unencrypted radio
> and satphone conversations. Yet it is no secret that these can be
> monitored by anyone with appropriate receivers.[107]

A number of informed outsiders shared Ball's frustration over the reluctance to divulge even such low-grade Sigint-derived information. But in this instance, despite the ramifications for those affected, the principle was still strictly upheld.

After a vote in favour of independence on 4 September 1999, violence broke out with anti-independence militia backed by the Indonesia military.

Within a few days, reports of widespread human rights violations led to a UN Security Council resolution mandating an Australian-led intervention. On 20 September, an Australian-led peacekeeping force (INTERFET) arrived, including signals squadron teams which were deployed with elements of the Australian land forces under Headquarters 3rd Brigade.[108] These teams provided timely and valuable reports,[109] and the 2004 Flood Report of the Inquiry into Australian Intelligence agencies notes that DSD's experience during the ADF's East Timor deployment in 1999 had been drawn on to develop enhanced support arrangements for ADF operations.[110]

Shortly after the East Timor intervention, allegations about the deaths of the five Australian journalists in 1975 resurfaced in a book suggesting a clandestine system of 'deceit and cover-ups' and challenging the official narrative. The book asked a series of loaded questions: 'Did highly-placed Australians secretly "sign off" on Indonesia's plan to invade its neighbour? Did they know that the newsmen were targets? Did they leave these young men at the mercy of the Indonesian Army?'[111]

The matter was closely scrutinised and the records re-examined once again as part of a detailed internal review. The review found that circumstantial evidence was assumed to point to a cover-up. That evidence, however, would be more accurately described as incomplete and untimely, although well intended, coverage of the situation by the intelligence community. Coverage was indeed incomplete. Translators translated material not in real time and analysts pieced together snippets thereafter. In reality, few could have predicted such a macabre turn of events. Pieces that may have looked like conclusive evidence in hindsight rarely did so beforehand. The Inspector General of Intelligence and Security, Mr Bill Blick, who investigated the allegations, concluded in May 2003 that they did not stand up to objective scrutiny.[112] A fuller account has yet to be made public.

Apart from the controversies over what was known or not known in advance, the operational experience in East Timor saw some notable successes. The tactical intelligence support provided to the deployed forces was similar to that provided to Australia's forces in Vietnam. This time, however, it was with a higher level of support covering Geoint (mapping and imagery), Humint (patrol and human intelligence collection team reports) and Sigint. The Sigint support was mainly provided through deployable elements from 72 Electronic Warfare Squadron, a part of the Australian Army's 7th Signal Regiment, which had a direct connection through lineage with the work of 547 Signal Troop in Vietnam and its roots in the special wireless units of the Second World War.[113]

The signal squadron teams were deployed with elements of the 3rd Brigade – the Australian Army formation (along with New Zealand, British, Canadian and other coalition partners) which after commencing in Dili, operated along the East Timor–Indonesia border. The squadron commander and his operations officer coordinated technical advice, while analysts provided reports that added to the force's understanding of the situation and supported the commanders' objectives. This unit worked closely with the brigade headquarters and unit staffs, coordinating tasking and requirements in support of the mission. It soon became clear that more integrated training was required at battalion level, since the light electronic warfare teams and the command elements were not conversant with battalion procedures as they had not worked together regularly on exercises for some time.[114]

Australian forces subsequently deployed to Afghanistan in 2001, Iraq in 2003, Solomon Islands in 2003 and elsewhere. The positive experience of tactical level intelligence support in East Timor created an expectation that deployed forces would receive intelligence reports that would directly contribute to operations – that is, they would receive timely, accurate and actionable intelligence.

Support to military operations in the Middle East

In the 1990s, ADF operations had gathered momentum, with contingents on peacekeeping missions in Namibia, Cambodia, Mozambique, the Afghanistan–Pakistan border, Somalia, Rwanda, Bougainville, and East Timor. By the late 1990s, intelligence agencies were refining their support to the ADF on such deployments. The agencies contributed as collectors and assessors, often working together and with allied counterparts. DSD liaison officers found themselves deployed on a range of these operations, and their 'kit' included a suitcase-sized deployed Sigint support facility (DSSF). Sometimes agency representatives were forward-deployed to ADF headquarters to provide maximum intelligence and operational support. One such example was the support provided for the INTERFET mission in 1999–2000.

In late 1999 and 2000, the operational experience in East Timor looked to be the most significant practitioners would see for a generation. But the events of 9/11 changed that. In short order, Australian troops deployed to Afghanistan alongside a US-led international coalition. Few realised that an initial deployment of a few months would morph into a conflict spanning two decades, with contingents from Australia's Navy, Army and Air Force committed on operations alongside US and other coalition partners, not just in Afghanistan but in Iraq and across the Middle East. Included in the mix was an important, but largely unheralded, cohort of intelligence practitioners from the three services, alongside civilian counterparts from agencies in the Australian intelligence community, who were there to provide detailed, timely and actionable intelligence support.

Increasingly through the 1990s, the challenge had been not just monitoring how nation states and their security apparatus exploited the new communications technology. It also involved remaining abreast of terrorist groups as they became better organised and more deadly. These groups were savvy adopters of the new technology, devising ways to turn mobile phone devices into timers for explosive devices, establishing networks through subscriber messaging services (SMS) and websites. This presented additional challenges for intelligence collection, processing and dissemination.

In his chapter in *Niche Wars,* Colonel Mick Lehmann outlines some of the practical workings of intelligence in the Afghanistan campaign. He notes the importance of actionable intelligence and the essential nature of the Five Eyes relationship. He explains how the intelligence support teams developed a level of proficiency that facilitated many tactical and operational successes, but also acknowledged some major failings.[115]

Tactically, intelligence contributed directly to operations. In the

Special Operations Task Group, for instance, each four- to five-man team is said to have included operators and a signaller (Electronic Warfare Operator or Signals Intelligence Operator). One operator described being taught how to monitor communications and to find technical or electronic devices that transmit signals. The Sigint operator, nicknamed a 'Bear', would operate a communications monitoring device to analyse data from enemy communications and provide threat detection and intelligence to the troops on the ground. This involved carrying equipment to establish line of bearing to phones and 'ICOM' two-way radios used by the insurgents. The techniques echoed the DF techniques used by their predecessors a generation earlier in the Vietnam War, albeit in a more compact and portable form. Sometimes, the 'Bear' would be placed with an overwatch patrol to monitor communications and pass on information to the supported patrols.[116]

Those captured or killed, we are told, would be subject to a sensitive site exploitation procedure, whereby contents of pockets are laid out and electronic devices identified and 'processed', and their phone and other communications devices would be examined, their details reported back to base, and the database scanned for matches. The base included a 'fusion and targeting cell,' where multiple-sourced databases could be interrogated, not only to prepare a 'package' outlining the details of a designated target, but to seek information for those out on patrol.[117] A 'touchdown', we are told, occurred when the base reported back, confirming the identity of the phone as one used by insurgents.[118] An unprecedented level of sophisticated tactical support was available.

Air Force Heron Unattended Aerial Vehicles were also deployed in Afghanistan in support of ADF contingents. Fitted with electro-optical, infra-red and radio frequency spectrum monitoring equipment, the Herons provided significant and timely support.[119] They were particularly in demand by forces out on patrols, and were considered an integral part of the spectrum of information sources used for their protection.[120] The Herons provided over 27 000 mission hours of high-resolution ISR support. In essence, the Herons provided information the soldiers needed to fight and to keep their people safe.[121]

As Australia's military commitment in Afghanistan wound back, Operation Okra was launched in September 2014. This mission, comprising hundreds of mostly Air Force and Army personnel, was in response to the advances in Syria and Iraq made by the so-called Islamic State in the Levant (ISIL) or the Islamic State in Syria (ISIS), otherwise known as Daesh. RAAF E-7A Wedgetail AEW&C aircraft contributed to the mission to 'support the Iraqi Security Forces in their ongoing fight against terrorism in Iraq'.[122]

Despite access to national and coalition technical means to deliver surprisingly accurate and timely insights, tactical successes were not matched by strategic breakthroughs. Fused intelligence, incorporating insights from Humint, Geoint and Sigint to provide a more complete picture of the tactical situation was a capability broadly acknowledged and widely appreciated by commanders of the deployed forces. The level, volume and fidelity of actionable intelligence supplied to Australian troops in Afghanistan surpassed anything supplied in earlier conflicts. Intelligence practitioners felt they were making a real difference in helping keep own forces safe and in ensuring operational successes. There was deep frustration, however, at not being able to contribute to a lasting legacy of coalition success in Afghanistan.[123]

This frustration indicates the limitations of excellent intelligence support, particularly if those arrangements are in support of a flawed strategy. Criticism of the strategy of Australia's involvement in the Middle East in the first two decades of the twenty-first century is widespread.[124]

Notwithstanding these challenges, as we shall see in the next chapter, there are positive developments to report concerning the reform of the intelligence arrangements at the national level, and the transformation of ASD from a secretive to a more accessible national institution with a mandate to enhance cyber security.

10
LEGISLATIVE REFORM & THE COMING OF CYBER

In the first two decades of the new millennium, the surge in ADF operations saw a corresponding increase in Sigint support. But another transformation was turning the once secretive national Sigint agency into a more accountable, visible and better understood entity. This chapter considers the leaks, revelations and reforms that transpired at a time of social transformation associated with the digital or information revolution, sometimes described as the Fourth Industrial Revolution.

Even a decade in to the new millennium, few would have guessed how crucial the Sigint heritage of ASD would be to managing the security challenges of a digital and web-enabled society. Perhaps the most significant reform was the *Intelligence Services Act* of 2001. Yet to understand its importance, one needs to look back further, to the time of Samuels and Codd.

Further inquiry, revelations and legislative reform

Some years before the intervention in East Timor in 1999 and the wars in Afghanistan and Iraq from 2001 onwards, and shortly after the move to Canberra, the Australian government had commissioned another inquiry under Justice Gordon Samuels and Michael Codd. Submitting their report in March 1995, their work would eventually lead to the *Intelligence Services Act 2001*, which, with subsequent amendments, regulates ASD to this day. The Act implemented recommendations of the Samuels and Codd judicial inquiry into problems that had been reported within ASIS, including the recommendation that Australia's foreign intelligence agency be placed on a statutory footing.¹ After some deliberation, the government decided the Act should also cover DSD, as it then was.² This decision was logical, as DSD, like ASIS, was an external intelligence collection agency, but there was some additional background to DSD's inclusion. The Act may have come into effect earlier, but other events led to a delay.

On 23 May 1999, a controversial cover story was aired on the *Sunday Nine Network TV* program. It featured allegations about the existence and operations of Echelon, a computer-automated satellite surveillance network under the UKUSA alliance. Specifically, it was alleged that Britain had used a similar agency in Canada to undertake surveillance for domestic political purposes. It was also alleged that Echelon intelligence had been used by the United States to obtain commercial advantages for domestic companies negotiating for contracts with Indonesia. The implication was that the network could be, or had been, used by the larger parties to gain information for their interests, potentially to the detriment of those of Australia.³ Some reports alleged that Australia became part of the Echelon program during the prime ministership of Malcolm Fraser. According to this view, the capability represented 'the next generation of surveillance'; one that 'would extend Pine Gap's capacity to listen in on domestic communications around the world'.⁴

Concern was expressed about the program, particularly the focus on commercial interests and the potential for breaches of privacy for Australian citizens. It raised a public interest issue regarding DSD and the fact that, prior to the enactment of the *Intelligence Services Act* in 2001, at least, 'its intrusive surveillance powers [were] not restrained by an act of parliament' and that its operations were not subject to parliamentary scrutiny.⁵

With the revelations in the headlines, in May 1999, the DSD director, Martin Brady, made a groundbreaking disclosure; he recognised formally what was by then widely known: that DSD cooperated with counterpart Sigint organisations overseas, including the USA's NSA, New Zealand's GCSB, the UK's GCHQ and Canada's CSE, under the UKUSA relationship.⁶

Further revelations in May 2001 in the European parliament attracted media attention,⁷ and led it to commission a study into 'The Echelon Affair'. In the end, the study, which emerged in 2014, observed:

During the second half of the 1990s press and media reports revealed the existence of the Echelon network. This system for intercepting private and economic communications was developed and managed by the states that had signed the UKUSA and was characterised by its powers and the range of communications targeted: surveillance was directed against not only military organisations and installations but also governments, international organisations and companies throughout the world.[8]

These revelations again underscored the need for a more transparent and accountable approach to managing Sigint. While legislation had been under consideration for several years, this added impetus towards a clearer legislative basis for DSD's actions.

Mindful of public concerns about lack of transparency on the one hand and surveillance on the other, the *Intelligence Services Act* made DSD's functions public in legislation for the first time, including the fact that DSD's intelligence gathering occurs outside Australia.[9] The Act also introduced a system of oversight and accountability. The director of DSD (now the director-general of ASD) and director-general of ASIS were required to obtain ministerial authorisation before undertaking intelligence collection or other activities affecting Australian citizens.[10] The Inspector-General of Intelligence and Security (IGIS) was given oversight powers in relation to ministerial authorisations and directions under this Act[11] and now the attorney-general must agree to them. The Act obliges the agencies to respect Australians' rights to privacy.[12] Ministers must provide written rules to ensure protection of these rights and the attorney-general and the IGIS are consulted in the development of the rules.[13]

The Act also established a parliamentary committee to oversee the administration and expenditure of the intelligence agencies involved in intelligence collection, namely DSD as well as ASIS and ASIO.[14] Initially, this was the Parliamentary Joint Committee on ASIO, ASIS and DSD (PJCAAD), but after the enactment of the *Intelligence Services Act* in 2001, the PJCAAD became the Parliamentary Joint Committee on Intelligence and Security (PJCIS), extending its coverage to include DIO, ONI and AGO. This expansion was matched by a proliferation of satellite and digital communications which added to DSD's Sigint collection facilities. If there were to be increased collection, it became increasingly apparent that there was a need for a commensurate increase in accountability and reporting. These mechanisms, the PJCIS, the IGIS and the legislative provisions of the *Intelligence Services Act,* went some way towards keeping alive Justice Hope's spirit of reform.

The reports by Philip Flood and by Bill Blick mentioned earlier were published in the years following the 1999 East Timor intervention and the so-called Global War on Terror in 2001. Much of this concerned how the military applied or misapplied intelligence, including Sigint.

The Darwin and Geraldton stations

In the meantime, DSD's facilities – which had been in Melbourne, with outstations in Darwin, Canberra and near Perth – were upgraded and reoriented, as a result of changing priorities, reallocation of resources, and new technologies which presented challenges and opportunities. This work was undertaken as the transition from reliance on analogue communications systems to digital was in full swing, setting the scene for a revolution in communications technology and for cyber that would follow.

From the earliest times, technological breakthroughs like the printing press, the discovery of electricity, the invention of the telegraph and the wireless radio have required intelligence to rise to the challenges and opportunities they present. The first two decades of the new millennium have been no exception. With the proliferation of satellite communications systems, the internet age, and the expansion of digital communications services, the reliance of the Australian government, as well as industry and individuals, on internet services has increased exponentially, as have the accompanying security risks.

Once satellite communications networks started to proliferate, it became evident that there would be value in setting up a ground station in Western Australia, along with the Sigint facilities at Shoal Bay near Darwin, co-located with the existing Naval HF receiving station which had been established in 1975 after being moved from HMAS *Coonawarra,* closer into the city of Darwin.[15] Desmond Ball described the Shoal Bay facilities as including two different signals interception systems. One he described as a circular antenna array used for interception, monitoring, direction finding and analysis of radio signals in the high frequency band. This monitored regional military communications systems and was primarily managed, maintained and operated by naval personnel. The second system, he said, was concerned with the interception of regional satellite communications systems, which was a collaborative effort between Navy and DSD until ASD assumed control in 2002.[16]

Australia also established the Australian Defence Satellite Communications Station (ADSCS) in 1993 at Kojarena, 30 kilometres east of Geraldton. Kojarena also faced public scrutiny when it was named as one of five satellite monitoring stations that formed Echelon – a system that journalists reported as intercepting tens of thousands of emails, telephone calls and faxes across the planet using a 'keyword' search program, or 'dictionary'.[17]

The Kojarena facility expanded in 2007, reportedly to include a US military mobile phone satellite communications system for its troops in the Middle East and Asia. The Kojarena location is said to allow for the interception of more than 100 geostationary satellites in orbit, 'including those controlled by Russia, China and Pakistan'. Reports suggested that similar stations in the United Kingdom, the United States and New Zealand ensure the world's communications satellites are covered by intelligence-sharing arrangements that form part of the UKUSA Agreement.[18]

By 2020, the government had upgraded the satellite ground station. This facility is reported as 'jointly operated by ASD and the United States as part of a SATCOM partnership'.[19] The ground station is reported to provide 'an anchoring capability to Wideband Global Satellite Communications satellites primarily located over the Indian Ocean'. The program was stated to include two military hosted payloads on commercial satellites, a military satellite as part of the United States Wideband Global Satellite-Communications system, ground infrastructure across Australia and deployed terminals with accompanying network management capabilities. This was seen to offer 'real-time operational and logistical information which is essential for command and control of deployed forces'.[20]

Australia paid for one of ten of the wideband global satellites in return for global access to the entire constellation's capabilities, giving the ADF greatly expanded communications functions.[21]

In the 1990s and 2020s, Kojarena demonstrated Australia's commitment to development and refinement of high-end Sigint capabilities. Collaboration with UKUSA partners has been seen as critical for access to the technology enabling this capability. The pace of technological change has reinforced the value of this arrangement.

Leaks and controversies

Claims about the Echelon system at the end of the millennium were controversial enough, but there would be further leaks and scandals, both domestic and international.

In 1995, there were media reports that a sophisticated joint US–Australian operation had been launched in the late 1980s to bug the newly constructed Chinese Embassy in Canberra. Reports suggested 'Australian intelligence officers and NSA technicians' covertly installed 'an elaborate system of fibre optic bugging devices throughout the embassy during its construction'. Apparently quite successful for several years, by the mid-1990s, its existence was becoming surprisingly common knowledge. Reports indicate Foreign Minister Gareth Evans issued a suppression order blocking publication of the story. Nonetheless, it eventually aired on ABC TV, to much embarrassment and Chinese chagrin.[22] The episode demonstrated the risk of compromise of audacious undertakings and the likely repercussions. Reports of a similar bugging operation in East Timor some years later, and its consequences, were another sharp reminder of the risks of such operations.

In September 2002, media reports suggested a 'spy sex scandal dossier' on the so-called DSD 'royal family', so named 'because of their elitist social activities'. Concern was expressed that DSD was 'in considerable decline', but others claimed this 'concern' was in reality insiders' resentment at missing out on promotions. Eager to manage the fallout from this story, Defence Minister Robert Hill declared there were no evident security breaches involved. The matter was examined by the Defence Security Branch, but an independent inquiry was not commissioned.[23] There is little to show in the public domain on how that investigation unfolded, but internal reforms followed, aimed at building a unified team from the disparate elements of DSD.

Australian intelligence agencies also faced problems from abroad. As mentioned earlier, an American contractor working in the US intelligence community, Edward Snowden, raised concerns that metadata collected on US citizens was being inappropriately exploited (metadata being the digital information on the number dialled, including the time, date and frequency of the call).[24] Dissatisfied with the responses to his concerns about the use and, in his view, abuse of metadata, Snowden decided to take the law into his own hands and defected in 2013.

Snowden took with him to Hong Kong electronic files containing intimate details of the working of the US-led Sigint architecture, to which Australia was a party. After hiding briefly in the Chinese city, Snowden moved to Russia and in 2022 was granted Russian citizenship. Some defended him as a patriot, exposing intelligence overreach. Others saw him as a traitor who gave away the 'plumbing' of the Sigint collection system.[25] Regardless, his actions caused great concern about the trustworthiness of national security institutions with vast and potentially intrusive powers. The Snowden revelations fed the views of conspiracists in the United States and elsewhere who were already railing against government excess. His revelations seemed to confirm the worst of their suspicions and fuelled international protests. The irony of Snowden seeking shelter in an authoritarian state such as Russia was overshadowed by concern about his secrets being divulged to Russian intelligence.

For Australia, the repercussions were severe, with documents indicating Australia had sought to monitor the mobile calls of then Indonesian President Susilo Bambang Yudhoyono and his wife. Indonesian Foreign Minister Marty Natalegawa was said to have reacted angrily, threatening to reconsider cooperation on issues crucial to Australia.[26] Reports also suggested 'clandestine facilities at embassies to intercept phone calls and data across Asia as part of a US-led global spying network'.[27]

Ball was philosophical about the news, declaring that 'knowing what our neighbours are really thinking is important for all sorts of diplomatic and trade negotiations'.[28] His was the perspective of one who had spent decades researching and writing about Australian Sigint arrangements. In response to the news and protestations, Prime Minister Tony Abbott told the Australian parliament that other countries should not expect Australia to reveal the kind of details they themselves keep secret.[29]

The Snowden revelations cast a long shadow. In addition to the groundswell of protests internationally, many states tightened their security measures. In the case of Australia's close neighbour, Indonesia, police and other military cooperation was temporarily suspended.[30] Although Australian support was important in the identification of terrorists and their networks, it would take some time for the relationships to recover.

The United States also made adjustments, with the US Congress passing the *USA Freedom Act* in mid-2015. This move validated, for many, the actions of Snowden – even though Snowden's steps went well beyond being a whistleblower. Nonetheless, the new legislation would constrain NSA surveillance and the bulk collection of Americans' phone records.[31]

In Australia, the demand for change was less strident. This was in large part because of the reforms of the preceding decades, starting with the work begun by Justice Hope, culminating in the introduction of the *Intelligence Services Act* in 2001. These reforms reduced the prospect of overreach that Snowden invoked in the US context. Additional Australian legislation would follow, but mainly to strengthen the powers of intelligence agencies, rather than constrain them.

Nevertheless, calls for greater accountability would grow, echoing similar developments in other Five Eyes countries.[32] That pressure continued until 2017 when the *Independent Intelligence Review* examined accountability and oversight measures for the Australian intelligence community.

Defence of the joint facilities

The revelations by Edward Snowden in June 2013 elicited a strong reaction around the world, including Australia. Stephen Smith was defence minister at the time. In late June, Smith went on the public record to defend the close ties between Australia and the United States.[33] He spoke about Australia's hosting of two shared facilities: the Joint Defence Facility Pine Gap and the Joint Geological and Geophysical Research Station, originally established in 1955 and located near Alice Springs.

Smith described the Joint Geological and Geophysical Research Station, jointly operated by Geoscience Australia and the US Air Force, as a 'seismic monitoring station' for earthquakes and nuclear explosions during the Cold War and associated with the Comprehensive Test Ban Treaty.[34]

Smith gave a strong endorsement of Pine Gap, noting 'it supports monitoring of compliance with arms control and disarmament agreements and provides ballistic missile early warning information as well as being central to Australia's security and intelligence relationship with the United States'. Pine Gap, Smith argued, 'makes a vital contribution to the security interests of both countries and reaffirms the high level of cooperation that has been achieved in Australia's closest defence relationship'. Through the information gathered at this joint facility, he said, 'Australia is able to access intelligence and early warning that would be unavailable from any other means and is unique in our region'. The facility 'delivers information on intelligence priorities such as terrorism, the proliferation of weapons of mass destruction, and military and weapons developments'.[35]

On nuclear weapons, Smith reminded Australians that 'the risk of the use of nuclear weapons continues'. As a nation that 'prides itself on playing an active role in the counter-proliferation of nuclear weapons', he observed, the 'value of the data obtained from Pine Gap cannot be underestimated'. In addition to the intelligence benefits, the facility 'provides Australia a world-class capability which we could not independently develop'. With this in mind, he said, 'Pine Gap will remain a central element of Australia's security relationship with the United States for the foreseeable future.'[36]

After the Snowden revelations and Smith's defence of the joint arrangements, in 2014 Ball was highly critical of the role Pine Gap played 'in American drone strikes', calling its work 'ethically unacceptable'.[37] An ABC News report declared that the facilities at Pine Gap provided 'detailed geolocation intelligence to the US military that can be used to locate targets for special forces and drones'.[38]

Notwithstanding robust government defence of the joint facilities, periodic but declining protests were a reminder that not all Australians were happy with this degree of connectivity with powerful partners. One of Ball's research collaborators, Richard Tanter, stated that through the 1980s and 1990s, 'Ball reluctantly supported retention of Pine Gap because, despite its role in nuclear war planning, its interception of Soviet missile telemetry was essential for verifying compliance with US-Soviet arms control agreements'.[39]

Kim Beazley, who had been defence minister under Bob Hawke, disagreed, arguing that 'Pine Gap became critical for us. We would be deaf and blind without Pine Gap.'[40] Beazley saw the facilities as 'deeply embedded in Australia's order of battle'. Noting the sunk cost, he argued their replacement 'would be not only unaffordable but technologically impossible', and that 'in a tight financial situation and a more complex regional security environment, they are invaluable'.[41]

At the time of writing, over 55 years since becoming operational, Pine Gap remains one of the most important United States intelligence facilities outside that country. It plays a vital role in the collection of a wide range of Sigint material shared between the United States and Australia. It is still a highly secretive organisation, but is understood to provide early warning of ballistic missile launches, targeting of nuclear weapons and battlefield intelligence data for United States armed forces operating in the Middle East and elsewhere. Reports indicate it supports United States and Japanese missile defence and arms control verification and contributes targeting data to United States drone operations.[42]

Australians now have access to all areas of the base at Pine Gap, except the US National Cryptographic Room, as well as to all of its 'product'; namely, its reports. Australians are employed in all of its sections. The number of Australians at Pine Gap was estimated to have increased from around 340 in the early 1990s to about 420 a decade later.[43] As a result, it could be argued that in these aspects of Sigint, Australia is better placed even than Britain.

Critics point out, however, that most of the tasking of satellites comes from the United States, reflecting its strategic priorities and ownership of the satellites and their infrastructure. They also contend that Australia's participation in the base's operations carries some responsibility for the consequences of those operations.[44] They object to the automatic integration of Australia into United States operations, arguing that Australia lacks the strategic and political control a country needs in order to maintain policy independence and accountability. The argument is made also that Pine Gap makes Australia inherently vulnerable in a conflict involving the great powers.[45] However, the joint facilities are recognised by practitioners and policymakers as conferring enormous benefits for Australia, including a strong and unique tie to a powerful ally. These benefits mean that the prospect of vulnerability is considered to be more than offset by the strength of insight, assurance of support and depth of deterrence it provides. In practice, therefore, to forgo the base, to ask the United States to withdraw, is almost unthinkable. And yet, if the Chifley government had not lost the December 1949 election, Australia's Sigint would have taken a different, probably more independent but less capable, course at least for a time.

With simmering disagreements in mind, it is incumbent on policymakers and intelligence practitioners today to weigh carefully the pros and cons of Australia's investment in the intelligence infrastructure shared with the United States. To date, that continues to be seen among policymakers, practitioners, parliamentarians and numerous academics as tipping strongly in favour of retaining and even deepening those ties in the pursuit of Australia's ability to understand and respond to the regional dynamics and to help deter acts of aggression. For those privy to the classified briefings relating to the joint facility, the case has been compelling. Successive Australian governments have considered this as an arrangement whereby Australia's common interests with the United States justify the investment in the facilities and the shared arrangements.

Some commentators have questioned whether in future, as an economy measure, the United States may wish to cancel the arrangements at Pine Gap. Such a step may be technically feasible as one which could capitalise on technology that enables networked global communications between satellites that, some would contend, makes Pine Gap technologically redundant. That may well be the case where space is not contested. However, as disputes over space increase and the importance of redundancy of systems grows, there remain considerable advantages to retaining Pine

Gap, even if largely for redundancy, to act in part as a backup for facilities that could be remotely operated from continental United States. Resilience and redundancy are not inconsequential factors. Either way, Australia is now heavily invested in the shared facilities and their benefits; including a greater understanding of regional security challenges, enabling more effective and nuanced responses to them.

Growth and statutory independence

In the years following its establishment in 1947, DSD's technical capability and capacity grew significantly, and its ability to provide support to government decision-making became more apparent. The insights provided were increasingly useful for policymakers responding to tricky international relations, as well as for those formulating defence strategy and policy. The experience in Vietnam and elsewhere had also demonstrated DSD's utility for military practitioners eager to provide the best situational awareness to troops in harm's way abroad. These capabilities, plus access to an expanding array of technology, demonstrated, for instance, by the joint facilities at Pine Gap, foreshadowed DSD's increasing importance and the inevitability that it would become a statutory authority. The digital era would see demand for its services grow exponentially.

For some time after the Vietnam War, however, when there was a significant lull in the military's operations, the imperative for a senior military appointment inside DSD seemed less pressing. In the post–Cold War years, however, with the proliferation of military operations around the globe, often on humanitarian assistance and peacekeeping missions, there was again a need for senior military leadership to be involved in coordinating direct support to military operations. That need intensified with the commitment to operations in Afghanistan and Iraq. As a result, a Director General Support to Military Operations appointment was created. This was filled by a succession of 'one star' officers, including Commodore Kim Pitt, Air Commodore (later Air Marshal and Chief of Air Force) Geoff Shepherd, Commodore Denis Mole, and Brigadier (later Major General) Stephen Day and Major General Steve Meekin.

By the mid-2010s, the national Sigint agency and the military Sigint units had a wealth of operational experience. And with cyber came further reform and expansion. Finally, 40 years after Justice Hope had recommended it, ASD became a statutory authority, with corporate reporting obligations; in particular, through meeting the requirements of the *Public Governance, Performance and Accountability Act*, ASD became answerable to the minister for defence, with its priorities set by the National Security Committee of Cabinet. The Parliamentary Joint Committee on Intelligence and Security has powers to review and oversee ASD's administration and resourcing. In addition, the Inspector-General of Intelligence and Security has enduring powers of a royal commissioner, with complete access to agency records and powers to require evidence.[46] This step up to becoming a statutory authority reflected ASD's expanded capabilities. It was also a consequence of the creation of the Cyber Security Operations Centre and its expansion into the Australian Cyber Security Centre.

Mike Burgess was appointed the inaugural Director-General at departmental secretary level (band four of the Australian Public Service's senior executive service).[47] Burgess had a military deputy appointment created and the Australian Army officer, Lieutenant General John Frewen, was appointed. Burgess was subsequently appointed Director-General of ASIO in September 2019 and Frewen was made Acting Director-General of ASD. This was the first time a military officer had held the appointment. But he only held it for three months and would not be replaced inside DSD at that three-star or four-star ranked level.

After acting in the position of ASD Director-General for several months, Frewen handed over the reins to Rachel Noble, the first woman to fill the role and someone with longstanding connections to DSD, being the daughter of a former DSD employee. For Noble 'Joining the then-DSD in 1994 was a bit like joining the family business' as her father, Tim Noble, and sister Rebecca Skinner worked at DSD.[48] Frewen would go on to head the ADF's COVID-19 Task Force and the national Covid Vaccine Task Force through to mid-2022.

Thereafter, the military function was downgraded with the new arrangements for ASD senior management including three civilian band three senior executive service (SES) deputies and one band four SES director-general. Under Noble's direction, the military's highest level representation was downgraded to one-star or two-star level, below the level of the band three SES deputies. This new management model left the services uneasy about the responsiveness of ASD to their operational demands. There is a view that this contributed to the decision to create in the Defence Department a new three-star-level senior military Chief of Defence Intelligence (CDI). A seasoned operational military intelligence practitioner, Lieutenant General Gavan Reynolds, was promoted to the position in 2020.

Cyber and accountability

In the late twentieth century, when the internet had taken hold, it seemed to present an existential threat to the traditional twentieth-century approach to Sigint – collected mostly from analogue signals passed over line, or over various forms of radio technology over the 'airwaves' or via satellites in space. The interconnected network of computers (or internet) started to transform the way society interacts and does business, as well as how military operations are planned and conducted. This online digital interaction gathered pace in response to the coronavirus pandemic in early 2020.

With society going from being web-enabled, to web-dependent and, in turn, web-vulnerable, investment in cyber has become critical to the rest of the Sigint domain. A key participant in the transformation in Australia was Stephen Merchant's successor as Director DSD, Ian McKenzie, a senior public servant who served as director for six years from late 2006 until he retired in 2013.[49] While he was director, DSD commenced a new chapter. As we have seen, the military had long thought in terms of land, sea and later air power, but cyber would come into its own in the opening decades of the new century and DSD would be in the vanguard.

In 2010, DSD established the Cyber Security Operations Centre (CSOC) to respond to security threats to critical Australian systems. On 7 May 2013, DSD was renamed ASD in recognition of its national security role. A year later, the CSOC became the Australian Cyber Security Centre (ACSC), which includes staff and secondees from the Attorney-General's Department, Australian Federal Police (AFP), the Australian Criminal

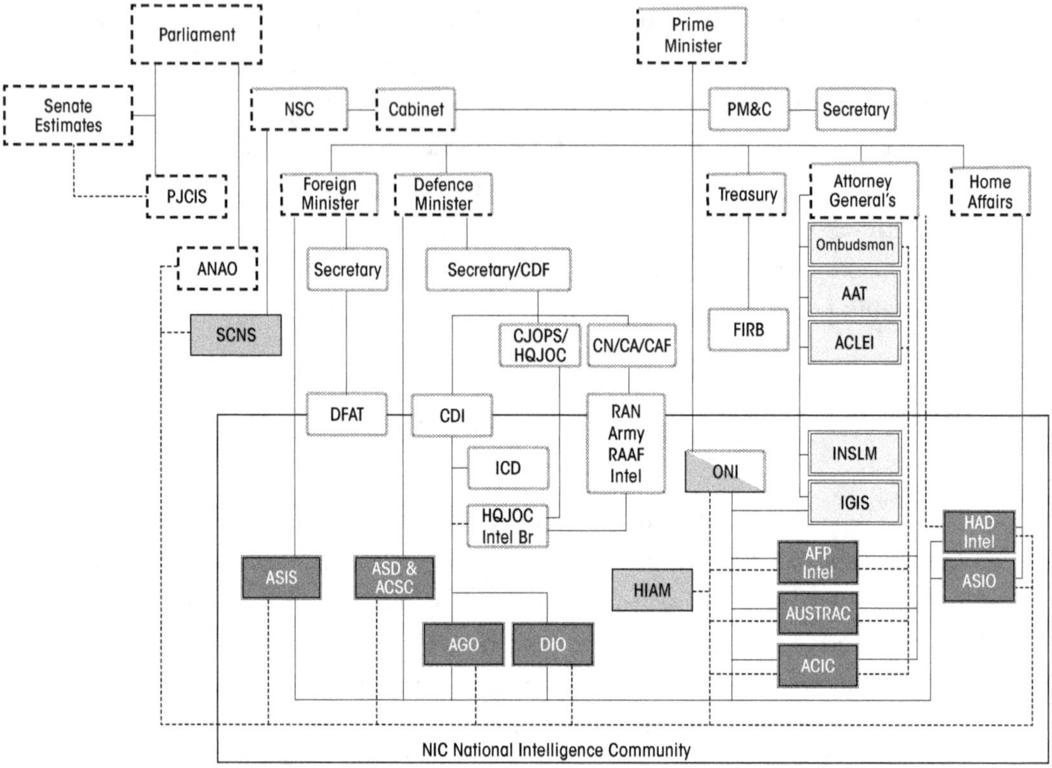

Executive Control

Parliamentary Accountability

Independent Accountability

Coordination

NIC Agency

Legend

AART	Administrative Appeals Tribunal
ACIC	Australian Criminal Intelligence Commission
ACLEI	Australian Commission for Law Enforcement Integrity
ACSC	Australian Cyber Security Centre
AFP	Australian Federal Police
AGO	Australian Geospatial-Intelligence Organisation
ANAO	Australian National Audit Office
ASD	Australian Signals Directorate
ASIO	Australian Security Intelligence Organisation
ASIS	Australian Secret Intelligence Service
AUSTRAC	Australian Transaction Reports and Analysis Centre
CDF	Chief of Defence Force
CDI	Chief of Defence Intelligence
CJOPS	Chief of Joint Operations
DIO	Defence Intelligence Organisation
DFAT	Department of Foreign Affairs and Trade
FIRB	Foreign Investment Review Board
HAD	Home Affairs Department
HIAM	Heads of Intelligence Agencies Meeting
HQJOC	Headquarters Joint Operations Command
ICD	Intelligence Capability Division (CDI)
IGIS	Inspector-General of Intelligence and Security
INSLM	Intelligence and National Security Legislation Monitor
NSC	National Security Committee of Cabinet
ONI	Office of National Intelligence
PJCIS	Parliamentary Joint Committee on Intelligence and Security
PM&C	Prime Minister & Cabinet Department
RAAF	Royal Australian Air Force
RAN	Royal Australian Navy
SCNS	Secretaries Committee on National Security

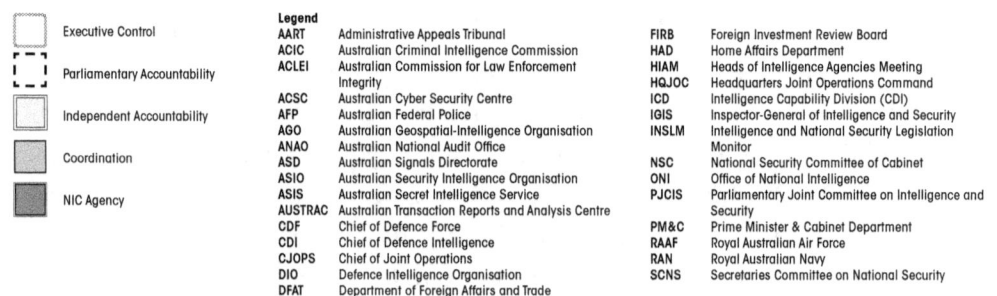

Figure 8. Australia's national intelligence community structure and accountability arrangements.

SOURCE John Blaxland.

Intelligence Commission (ACIC), ASIO, DIO and other cyber security experts under ASD's leadership.

Figure 8 (NIC structure & accountability arrangements) demonstrates the arrangements that have evolved following the reviews dating back to Justice Hope in the 1970s and 1980s. ASD and ACSC appear in the context of a broader national intelligence community, with each agency having separate but complementary responsibilities and lines of authority, mandated by legislation accumulated over more than half a century – legislation which has proliferated in recent years. In addition, there are independent oversight and executive controls, as well as parliamentary accountability lines of responsibility. The system has its limitations, and periodic adjustments are required to make sure that oversight and accountability keep up with the community's growth and expanded powers. On balance, however, it has provided Australia with perhaps the most robust mechanisms overseeing the intelligence agencies of any liberal democracy. Perhaps it is for this reason that ASD has been able to take the lead in raising cyber security awareness across the nation.

The public is keenly aware of the cyber threats posed by ransom attacks, corporate espionage, threats to the grid, election interference and others. American international relations professor Joseph Nye lists the special characteristics of the new cyber-domain as the erosion of distance, the speed of interaction, the low cost, and difficulty of attribution.[50] Notwithstanding the need for ASD's capabilities to counter cyber threats, there were and are fears in some quarters of 'mass surveillance' of Australians.

So far, no one has died in a cyber attack in Australia, and there has been no cyber Pearl Harbor, although Medibank could face a $1 billion compensation bill from a cyber attack affecting 10 million customers.[51] Good governance of cyberspace may be the greatest challenge the world has faced so far. The world has experienced cyber attacks since the 1980s, but the attack surface now includes everything from industrial control systems, to cars, to personal digital assistants (mobile phones).[52] The US Administration under President Joe Biden has complained that the scale and duration of some Chinese and Russian cyberattacks have moved them beyond normal spying. History has shown that societies take time to learn to respond to major disruptive technological changes and to put in place rules to make the world safer from new threats. As Nye points out, it took two decades after the United States dropped nuclear bombs on Japan for countries to agree on the Limited Test Ban Treaty and the Nuclear Non-Proliferation Treaty.[53] We should expect the rules of the road for cyber to take some time to be agreed.

Building on the legacy of resourcefulness and enterprise of the two world wars, Australia has become known as a nation of inventors and entrepreneurs, with innovations that include the 'black box' flight recorder, ultrasound, multi-channel cochlear implants, wi-fi and Google Maps. Still, there is no room for complacency and there is more work to be done. But the combination of these skills and qualities and the legislative framework created by Hope and his successors have prepared Australia well to navigate the fifth dimension of cyber.[54]

Australia's Cyber Security Strategy

In Malcolm Turnbull, Australia had a prime minister who was genuinely interested in the new technology, and knowledgeable about it. Drawing on ideas from a classified cyber security review, in April 2016, Turnbull announced Australia's Cyber Security Strategy. He explained how the government proposed to meet 'the dual challenges of the digital age—advancing and protecting our interests online.'[55] The Department's Special Adviser on Cyber Security would lead implementation of this strategy. It was pitched as something ASD and the ACSC would facilitate, while relying on broader social engagement. The strategy declared that 'all of us – governments, businesses, communities and individuals – need to tackle cyber security threats to make the most of online opportunities'.[56] This was a more public role for ASD than its founders had ever imagined possible.

Turnbull stressed that the 'need for an open, free and secure Internet goes far beyond economics', with the caveat that 'cyberspace cannot be allowed to become a lawless domain'. He went on to say that both government and the private sector have vital roles to play, that Australia and Australians are targets for malicious actors, including serious and organised criminal syndicates and foreign adversaries. Turnbull promised that Australia and its allies would work together to promote norms of behaviour consistent with a free, open and secure internet. These norms include that states should not knowingly conduct or support cyber-enabled intellectual property theft for commercial advantage. Turnbull then turned to a range of other threats, including online propaganda inciting extremist and terrorist violence. He mentioned the Snowden case as demonstrating that often the most damaging risk to government or business online security is not 'malware' but 'warmware'; that is, trusted insiders who have the potential to cause massive disruption to a network or to use legitimate access to obtain classified material, and then illegally disclose it.

The circumstances that led to the Snowden case were specific to US Sigint and cyber security practices. But there are lessons for Australia if such a security breach is to be avoided in future. With this in mind, Turnbull declared that 'technical solutions are important but cultural change will be most effective in mitigating this form of cyber attack. As businesses and governments we must better educate and empower our employees to use sound practices online.'[57]

Turnbull's announcement was a major step, with cyber now centre stage of national security and economic development. Cyber had moved beyond the remit of one single government agency like ASD. Sigint and cyber would no longer be left to a middle-level Defence bureaucratic appointment. It was to be controlled through the Prime Minister's Department, by a senior head of agency operating at the level of a secretary of a department.

To clarify ASD's new functions and powers, the 2017 *Independent Intelligence Review* paved the way for ASD to be established as a statutory authority with national powers concerning cyber.[58] Under this arrangement, ASD has a formal legislative mandate for its role as the national information and cyber security authority, including combating cyber crime and advising the private sector on cyber security matters.[59]

The cyber announcement was followed in 2017 by the creation of an Information Warfare Division in the Defence Department's Joint Capability Division, separate from ASD, which also had a stake in the cyber domain.[60] With ASD a statutory authority, there were concerns about its responsiveness to the needs of the military, particularly with competing demands from other parts of government and society.

The Australian Cyber Security Centre

Those competing and growing demands were behind the establishment of the ACSC. By 2018, the ACSC custom-built facility opened in Canberra, with classified and unclassified sections to allow engagement with the private sector and collaboration with industry. It represented the government's 'commitment to the online safety of all Australian families and businesses'.[61] In 2018, Malcolm Turnbull observed that, since the release of the 2016 Cyber Security Strategy, the cyber threat had 'shifted and evolved dramatically', with criminal networks and malicious foreign actors using sophisticated technology to steal intellectual property and interfere with institutions. Attempted attacks were seen as

> occurring every day. Billions of cyber events orchestrated by criminal, and indeed nation state actors are aiming at the very heart of the Australian Government, business and our public life. It is a global threat. Nation states are among the worst and most consistent offenders, persistently targeting Australian institutions across government and business.[62]

The government further revealed that 33 Australian universities had been targeted by an Iran-based spear-phishing campaign, in an attempt to steal intellectual property and academic research. Meanwhile, Russian-backed hackers had infiltrated Western political institutions in a bid to influence elections. In addition, cyber criminals continued to target bank balances and personal data of vulnerable members of the community. Estimates of the global cost of cybercrime in 2018 were at around $600 billion, of which the estimated cost to Australian business and individuals was $7 billion.[63] Reflecting the changing dynamics and the worsening security situation, in 2018, the *Intelligence Services Act* was amended.[64] Yet this was done without providing a clear and transparent legal framework for the ACSC.[65]

This delay in legislative reform has been caused in part by the blurring of the distinction between law enforcement, intelligence and the field that spans the two – disruption. Increasingly, cyber-related intelligence collection has contributed to tracking and monitoring criminal and espionage-related activity. With a growing expectation of a no-fault approach to policing and intelligence, Sigint and cyber have been used to help track and disrupt potential actions by malevolent individuals and groups in anticipation of an act, rather than as a result of one. The result is a category of response that lies between old-style intelligence collection and collection of evidence by police for prosecution. That category involves disruption in anticipation of a crime. This is a departure from traditional intelligence collection operations (where material is collected for assessments of likely future developments) and conventional policing (where evidence is collected of crimes in the recent past for use in judicial prosecutions).

This also involves close cooperation and deconfliction, with state police and related security entities involved in managing cyber threats. In New South Wales, for instance, tens of millions of dollars have been invested in establishing a cyber security centre to protect the NSW Police network from online security threats in real time.[66] The major banks have invested on an even greater scale to ward off malicious cyber actors.[67] Much of this also reflects a surge in ransomware attacks.[68]

Geoeconomics, Sigint, cyber and 5G

While fighting cybercrime monopolised the attention of the ACSC, globalisation became the focus of public commentary.[69] In 2018, agents of ASD played a digital war game. As Simeon Gilding, a former senior Australian cyber security official, explained:

> We asked ourselves, if we had the powers akin to the 2017 Chinese Intelligence Law to direct a company which supplies 5G equipment to telco networks, what could we do with that and could anyone stop us? ... We concluded that we could be awesome, no one would know and, if they did, we could plausibly deny our activities, safe in the knowledge that it would be too late to reverse billions of dollars' worth of investment. And, ironically, our targets would be paying to

build a platform for our own signals intelligence and offensive cyber operations.[70]

In Gilding's view, the 'fundamental issue is one of trust between nations in cyberspace', and the Chinese Communist Party had 'destroyed that trust through its scaled and indiscriminate hacking of foreign networks and its determination to direct and control Chinese tech companies'. Allowing Huawei to build the country's fifth generation (5G) network would have been like 'paying a fox to babysit your chickens'.[71]

In their 2019 book *Six Faces of Globalisation*, Anthea Roberts and Nicolas Lamp make the case that 'Instead of assessing gains through an economic lens alone with a win-win assumption that all countries can benefit from economic globalization', 'great-power rivalry, strategic concerns, security threats, and technological competition' are now dominant.[72] Another perspective, found in *AI Superpowers,* focuses on Sino-US competition, extolling the positive potential of advanced computing and assertive, expansionist state-backed business enterprises, such as Huawei.[73] Defenders of Chinese-sourced 5G technology argue that there was no evidence of improper use of the technology. This work highlighted 5G mobile cellular phone technology in the geostrategic competition.

Central to the Australian government's response to this challenge are ASD and the ACSC. In Australia, the debate focused on the economic, technical, geopolitical and strategic considerations of the departments and agencies concerned. The consensus was that it was no longer enough to say there was no 'smoking gun'. The absence of smoke no longer meant the absence of a loaded weapon.

In the end, armed with these insights, Prime Minister Malcolm Turnbull banned 'high risk vendors' from the country's 5G networks, thus excluding Huawei from Australia's 5G rollout.[74] At first, Australia's landmark 5G decision came in for criticism from those Five Eyes partners eager to capitalise on Chinese investment opportunities. Later, however, the Australian decision to block Huawei and ZTE from rolling out their 5G infrastructure was seen by many as the right one. In particular, the United States came to see 5G technology as part of the competition for global technological primacy. Australia's lead was followed by several like-minded countries.⁷⁵

The Richardson Review

Meanwhile, public debate about the powers of ASD continued, reflecting the grey zone of onshore disruption. For example, a *Daily Telegraph* article in April 2019 reported an alleged plan for ASD to 'monitor Australian citizens'. According to this plan, ASD would be able to see emails, bank records and text messages of Australian citizens, 'proactively disrupt and covertly remove' onshore cyber threats by 'hacking into critical infrastructure' and be given 'step-in' powers to force government agencies and private businesses to 'comply with security measures'.⁷⁶

Police raids followed in June 2019; these were seen by critics as catalysing 'the growing drift towards a surveillance state and secrecy state'. As a consequence, there has been heightened concern about 'encroaching dangers to journalism and free speech'.⁷⁷

These were among the issues considered in the *Comprehensive Review of the Legal Framework of the National Intelligence Community*, the most comprehensive review since the Hope Royal Commissions, with Dennis Richardson, the former head of ASIO and former secretary of the Department of Defence, as the principal reviewer. Richardson and his team found that, on the whole, the legal framework had been well maintained and was largely fit for purpose.⁷⁸ They noted, however, that even Justice Hope did not anticipate the need for an intelligence agency responsible for cyber security.⁷⁹

Richardson's report acknowledged that the legislative framework, under which ASD, or rather the ACSC under ASD's leadership, must rely on foreign intelligence warrants to collect information about cyber security threats, may not be fit for purpose. For example, if an Australian hacking group launched a major cyber attack from inside Australia, then a foreign intelligence warrant would not be available or appropriate.

Interestingly, in light of the *Daily Telegraph* article, ASD was apparently not seeking additional powers for itself in this area. Richardson and his team noted that ASD had been 'explicit that its current powers are adequate and that additional powers are "not necessary" for cyber security purposes'. The Review's recommendation on this point is classified.[80] However, the Review's consideration of the clarity and transparency about necessary activities noted that:

> We consider that Australia's lead cyber security authority should have the benefit of a legal framework that clearly and accurately reflects its role and responsibilities and that would not require it to rely on a 'patchwork' of warrants and authorities held by other agencies to investigate and combat cyber security incidents.[81]

Momentum was growing for the ACSC to be subject to the level of legislative control and oversight provisions that other arms of the National Intelligence Community had come to expect.

'Exceptional circumstances' and onshore offensive cyber capabilities

The Richardson Review also noted that, given the centrality of cyber, at some point 'exceptional circumstances' – for example, requiring the use onshore of offensive cyber capabilities – would be likely to arise. Mindful of this, it noted that the legal framework determining what a government can do in exceptional circumstances should be developed in the cool light of day, rather than in response to a crisis.[82]

At time of writing, the ACSC has been tasked to 'provide advice and information about how to protect Australians and their businesses online'. Its website states that the ACSC 'leads the Australian Government's efforts to improve cyber security', monitoring cyber threats across the globe on a continuing basis and then alerting people and organisations on how to respond.[83] The ACSC maintains 'A national footprint of joint Cyber Security Centres where we collaborate with nearly 200 businesses, government and academic partners on current cyber security issues. We also work with law enforcement authorities to fight cybercrime.'[84]

In the context of responding to online criminal activity, the Review records the Home Affairs position that criminal activity through the use of the 'dark web' (that is, the realm of unregulated and largely illegal online activity) represents a significant threat, and that Australian law enforcement responses need to evolve accordingly. This applies, especially, regarding online child exploitation offences.[85]

Richardson and his team considered that the role of cybercrime disruption is one for the Australian Federal Police, with technical assistance and support from ASD, but they did not agree that the AFP needed additional disruption powers to do the job. Indeed, they expressed concern that 'there appears to be a tendency for some officials to consider that "zapping" computers being used in child exploitation is a reasonable response'.[86]

The challenges the Richardson Review observed and analysed have increased, notably as a result of phenomena associated with the so-called Fourth Industrial Revolution. The first three industrial revolutions are said to have involved the steam, electric and nuclear ages. The fourth, the information age, is built not so much on mechanical and analogue technologies, but on advanced digital technologies.

The growing power of robotics; machine-assisted learning; unattended air, sea and land vehicles; sophisticated and personalised algorithms; quantum computing (including quantum cryptography) and so-called 'artificial intelligence' (ranging from assisted, semi-autonomous to fully-autonomous systems) means that providing effective cyber security and Sigint has become much more challenging. This applies to both of ASD's mandates – those being, 'to reveal their secrets and to protect our own'.

The challenges in providing adequate computing power for these needs have continued to grow. This applies to effective encryption to protect sensitive intelligence and security communications networks, and the myriad of other governmental, industrial and educational networks that look to ASD for advice and assistance. The challenges apply to both mandates.

We probably have not heard the last of the debate about extra powers for cyber security purposes. The Morrison government disagreed with part of this recommendation of the Richardson Review and introduced the Surveillance Legislation Amendment (Identify and Disrupt) Bill, intended to give the AFP and ACIC (but not ASD) power to 'collect intelligence, conduct investigations, disrupt and prosecute ... the most serious of crimes'.[87]

In response to the Review, the Department of Home Affairs published a discussion paper in late 2021 on *Reform of Australia's Electronic Surveillance Framework*.[88] The paper argued that with fast-paced change in law enforcement, intelligence collection and cyber (in part due to the rapid advance of technological change associated with the Fourth Industrial Revolution), Australia's legislation has struggled to keep pace. The paper stated that law enforcement agencies, including integrity and anti-corruption bodies and ASIO, 'at times require access to specific information and data to protect the community from serious crimes and threats to Australia's national security'. It claimed that without access to such information, 'law enforcement agencies could not prevent and prosecute the most serious criminal activities such as child sexual abuse, organised crime and cybercrime'.[89] The 'prevention' to which the paper refers involves attempts at disruption and proactive engagement that have caused unease and provoked media criticism with the expansion of intrusive intelligence and policing powers. What is clear is that there is a continuing need for adequate checks and balances, for solid safeguards to hedge against misuse and abuse that may arise from the accretion of additional powers, risking the erosion of public trust.[90]

The legacy of Australia's Sigint history in offensive cyber operations

In the meantime, with signs of a growing global cyber security threat, ASD teamed up with the Australian National University in an attempt to recruit and educate additional cyber security specialists, or what the *Canberra Times* described as 'the next batch of cyber spies'. According to the then Assistant Defence Minister, Andrew Hastie, 'the cyber battlefield, more than ever, needs those Australians with a rare mix of specialist skills, adaptability and imagination to defend the nation against the most sophisticated adversaries.'[91] While outright conflict or war has been avoided so far, competition and contestation, notably in the cyber domain, has become more challenging, suggesting it will attract a growing number of the next generation of cyber enthusiasts.

In an unprecedented and fascinating account to the Lowy Institute on 27 March 2019, Mike Burgess, then Director-General of ASD, described the role of ASD's offshore offensive cyber operatives. In the Middle East, at the height of a coalition battle with Daesh, ASD cyber operators in Canberra degraded Daesh communications within seconds. Terrorist commanders were unable to connect to the internet or communicate with each other. The terrorists were in disarray and driven from their position.[92] We see these ASD cyber operators as the professional descendants of the Australians in the world wars who yielded to none in the art of traffic analysis, which was decisive in many an emergency.

Burgess gave another equally gripping example, which also echoes the past. This concerned a highly trained young woman, a science graduate turned ASD covert online operator, who led a team that tracked down a man who had been radicalised and was overseas trying to join a terrorist group. Pretending to be a terrorist commander, she used a series of online conversations to gradually win her target's trust. Eventually, she persuaded the aspiring terrorist to abandon his plan for jihad and move to another country where ASD's partner agencies could ensure he was no longer a danger to others or himself.[93]

In professional terms, this young woman is the direct descendant of unsung women Siginters, including Mrs Mac and her pupils in the three armed services, the women of FRUMEL and Central Bureau.

REDSPICE

Momentum for a more comprehensive and better resourced approach to cyber challenges had been building for some time. On 29 March 2022, then Treasurer Josh Frydenberg announced by far the largest investment in ASD in its 75-year history.[94] The contrast between the $9.9 billion for the ten-year Project REDSPICE and the organisation's parlous resourcing situation in its early years is truly remarkable. As we have seen, as late as

1977, one of Justice Hope's few criticisms of what was then DSD was its tendency to reduce demands for funds and staff to levels it felt would be unexceptionable to Canberra. These attitudes had resulted in budgetary deprivations and low staff classifications.[95]

Hope went on to say that:

> Anyone who examines the recent history of budget cuts and staff ceilings imposed, arbitrarily, on DSD, and anyone who visits its ramshackle buildings, would see that, over the years, DSD has indeed been modest, not to say frugal, in its requisitions on government. In fact, it is only in the last few years that DSD's crying need for proper and secure accommodation has been recognised and acted on.[96]

Hope argued that these indicators all pointed to a perception by DSD of the indifference of outsiders to its work, leading to the agency's determination to cut its coat to the cloth it felt was available, rather than to seek the optimum to meet all the national intelligence requirements that might be placed upon it.[97] In one of the many examples of his prescience, Hope stated that 'this attitude will not suffice for the future. DSD will need to make it quite clear to Canberra that the resources it receives are directly related to the intelligence it will be able to produce.'[98] ASD has apparently done just that, in the years following, becoming the assertive organisation Hope wanted to see. The unashamedly brash name of REDSPICE seems to confirm the sea change in this once modest and self-effacing organisation. REDSPICE stands for Resilience, Effects, Defence, Space, Intelligence Cyber and Enablers.

It is instructive to remember here that, after extensive public consultation, the 2020 Cyber Security Strategy stated that 'we deny and deter, while balancing the risk of escalation. Our actions are lawful and aligned with the values we seek to uphold and will therefore be proportionate.'[99]

In 2021, increased threats to Australia's critical infrastructure led to the development and release of Australia's International Cyber and Critical Technology Engagement Strategy. Again, there was extensive public consultation. This 2021 strategy repeated the 'deny and deter' principles, and mentioned that as part of its 'cyber deterrence posture', Australia would provide messages to demonstrate willingness and ability to 'impose costs' on those who carry out malicious cyber activity. Australia's responses to malicious cyber activity could comprise law enforcement or diplomatic, economic or military measures, and responses would not always be public.[100]

Mike Burgess and his successor Rachel Noble, have been more forthcoming than their predecessors about aspects of ASD's capabilities and activities, including offensive ones. ASD's website indicates that REDSPICE was designed in response to deteriorating regional strategic circumstances, characterised by rapid military expansion, growing coercive behaviour and increased cyber attacks. The REDSPICE blueprint sets out plans to triple existing offensive capacity, double persistent cyber-hunt activities, quadruple its global footprint, create 1900 new jobs and capitalise on advanced AI machine learning and cloud technology. This is presented as an initiative that will harden computer networks against cyber attacks and enhance intelligence capabilities. Increasing ASD's national and international footprint is also portrayed as a move that will improve core ASD resilience.

The five broad goals of REDSPICE are listed as:

- scaling cyber effects capabilities, tripling the offensive cyber effects capability to support the ADF
- developing new intelligence capabilities
- enhancing Australia's cyber defence
- increasing resilience and redundancy
- improving foundational technologies.

Questions remain, however, about what ASD's offensive capabilities and 'persistent cyber-hunt activities' might be, and which AI machine learning and cloud technologies will be used to enable the tripling of current offensive capacity and doubling of persistent cyber-hunt activities.[101]

Indeed, we are left wondering what this unprecedented spending on ASD really means. REDSPICE reportedly came at the expense of a $1.3 billion 'Sky Guardian' drone project and money reallocated from elsewhere in Defence.[102] Along with the dramatic changes in ASD's resourcing and culture, the question remains as to whether the surge in resourcing means that a change in Australia's cyber strategy is also being contemplated or has indeed already occurred.

REDSPICE, of course, was not announced in a vacuum. It was the centrepiece of the former government's defence budget speech. It also came nine days after the passing of the *Security Legislation Amendment (Critical Infrastructure Protection) Act 2022*. This legislation was passed in response to increasing cyber security threats to essential services and businesses, including cyber attacks on federal parliamentary networks, logistics, the medical sector and universities. Its declared purpose was to improve the security and resilience of Australia's critical infrastructure.[103]

The announcement also followed a meeting of the Quad Senior Cyber Group (Australia, India, Japan and the United States) in Sydney which convened to discuss cyber security cooperation and to boost the resilience of the group's critical infrastructure.[104] The United States, of course, is not only a fellow Quad member but the most powerful of the Five Eyes partners. For a long time, as we know, its Sigint relationship with Australia has been particularly close and this is evident in some of the common cyber strategy terminology used by Australia and the United States.

In their book on the age of cyber threats, *The Fifth Domain*, Richard Clarke and Robert Knake noted that during the Cold War, the best strategies could be summed up in a couple of words; namely, containment and deterrence. For cyber, they proposed resilience, citing the psychologist Judith Rodin's definition as the capacity of any 'entity to prepare for disruptions, to recover from shocks and stresses and to adapt and grow from a disruptive experience.'[105] Australia, like the United States, refers to 'imposing costs' on cyber opponents. Exactly how this might unfold and what follow-on repercussions it might prompt has not been explained, leaving a certain unease about taking on an offensive cyber challenge, the scale of which may not have been fully fathomed and the repercussions of which could be profound.

Since the Trump presidency, the United States' cyber strategy has been an assertive one, known as 'persistent engagement' or 'forward defence'. NSA director, General Paul Nakasone, is also commander of US Cyber Command, and both are close partners of ASD. He confirmed that the

United States had engaged in offensive cyber operations in support of Ukraine. The actions described are consistent with the US doctrine of persistent engagement, which calls for the use of offensive and defensive capabilities in partnership with allies to disrupt hostile cyber actors before they can threaten US and allied networks. This policy gives US cyber teams significant latitude to operate on hostile networks and disrupt threat actors in their own digital staging areas.[106] It also contributes to a blurring of the threshold of war – something usually associated with the application of lethal force. Operating short of that kinetic threshold has become a crucial part of the art and science of cyber operations.

Nakasone himself characterised persistent engagement as 'centred on the construct of both enable and act'. 'Enable' means sharing threat indicators, pooling resources and providing insights. 'Act' entails 'hunt forward' – that is, proactively identifying security vulnerabilities in partners' networks overseas – with permission, as well as offensive operations and information operations. Rather than cyber deterrence, Nakasone and other officials speak of 'imposing costs' on adversaries via persistent engagement.[107]

The 'persistent engagement' doctrine is the subject of a healthy debate in the United States. It has strong critics, including Jason Healey, who points out 'an overarching worry is that Cyber Command does not appear to see this approach as fundamentally risky'. A more engaged forward defence might not result in 'negative' feedback – reducing conflict by bringing it back to the historical norm – but instead 'positive' feedback, exacerbating the conflict and adversaries may see the new US vision as a challenge to rise to, rather than one from which to back away.[108] In an era of heightened geostrategic competition, this is a robust and forward-leaning posture. For Australia, with its different geostrategic circumstances, such an approach would be more audacious than any previously adopted by Australia's national Sigint enterprise. Not surprisingly, therefore, there is a clear need for Australia to be able to calculate its own risks and weigh up the consequences of acting offensively, if not kinetically, in the cyber domain.

It may be that ASD's REDSPICE blueprint should be taken at face value, using a generous investment to continuously improve ASD's offensive and defensive capabilities to deal with deteriorating circumstances.

If it is more than that, and a change to Australia's cyber strategy is being contemplated or, as seems to be the case, it is already happening, we would still hope to see, albeit belatedly, the same level of public engagement and consultation that took place before previous cyber strategies were adopted. Among the issues debated might be whether any new strategy would be subject to the same extent to legal oversight as its predecessors, and whether it would be as consistent with Australia's support for the international rules-based order and obligations under international law.

The accelerating tempo of change

In the early days of computers, Siginters worried about loss of access to the analogue means of radio and cable telecommunications. Little did they realise that the transition would lead to a flood of new digital data sources that would require an entirely new way of managing Sigint collection and analysis. That has led to a new cyber domain and substantial growth in the Sigint network in response to the demand for additional security measures.

Since the establishment of the joint facilities in the mid-1960s, arrangements for full knowledge and concurrence have disarmed much of the criticism of what some described as a nuclear target in the middle of Australia. In an interconnected, digitised and satellite-dependent world, the value of Pine Gap's shared intelligence and security arrangements has increased.

Meanwhile, Australia's ties with New Zealand have been repeatedly on display (in Bougainville, East Timor, Solomon Islands and elsewhere), with combined contributions to stability and security in the region demonstrating the value of closely aligned intelligence efforts and strategic outlooks.

The East Timor experience for Australia proved traumatic, with highs and lows. The controversies over events in 1975, in particular, deserve a more detailed public explanation than is offered here, but there are understandable sensitivities in how Australia manages its relationship with not only East Timor but also Indonesia. The crisis of September 1999, which triggered the UN-mandated and Australian-led international force, had a cathartic effect on Australian Defence practices and procedures. While it set back bilateral relations with Indonesia, it had a positive effect, not just on the ADF but also on the Australian intelligence community. From September 2001 onwards, the ADF would face two decades of continuous deployments, particularly in the Middle East, for which intelligence support was crucial. Closer to home, the bombings in Bali and Jakarta in the early 2000s presented an opportunity for Australia to lend a hand to Indonesia and, in so doing, start to rebuild the important relationship between the two countries.

There were also technological advances and significant reforms. In the end, the defence force EW and Sigint units, working closely with the national Sigint agency in Canberra, as well as the collection sites in Darwin, Geraldton and Pine Gap, provided timely, accurate and 'actionable' intelligence. The focus on such tactical support came, to a certain extent, at the expense of careful and focused management of Australia's regional security priorities.

The pace of technological change, coupled with leaks and revelations, and an increasingly contested international security environment has meant that demand for Sigint services has not diminished and is likely to increase. The Five Eyes partnership has given Australia access to an unrivalled range of technology, and the burden of collection and analysis is shared. Membership of this Anglophone club, however, appears to have contributed to Australia being seen as separate from its Asian and Pacific neighbourhood, where many states would welcome greater US engagement economically and in regional security measures. Ironically, the perennial challenge of reconciling Australia's place in the world (its geography) with its sense of identity (its history) is played out most obviously and tangibly in its Sigint arrangements and partnerships.

CONCLUSION & LOOKING AHEAD

In the intelligence business, the secret of success lies in keeping one's successes secret. After all, were a target or adversary to become aware of one's ability to eavesdrop, they would be expected to alter their practices, making it that much harder to replicate the success that led to the boasting in the first place. History is littered with examples of such risky boasting and several are captured in earlier chapters. Notwithstanding this truism, there is a place for a judicious but honest account of the place of intelligence, and, in the case of this study, Sigint, to enable us to understand the past and shed some light on the future. This applies particularly to the use of military force and Sigint, as well as cyberspace, in the pursuit of Australia's national interest. Intelligence, and specifically Sigint, is intimately linked to the function of government.[1]

The available literature on military and national security affairs since Federation indicates that some historians have been reluctant to write about Sigint because they do not, or fear that they may not, have access to all the facts. Others, who may have spent time working as trusted insiders, with privileged access, may fear that what they say may constitute a security breach. Others still, such as official historians of the Second World War, were prevented from including material on Sigint, if indeed they were ever privy to information about it. While we were working on the official history of ASD, we had such access. It has informed this history of Australian Sigint up to and including the Second World War. This book opens the door to a deeper understanding of Australia's role in world history, revealing how Sigint influenced, sometimes determined, major events. The decryption of the Zimmerman telegram was key to the United States' decision to enter the First World War and contributed to the Allied victory. Any number of otherwise accurate, authoritative accounts of events like this that leave out the role of Sigint are leaving out something vital.

In his speech at the Lowy Institute in 2019, in addition to giving an unprecedented public account of ASD's use of offensive cyber capabilities, Mike Burgess spoke about transparency and helping to dispel myths.[2] He touched on capabilities that developed with the first computers at the height of the Second World War.

The Director-General of ASD at time of writing, Rachel Noble, has made public mention of technological challenges and opportunities and how, through the decades, Australian parliaments have made legislative changes to enable ASD to do its job within strict parameters while keeping pace with technological change.[3]

Over the course of history, secret activities have gone from being at the whim of a ruler to being controlled by the executive arm of government, without reference to the legislature. In the words of Dennis Richardson, what began as an assumed bargain between the individual and the state has evolved into an increasingly explicit bargain, the terms of which are made clear through democratic debate and then set down in legislation.[4]

We look forward to that democratic debate continuing and hope that this book will contribute to it, while encouraging personal reflection on what the history of Australian Sigint and the advent of cyber means. As the traitor unmasked at the end of the late John le Carré's *Tinker Tailor Soldier Spy* declares, 'secret services were the only real measure of a nation's political health, the only real expression of its subconscious'.[5] As Alex Younger, former head of MI6, said 'You can tell a lot about the soul of a country from its intelligence services'.[6]

Sigint qualities

If the secrecy of the services is the key to them being the real expression of a nation's subconscious or a means of penetrating the soul of a country, these observations apply particularly to Sigint, the most secret part of the intelligence services. Indeed, while researching and writing this book, we recognised in Australian Siginters strongly developed qualities, which are often thought of as quintessentially Australian. One of those qualities is resourcefulness.

In the words of the wartime British Sigint chronicler Nigel de Grey, 'it is impossible not to admire Australian enterprise and initiative in the speed with which they got into action'.[7] Central Bureau had 'grown up on its own',[8] having little early contact with other centres and yet 'it controlled its field organisation, ultimately not only Australian but also American, even more closely than was done in the last phases by the Sigint Directorate in South East Asia Command'.[9]

Another quality is diffidence, perhaps associated with an early lack of national self-confidence stemming from Australia's junior partner status with the United Kingdom and the United States. A vivid illustration of it, as we have seen, is found in a part of the secret Holden Agreement, signed by GC&CS and the US Naval Security Group on 2 October 1942.[10] The Agreement described the Australian naval Sigint unit, later known as D Special Section headed by Eric Nave, as a British–Australian Unit that was turned over to the United States. It also singled out Nave by name for recall to the United Kingdom. Fortunately, both moves were skilfully countered by Australians who recognised the importance of preserving Australia's national Sigint enterprise.[11]

Faced with a government that refused to spend on defence, as early as 1921 the Australia Navy did what it could 'on the smell of an oil rag'. It distributed Japanese Morse code manuals to all of its naval telegraphic stations and major units, and ordered all ships and stations to practise taking down Japanese Morse code.[12]

Larrikinism is another quality that comes through. In 1914, Australians were not above duping the German master of the *Hobart*. Similarly, in 1932, one Lieutenant Paymaster W.E. McLaughlin, then Liaison Officer aboard HIJMS *Asama*, seized the opportunity to copy the frequencies and schedule used by the Japanese squadron and provided them to the RAN escort ships.[13]

Larrikinism may have gone hand in hand with a certain disregard for security, which so dismayed the British and the Americans, culminating in the crisis in 1947 when decrypts revealed an Australian was passing information to the Soviet Embassy in Canberra. Remembering the war when the Soviets had fought bravely on the right side and largely ignorant of aspects of that regime that they would have found unacceptable, the

Chifley government had been too friendly with the Soviet Union for the liking of the Americans as well as the British. Australia was only readmitted to the cryptologic exchange after that government's electoral defeat in 1949. Australia's readmission was more than justified by the well-managed defection in 1954 of Vladimir and Evdokia Petrov, a timely intelligence coup for Australia.

So, what is it about Australian Sigint that would encourage the qualities we have mentioned? The penetrating observation by Justice Robert Marsden Hope on the early ASD, then known as DSD, may shed some light:

> The twin constraints of specialisation and secrecy combine to develop in DSD a degree of exclusiveness and inward looking that could prove dangerous for morale. So far as I can tell, that has not been the effect on DSD's morale. On the contrary, the management has contrived to engender a strong feeling of *esprit de corps* based upon these same constraints.[14]

The powerful twin constraints created a cloistered environment, not just for DSD, but for Siginters in the armed services, which protected and encouraged those qualities we tend to think of as quintessentially Australian as well as the strong esprit de corps, which Hope observed.

This book has traced the impact over the years of a series of gifted individuals, such as James McCay, Defence Minister and then Commander of Australia's Intelligence Corps in the early 1900s, as well as linguist and codebreaker Captain Eric Nave. Others include Lieutenant Colonel 'Mic' Sandford, the skilful Central Bureau diplomat and eccentric; 'Ath' Treweek, the Sydney University scholar in ancient Greek who taught himself Japanese; Ruby Boye-Jones, the intrepid Coastwatcher; Jack Ryan, the young wireless operator on the HMAS *Sydney*; Florence McKenzie, the inspirational trainer of wartime women signallers; and Miss Lake, the mysterious teacher of Japanese. These people tend to appear more than once in this story – because of the special skills and qualities involved, it is a small world.

The words 'We also served' were inscribed on the back of the commemorative badge belatedly given to those involved in Sigint in the Second World War by British Prime Minister David Cameron. This was in recognition of the unsung work of many who had for most of their lives kept the oath of secrecy they pledged at the height of the war. Australians involved in the wartime Sigint enterprise were included. Appropriate recognition for the work they performed and what it meant to the nation is long overdue. We hope this book will encourage such recognition.

A record of the past becomes all the more important if, as some contend, the future is looking more uncertain than it has for generations. This book covers some of the story, but there is more to be said about what has gone before in order to make balanced judgements on where Australia should stand internationally; whether the balance is right on accountability and oversight arrangements; and how important Sigint and cyber security to Australia's place in the world today and in the future.

History rhyming?

There was a time when Sigint served the needs of a small clientele in the National Security Community in Canberra near the end of the Cold War. Those were the days when, in response to the question 'who is your (Sigint) customer?', the answer was 'Christ knows!' It was with this other-worldly mindset that outside critics would facetiously describe its product as 'Sigint for Jesus'. In other words, the Sigint was written with little regard for the needs of policymakers. Those days have gone. The technological upheaval, combined with great power contestation and competition underscore the need for Sigint and cyber to be effective, relevant, focused on priority concerns and better explained to the Australian people.

The lessons of the past may provide some useful pointers. That becomes more important if, as some suggest, Australia and its place in the region is at a turning point. Since the Snowden revelations of the second decade of this century, Sigint has faced far more exposure. More secrets have been revealed. The increase in great power contestation has called into question the hitherto remarkably successful arrangement of intelligence 'ties that bind' Australia with the United States and the other UKUSA partners. A seemingly permanent feature of Australian security, after all, remains a contract between nations and their intelligence agencies that is subject to review. These arrangements cannot be taken for granted. Notwithstanding the strong criticism of some, to date the arrangements have been widely considered, on a bipartisan basis, as being in Australia's interests. The authors continue to hold this view as well.

At the outset, we wrote about EW and Sigint revealing secrets and protecting one's own. This is a distillation of procedural and technological complexity into apparent simplicity. In practice, many aspects of the Sigint and cyber arrangements we take for granted are being buffeted by the gathering pace of change. This applies to technology, accountability, adaptiveness, social licence, integration, and foreign threats.

Technology has been transformed by satellite communications, the switch from analogue to digital, the invention of fibre-optic cables, advanced computers, the internet, mobile phones, personal digital assistants, not to mention quantum computing and autonomous systems. The days of listening in closely to HF radio transmissions may not be completely over, but they are but a fraction of the spectrum of technological challenges faced by the modern Sigint and cyber enterprise.

Accountability has advanced, with a shift from intelligence being authorised by executive mandate at the discretion of a minister or prime minister, to being mandated and articulated in detail through legislation scrutinised by the Australian parliament before being enacted into law. Today, accountability mechanisms include parliamentary, executive and independent judicial oversight, as well as audit and review mechanisms. The Australian people expect this level of accountability, especially given the intrusive surveillance powers with which intelligence agencies are entrusted, and this will apply to the REDSPICE initiative as well.

Adaptation is the way the national Sigint and cyber agency responds to changing technologies and threats. The national Sigint agency was born with a veiled title and designed for limited communications methods from the era of analogue radio. Over the years, it reinvented itself. In addition, it was transformed from an organisation so secretive even its name was a cover, to having shopfront offices in state capital cities across the nation and a substantial online presence, including social media. Along the way, it has developed cyber capabilities to defend not just government facilities but also industry and society's networked infrastructure.

Integration means incorporating Sigint as part of the broader apparatus of state to protect the nation against subversive domestic and foreign threats. Integration also means ensuring that intelligence is not a 'self-licking ice cream'; that is, to ensure that intelligence collection, analysis and reporting is fit for purpose, suitable, relevant, accurate and timely for government decision-makers and relevant and helpful to the broader community. Increasingly, the National Intelligence Community, including the cyber components, are having to work not just across government agencies at the federal level, but also with counterparts in state and local government, business, industry, civil society and the broader community.

The democratisation of intelligence and demands for greater accountability have made the task more challenging. In the early days, Sigint was authorised as a ministerial directive; since then, there has been growth in public awareness, thanks in part to the reviews and royal commissions, and the proliferation of legislation on how and why Sigint and other forms of intelligence can be collected, analysed and reported. These reviews and laws have led to a degree of public exposure likely to make the traditional introspective intelligence practitioner uncomfortable.

Revelations, particularly from abroad, of intelligence overreach, are now common knowledge in Australia. Governments stand or fall on voters' trust in them. If they lose that trust, they lose office. Some details of course must remain hidden, as it would be unsafe or inappropriate to divulge them. Nonetheless, with heightened competition over narratives and a surge in misinformation, greater effort is required to ensure that the Australian people are given the clearest and fullest possible explanation of the rationale and workings of the Sigint and cyber enterprise.

The Fourth Industrial Revolution, or the information revolution, brings with it web-enabled digital systems, including robotics, autonomous systems and quantum computing, including precise and pervasive surveillance. History never repeats, they say, but it does rhyme. These new developments, however, foreshadow a different future for Australia.

The information revolution has increased the demand for transparency and accountability. This has added a degree of vulnerability to the social contract Australians have with their Sigint agency. The same can be said for the Sigint agencies of Australia's Five Eyes partners. Controversies over leaks, incompetence and malpractice feed conspiracy theories, undermining the authority of intelligence organisations and public confidence in them. A no-comment, or refusal to confirm or deny, let alone engage in a meaningful way, can be as damaging as a leak.

While secrecy is key to the practice of Sigint, it has become increasingly difficult to maintain. Revelations have reportedly affected collection capabilities and limited the ability of the intelligence community to understand and report on the region with confidence. Collectors and analysts feel the pressure of reporting deadlines, as well as the demands for 'actionable' intelligence. And while there can be refinements and enhancements to the technology and analytical techniques of those working with Sigint, the future remains unknowable.

Demands for greater insight are coming at the same time as calls for more accountability and denunciations of the surveillance powers of the state. Sigint has been overwhelmed by a data deluge that makes prioritisation of information collected for further analysis very challenging – even with the aid of advanced computing power to help 'sort the wheat from the chaff'. As Major General Steve Meekin observed, 'some things will not change – reliance on smart people, good ideas, nous, courage and initiative (in the context of risk), innovation, linguists and target knowledge'.[15]

Some outsiders imagine analysts loitering over salacious sound grabs or pictures. The reality is one of enormous pressure to focus exclusively on priority issues and targets, monitored by powerful oversight bodies such as the IGIS which is empowered to strictly enforce legislation, with legal consequences for those who abuse the trust placed in them. This is the notion of protecting one's own citizens from the malign use of intrusive Sigint capabilities.

What is more, some would say that practitioners accustomed to working in secrecy and shunning the limelight have not provided convincing answers to the conspiracists' claims. One explanation is that many intelligence practitioners learnt their business from highly classified internally-sourced documents and procedures. As a result, all they know is classified secret and they are unaccustomed to, and uncomfortable with, distinguishing between classified and unclassified matters; what can be talked about with the media and others and what cannot. Such risk-averse practitioners avoid saying anything in public at all. Often enough, a grain of truth that reaches the public is exaggerated, and the 'conspiracy' can be explained by simple human error.

Conspiracy theories are encouraged by the drift away from the model of separation of functions established by Justice Hope. According to his model, separate foreign intelligence collection functions were to be performed by separate agencies (ASD being tasked with Sigint and cyber), analysis was to be separate from the collection of intelligence (principally by ONI and DIO), and intelligence analysis separate from the policymaking process. The policy/analysis divide was seen as important also, because policymakers were more directly subject to the demands of elected political masters. Justice Hope recognised that for intelligence analysts to be able to report critically, accurately and authoritatively, they needed to be managed separately. They had to be free of the pressures of the 24/7 news cycle and the politically-charged demands of senate inquiries and parliamentary question time for which policymakers are required to prepare their ministers.

Hope's model is also threatened by a new breed of intelligence agencies that, for some time, were under the umbrella of the Home Affairs Department. These include the Australian Criminal Intelligence Commission (ACIC), Australian Transaction Reports and Analysis Centre (AUSTRAC), Home Affairs Intelligence Branch and the intelligence arm of the Australian Federal Police, as well as financial oversight responsibilities for the management of the domestic security intelligence organisation, ASIO. Demands have grown for access to Sigint for these agencies. Hope's recommendation, endorsed by Dennis Richardson in his 2019 review, was for ASD to keep its focus on collection of foreign, and not domestic Sigint. Domestic security issues were the remit of ASIO, but with the Home Affairs Department's suite of intelligence functions, the demarcation became blurred.

At the time of writing, it was still too early to comment authoritatively on the effect of the Albanese government's decision to shift the ACIC, AUSTRAC and AFP back from Home Affairs to being under the remit of the attorney-general. But this is a move taken to strengthen community confidence in the government by showing that power was not being further concentrated into the hands of fewer, more powerful and apparently less accountable figures. The need for increased accountability and reassurance to the community seems to have weighed heavily in the decision to shift these bodies from under the Home Affairs portfolio back to being under the Attorney-General.

Demands have also grown for the new intelligence agencies to collect their own intelligence, including Sigint and cyber. Provisions are before parliament which would give these agencies new legislative cover for their tasks. Most of what these agencies do may be unquestionable and wholly legal, but there are concerns that the blurring of Hope's demarcation may compromise the integrity of the system, damaging the confidence of Australians and international partners. Additional scrutiny and oversight have been necessary to mitigate the risk of exposure of sensitive collection systems, targets and methodologies and more may be required.

A clear explanation of what is being done and why it is being done is critical to ensure that the national Sigint and cyber instrumentalities of state maintain their social licence to operate – a licence that comes through an elected parliament, which depends on elected officials and, above all, on the Australian electorate believing in it. People need to understand what it is that the national Sigint and cyber agency does for the nation. This requires an authoritative historical account, alive to national security sensitivities. By now it should be clear that there is a fine balance to be struck between openness and the secrecy necessary to maintain operational effectiveness.

Yet another challenge to established norms with regard to Sigint and cyber management relates to technological solutions that would require a realignment of resources. Nowadays, intelligence agencies are not the sole repositories of impressive computing power, nor of open-source intelligence analysis and reporting. There are several examples of companies delivering accurate and timely reporting that once was the preserve of the intelligence agencies, but drawing on open-source intelligence. None of these companies, however, have the computing power, global reach and benefits of other significant investments (including artificial intelligence and quantum computing) of the government intelligence agencies. This means that what the companies have to offer will remain bespoke functions, likely to be increasingly useful without posing a direct threat to the state and its management of its classified intelligence. Commercial service providers are ready to supply customers with satellite imagery and open-source intelligence analysis that would have been unimaginable a generation ago. Closeted in secure compartmented intelligence facilities, air-gapped from the rest of the world and the internet, intelligence collection and analysis can sometimes lose track of fast-moving events. Tech-savvy outsiders, with mobile devices and instant news updates on apps, can monitor what is happening without access to secret intelligence sources and could sometimes leave the professional intelligence practitioners behind.

The question then arises: can more of the Sigint function be outsourced to commercial service providers operating outside such strict provisions and drawing on publicly available data? In addition, can the remaining resources be directed towards acquiring only what needs to be collected above and beyond what can be gathered from other open sources? Our view is that a reassessment is required of which functions are critically important and need to be undertaken under the strictest secrecy provisions and which require less restrictive management. This reassessment needs to happen with an eye to the difference between facts and fabricated information – that is, fake news or deliberate and malevolent misinformation and disinformation.

Australia's Sigint arrangements are more advanced and sophisticated than ever, and the strong relationships with UKUSA partners are key to this. That seems unlikely to change. Indeed, many of the cyber security measures that help protect critical infrastructure are made possible with the help of the ACSC, with its links to trusted partners, notably those which are part of or closely connected to the UKUSA and Five Eyes network, although there are also advanced options available from other close security partners beyond the Five Eyes network.

To date, the security mechanisms of the past have served the nation well. But in view of damaged public confidence thanks to disturbing revelations in recent years and accelerating technological and geostrategic change, there is a need for a broader, continuous conversation with the nation in order to maintain and protect the social licence to operate.

Calls for additional self-reliance in Australia raise the question: how much is to be gained by maintaining close ties with UKUSA partners to enhance Australia's ability to pursue its own national interests and strengthen its own defence? To date, the answer has been overwhelmingly in favour of maintaining, strengthening and deepening alliance ties. The argument is based on the idea that enhanced self-reliance, curiously enough, comes most effectively through enhanced alliance cooperation and sharing. Perhaps, in future, that self-reliance could be more inclusive of our neighbours in Southeast Asia and the Pacific.

Australia's national Sigint agency, ASD, is already adjusting to the new dynamics with the expansion of the ACSC. With its branch offices in state capital cities, the ACSC is engaging members of the Australian community to a degree unimaginable in the days of the Cold War.

Indeed, cyber is taking on a life of its own. With its origins in Sigint, cyber is digitally and not analogue based. Indications are that, in future, the scope and influence of cyber will eclipse the traditional Sigint functions. Understanding how cyber was born from Sigint will inform judgements we make about the future.

Many challenges of today are about a future which cannot be predicted with certainty. Who could have foreseen that in 2022 Elon Musk's Starlink satellite internet service would change Ukraine's war with Russia, not only keeping Ukrainians online, but helping troops to communicate on the battlefield and enabling drones and weapons systems to stay operational.[16] But, at the same time, there are aspects of today's technological, political and social challenges which echo the past. As historians working in strategy, it seems self-evident that intelligence agencies cannot afford to operate in a way that is, or appears to be, unaccountable. They must negotiate the boundaries with their stakeholders – especially the main one, the Australian people.

With demands for more help to 'protect our own', there is no room for complacency. We hope that this consolidated and unofficial history of Australian Sigint will contribute to public awareness, understanding and engagement on these issues. The first recommendation of the Richardson Review mentioned at the beginning of this book is that National Intelligence Community agencies ensure that their induction and training address the history, background and principles that underpin their legal frameworks.[17] The pace of change and unprecedented demands on Sigint add weight to that recommendation.

We would take this one step further and recommend that all

Australians, not only those who work for National Intelligence Community agencies, take a renewed and personal interest in their intelligence history, and especially in what its most secret aspect – the history of Sigint, and now cyber – means for them as Australians. Cyber will grow in importance in society, in government, schools and industry and in the armed forces and emergency response agencies. With so much at stake in protecting our own, reliable and timely intelligence to support decision-making will be more important than ever. That means that apart from being more engaged with society than ever before, particularly on cyber issues, the nation's Sigint enterprise will need to continue being effective at revealing secrets.

Appendix A
CENTRAL BUREAU, BRISBANE, WORK FLOWCHART

The flowchart below illustrates the Central Bureau (CB) processing of Japanese encoded traffic from interception through to the dissemination of decrypts to recipients and/or destruction. It is derived from the Central Bureau Technical Signals Intelligence Report No. 2, and is a supplement to that report. The version below is file HW 52 87 held by the UK National Archives. As described in the pages that follow:

- CB Field Station intercepts the Japanese messages and distributes three copies, either by teleprinter link or safe-hand bag.
- Once intercepted, messages are counted and sorted at CB, copies are sent to the Signals Intelligence Service in Washington, relevant advanced elements of CB (e.g., Philippines), and a copy processed further by CB in Brisbane.
- The 'To' and 'From' addressees on each message are then decoded.
- The 'discriminant' for each message is then decoded. The 'discriminant' is a set of figures or letters incorporated into a message that tells the recipient which encoding system is being used for this message.
- Based on the 'discriminant', messages can then be separated into:
  » 'unsolvable systems' for filing. In this flowchart they use OTT as an example system – OTT stands for One Time Table.
  » unwanted systems, because they are of low importance or processed by centres other than CB.
  » 'readable' (i.e., ones that can be decoded) messages for further processing by CB.

- A worksheet is created for each 'readable' message using CB's IBM suite. This stage might engage the IBM-036 punch card generator, the IBM or Hollerith sorter, and an IBM-405 tabulator.
- The worksheet is used for some initial hand analysis, which may involve identifying the 'indicator' group. The 'indicator' is a group that flags to the recipient what setting, additive or other type of crypto variable was used specifically to encrypt this message.
- On the basis of the 'indicator' and 'discriminant', a decision will be made to send a message either for processing/decoding by hand, or back to the IBM suite for machine decryption. Machine processing relies on much of the relevant code system having already being broken (or 'recovered').
- A quick scan of each decrypted message is performed to determine its importance:
  » 'unimportant' messages will be filed
  » messages containing technical information regarding Japanese communications practices will be translated and the knowledge used to improve Sigint against the target
  » 'important' messages will be fully translated.
- Translations of 'important' messages will be edited and disseminated to a range of users and filed for future use.
- An Order of Battle Section also uses decrypted messages and a range of other inputs to maintain an up-to-date picture of the relevant Japanese service structure, movements, location, strength etc.

TOP SECRET
ULTRA

DESCRIPTION OF PROCESSES CARRIED OUT AT C.B. BRISBANE AS AT APRIL 26, 1945.

1. Field Sections send all raw material to Central Bureau by Safe Hand Bag. However, traffic with clear discriminants or specially designated, is first sent by radio teletype when such facilities exist. Usually three copies of each message are made by the field sections.

2. After counting and sorting for clear discriminant traffic one copy is sent to the SIS signal office for transmission to Washington, one copy is sent to Advanced Central Bureau and one copy sent on for processing.

3. The message addresses are first decoded, if this has not already been done at the field section.

4. The discriminants are next decoded.

5. This process is followed by sorting in order to separate the readable messages from those which cannot be solved (e.g. one time table messages, etc.) and unwanted signals known to be of minor importance or dealt with by other Centres.

6. The messages are next sent for IBM processing and a work sheet is prepared, the original copies being returned and filed.

7. The messages are now distributed according to their systems to appropriate solution departments where preliminary work on indicators is done before passing on to the next stage, or the message may be processed entirely by hand.

TOP SECRET

ULTRA

8. The IBM stage decodes the message and produces a new work sheet for the translators.

9. Messages are first scanned by translators in order to separate those of an unimportant nature.

10. Messages are next indexed using the logging system employed in the Japanese Signal Office, and at this stage consecutive parts of messages are found and associated. The unimportant signals are filed, and messages of a technical nature, especially those dealing with ciphers or cipher instructions, are separated and passed to special translators who produce a weekly report.

11. The translators next receive the messages for translation and return them to the loggers who pass them on for editing.

In the Editing Section many parts still missing are located and put in their right places, obvious errors in translation are corrected and normal wording and designations ensured. The messages are now duplicated, the original work sheets filed and copies sent through appropriate channels to authorised recipients.

Copies on local distribution are finally destroyed.

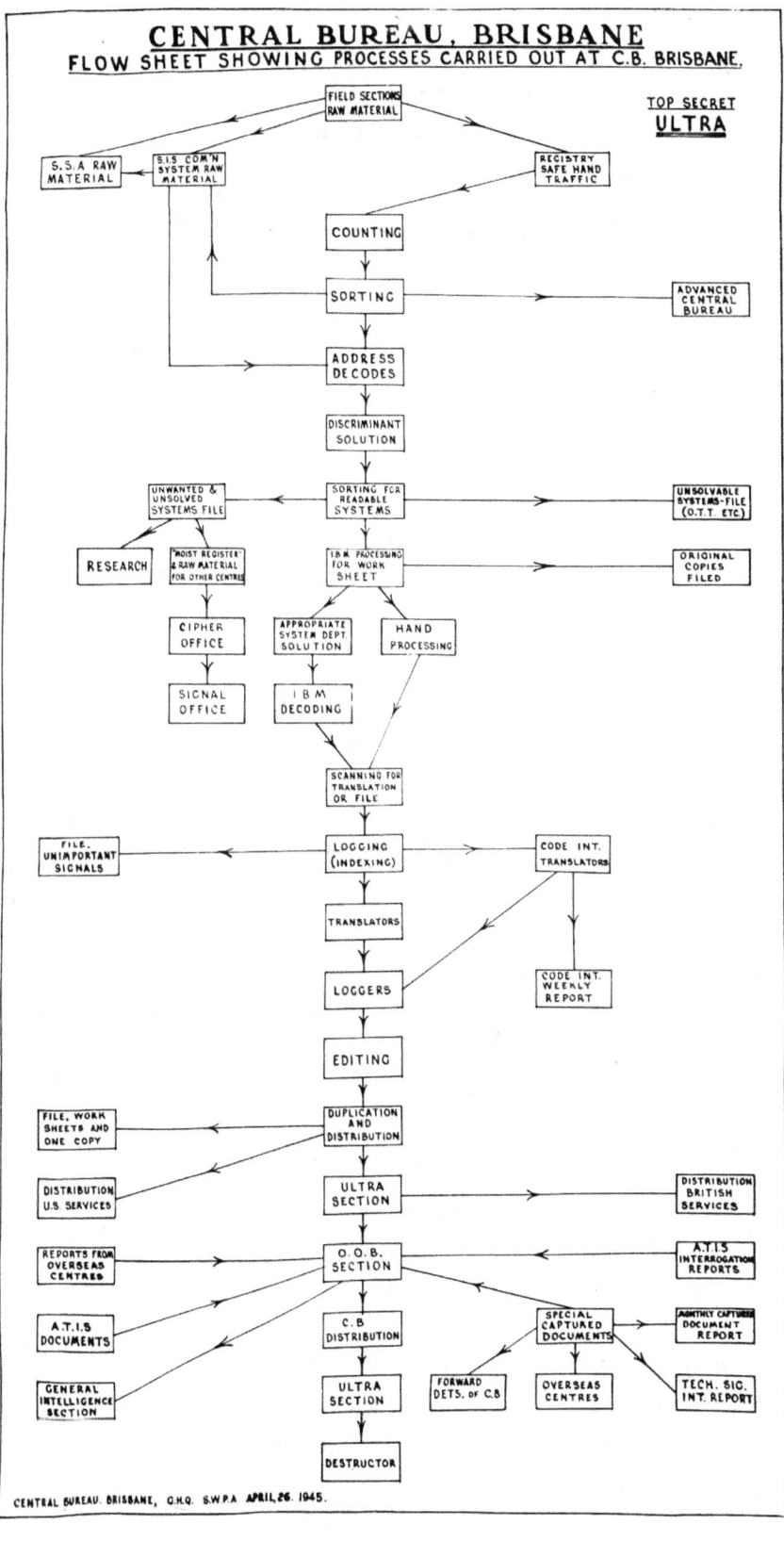

Appendix B
ARMY & AIR FORCE WARTIME SPECIAL WIRELESS UNITS

This appendix is drawn from material found in the following documents: Report, 'Central Bureau Technical Records. Part J – Field Sections', (undated) NAA: B5436 Part J; and Report, Lt. Colonel J.W. Ryan AIF to the Director of Military Intelligence and Director of Intelligence RAAF, 'Report on Special Wireless Units (Signals) 1940–45', 19 December 1945, NAA: A10908, 2.

ASWS teams included two groups. The first were Signals personnel – officers, operators, linesmen and administrative personnel. The Signal officer-in-chief was charged with preparations, training, appointments and supply of qualified personnel and equipment. The director of military intelligence was responsible for provision of personnel for the second group, the wireless intercept (WI) team, otherwise known as 'Y', Intelligence personnel. The 'Y' personnel were provided and directed by Central Bureau and organised separately as Australian Special Intelligence Personnel Sections (ASIPS).

51 Aust. Special Wireless Section
Raised in March 1942 following the return of the Special Wireless personnel from the Middle East, the section operated in the Darwin area until May 1945, when its functions were taken over by the 1 Canadian Special Wireless group. Its function was primarily interception of enemy army-air and naval-air wireless radio activity.

52 Aust. Special Wireless Section
Formed in June 1942 at Ferny Creek, Victoria, the section moved to Bonegilla where it worked with the Headquarters Group. In May 1943, it

moved to an independent camp site at Mornington race course, Victoria, where it focused on the interception of enemy and neutral diplomatic traffic.

53 Aust. Special Wireless Section
Formed in October 1942, the section was located with the Headquarters Group until November 1943, when it moved to Finschhafen, New Guinea. In August 1944, it moved again to Hollandia until disbandment in May 1945. Its focus was on interception of enemy army and army-air administrative radio links.

54 Aust. Special Wireless Section
Formed in October 1942, the section was located with Headquarters Group until September 1944 when it moved to Darwin, operating alongside 51 Section and then the Canadians until it was withdrawn in June 1945. Its function was to intercept enemy army and army-air administrative radio traffic.

55 Aust. Special Wireless Section
Formed in July 1942, the section moved to Port Moresby in September until it was withdrawn to the Headquarters Group location in March 1944. In the meantime, detachments had operated at Wangela, Milne Bay and Karema (in late 1942 and 1943), as well as Wau (July 1943) and Nadzab (October 1943 to January 1944).

56 Aust. Special Wireless Section
Formed in October 1942, the section was manned principally by AWAS personnel and focused on interception of enemy army and army-air administrative activity. Located alongside Headquarters Group until April 1943, it then moved to Perth, where it was attached to Headquarters 3rd Australian Corps. In October 1944, it moved to Brisbane, where it remained until disbandment in May 1945.

58 Aust. Special Wireless Section
Raised in December 1943, the section, like 56 Section, was operated by AWAS personnel. It focused on interception of enemy army, army-air

administrative and naval shore-based wireless telegraphy activity. It was located alongside Headquarters Group until disbandment in May 1945 when remaining personnel transferred to 96 Aust. Special Wireless Section.

59 Aust. Special Wireless Section
Raised in May 1944 and tasked with intercepting and publishing enemy and foreign news and propaganda transmissions, the section also published daily reports on Japanese language broadcasts.

96 Aust. Special Wireless Section
Raised in May 1945, due to the disbandment of several other sections, the section was collocated with HQ ASWG in Brisbane. Perhaps the largest ASWS, it included 51 men and 84 women. Tasked with interception of enemy army, army-air administrative and naval shore-based transmissions, it was disbanded in September 1945.

RAAF Wireless Units
RAAF's Wireless Units (WUs) were numbered 1 to 7. These were located at various times in Melbourne, Brisbane, and Townsville; and offshore in New Guinea at Port Moresby, Madang, and in the Netherland East Indies (now Indonesia) at Hollandia, Merauke, Biak, Morotai, and Labuan (now in Malaysia). By June 1945, WUs 3 to 6 were in operation.

US Army Special Reconnaissance
The US Army also established an intercept and DF site at Adelaide River (NT), as well as US Army Special Radio Intelligence (SRI) companies numbers 111, 112, 125 and 126, forward-deployed from 1945, in the Philippines. There was considerable overlap and exchange between the services and between the national contingents. This would have a longlasting spill-over effect because the close and trusted ties at the working level built during the shared trials and tribulations would outlast the war.

ACKNOWLEDGMENTS

We would like to acknowledge the select and highly qualified team that worked enthusiastically on this project, particularly while it was part of an official history project sponsored by ASD until the contract was cancelled in 2020. They are Major General (Ret'd) Steve Meekin, Jonnine Negus, Vikki Jakobson, Guy Hansen and Stewart Skelt.

Rochelle deserves a shout-out for her help and understanding. Kelly, Mel and Felicity deserve a shout-out as well for their administrative support as well as former Director of DSD, Steve Merchant. Bob Hartley, Petr Dunn OAM and Ian Pfennigwerth graciously allowed us to use images and diagrams from their publications. Thanks also for the support provided by archivists at the National Archives of Australia, the Australian War Memorial, the National Archives of Kew (UK) and the NSA Cryptologic Museum.

We also thank Air Commodore (Ret'd), Rick Keir and Captain (Ret'd) Gordon Andrew, RAN, the RAN Seapower Studies Centre, as well as the ANU's Judi Burtenshaw, Michelle McGuinness, Professor Anthea Roberts, Professor Emerita Joan Beaumont and the late Professor Brendan Sargeant.

This publication would not have been possible without the support of SDSC and colleagues at the ANU who helped us secure the intellectual property rights for the draft up to 1945. Special thanks to Judi Burtenshaw, Christine Sullivan and Professors Toni Erskine, Stephan Frühling and Brendan Taylor , as well as Dr Greg Raymond. In addition, mentor, friend and colleague, Emeritus Professor David Horner, warrants particular thanks for his support and advice. It would be remiss not to also acknowledge the contribution to our understanding of Sigint and national security policy of our late friend and colleague, Emeritus Professor Desmond Ball. Without the ground-breaking research and writing he undertook over many years on Australia's intelligence arrangements, much of this may never have been written.

The staff at UNSW Press were terrific to work with. In particular, we would like to thank the Executive Publisher, Elspeth Menzies, as well as editors Paul O'Beirne and Briony Neilson and designer, Josephine Pajor-Markus, who helped us make sure the final product came out polished! John is also indebted to his wife, Judith Steiner, and the family, for their support.

Finally, we would like to acknowledge and dedicate this work to the memory of those who could say 'we also served', having maintained their oaths of secrecy for many years – in most cases until their passing.

GLOSSARY

Unless otherwise annotated, this is derived from, and using the original spelling found in, a glossary compiled at HW 43/49, 'GC&CS Army and Air Force Sigint Volume VI: The Organisation and Evolution of British Sigint II' by Lt. Cmdr A. Griffiths RNVR.

Additive: A group added non-carrying to a code group in order to cipher it.

Additive table: A table or book of figure groups used for ciphering code groups or plain language addition.

Break: To establish the plain language equivalent of the groups of a code book or to reduce a cipher to plain language; to reconstruct (any code or cipher system).

BRUSA: Sigint exchange circuit, linking GC&CS, Washington, Colombo, Pearl Harbor, Melbourne and Guam.

Call sign: A group, usually of three or more letters and/or figures, sent either in clear or in cipher, either in the preamble or in the body of the message and serving to identify the sender and/or the recipient.

Cipher (n): Any system whereby the individual letters, figures, punctuation marks, etc., of plain language or the individual letters, figures or other symbols of an encoded message are rearranged among themselves (transposition), or with an admixture of other figures or letters (dummies), or replaced by different letters, figures, etc. (substitution), or hidden in seemingly innocent material (concealment), with a view to making the message unintelligible to anyone not in authorised possession of the knowledge or apparatus necessary to reverse the systematic process and to restore the order of letters, figures, etc. of the original plain language or encoded message.

Cipher (v): To change plain language or code groups into cipher by transposition, substitution or concealment, either by hand or by machine.

Comint: Communications intelligence, a subfield of Sigint concerning voice or written communications (other than material publically broadcast) which was collected for use (often including translation and decryption) by someone other than the intended recipient.

Control: The group or groups which indicate with what key-group the indicator has been enciphered.

Coverage: Provision for intercepting wireless signals.

Cryptanalysis: The art or science of ascertaining the essential nature of codes and ciphers and reconstructing the system and operations used by the encoders, and the encipherers, or enough of these, to enable the message to be read.

Cryptanalyst: A person engaged on cryptanalysis.

Cryptanalytic: Of, or belonging to, or employed in cryptanalysis.

Cryptographer: One who encrypts or decrypts messages or has a part in making codes or ciphers.

Cryptographic: Of, belonging to, or employed in cryptography.

Cryptography: The making and use of codes and ciphers.

Cyber: Of, relating to, or involving computers or computer networks (such as the internet).

Decode (n): Plain-language message obtained by decoding a code message.

Decode (v): To substitute the plain-language equivalents for the code groups of a coded message, to reduce to plain language thus.

Decrypt (n): The plain-language version of a coded and/or cyphered message.

Decrypt (v): To decode and/or decypher a coded or cyphered message.

Decipher (n): To change back into plain language or code groups a text that has been enciphered.

Depth: Two or more messages in a machine or similar letter substitution cipher that have been encyphered on the same machine setting or on the same key.

Discriminant: A group (or, rarely, two groups) placed normally in front of the text of a cypher message, indicating – in the case of cyphered code – the particular code book and cyphering key, and/or the cypher procedure, used, or – in the case of machine cyphers – the particular set-up used, and so serving to indicate the degree of secrecy of the message or to distinguish one type or section of traffic from another.

Elint: Enemy electronic intelligence.

Encode (v): To substitute code groups for the plain-language units of a text according to a system embodied in a code book.

Encrypt (v): To convert a plaintext message into unintelligible language by means of a cryptographic system.

Encypher (v): To change plain language or code groups into cypher either by hand or by machine.

Frequency: Classification of wireless waves according to the number of cycles or complete undulations completed in a second.

G2: Intelligence Branch of the US Army.

Group: Any number and combination of letters, figures, or symbols transmitted or set out consecutively so as to form one entity.

High-grade (of a code or cypher system): Designed to provide security – that is, resist breaking – for a comparatively long period or indefinitely.

Indicator: One or more letter or figure or letter-and-figure groups (either sent in clear or enciphered on a separate system and placed at the beginning and/or end of a message, or in the body of it) indicating the key or table used, or in the case of a long subtractor the starting-point or starting-point and finishing-point.

Intercept (n): A signal obtained by interception.

Intercept (v): To listen and record W/T signals, especially those not intended for one's reception.

Kana: A system of basic syllables by means of which all Japanese words can be spelt in contrast to characters.

Key: In cryptography, a symbol or sequence of symbols applied to text to encrypt or decrypt.

Key recovery: The cryptanalytic reconstruction of a key.

Machine cypher: A cypher system in which the encyphering and decyphering are performed by means of a machine.

OP-20-G: The sub-division of OP-20 (Department of US Naval Communications) which was charged with the work on enemy communications intelligence.

Plain language: Language intelligible without the assistance of a decode (or decypher).

Preamble: Introductory part of cypher or code message, including, for example, call sign, date, time, serial number, number of groups, address, indicator etc., sent either in clear or before the cypher text, or partly in clear and partly cypher, or wholly in cypher at beginning of text.

Recovery: A code or key identification, or the action of producing it.

Re-encode: To encode (the same clear message) a second time, especially with a different code or system.

Re-encrypt: Re-encode and/or re-encypher.

Sigint: Any intelligence from the Sigint Service.

Sigint Service: General term comprising all those elements which in combination or individually produce intelligence from a study of all communications and radio transmissions.

Sigsaly: Encyphered voice communications telephone.

Special intelligence: Intelligence resulting from the solution of high-grade codes and cyphers, or those for security reasons placed in this category.

Strip: To remove key, especially from encyphered code.

'Thumb': Security prefix for all information derived from the study of wireless communication short of decoding.

Traffic: Messages in cypher, code or clear text transmitted, or prepared for transmission, by electronic means.

Traffic analysis: The study of the external characteristics of enemy W/T traffic for the purpose of making deductions useful to cryptanalysis and/or intelligence.

Transposition: An encyphering system or process whereby the plain-language letters (figures or letters, etc., of an encoded or encyphered version) of a message are re-arranged among themselves according to a key before transmission.

True (of the figures or letters of additives, code groups, etc.): (Requiring no further correction to make them) the same as those actually used by the cypherers; (opposed to provisional or untrue).

Typex: British cypher machine, basically similar to the German Enigma machine.

'Ultra': Code word for Special Intelligence.

W/T: Wireless/Telegraphy.

Y: Interception also used to include the preliminary analysis of intercepted signals frequently undertaken by Y stations. Normally Y is exclusive of cryptanalysis, but in such terms as 'Y Afloat' or 'Y in the Field' it also comprises the reading of low-grade codes undertaken at the intercepting point.

Y inference: Intelligence deduction from the study of the external characteristics of enemy W/T traffic.

Y station: Wireless intercepting station.

ACRONYMS

1 ATF	1st Australian Task Force
1 RAR	1st Battalion Royal Australian Regiment
3 RAR	3rd Battalion Royal Australian Regiment
3TU	3 Telecommunication Unit (RAAF)
ABC	Australian Broadcasting Commission/Corporation
ABDA	American–British–Dutch–Australian command
ACIC	Australian Criminal Intelligence Commission
ACNB	Australian Commonwealth Naval Board
ACSCS	Australian Cyber Security Centre
ADF	Australian Defence Force
ADSC	Australian Defence Satellite Communications Station (Geraldton)
AEF	American Expeditionary Force
AFP	Australian Federal Police
AFSA	Armed Forces Security Agency (USA)
AGO	Australian Geospatial-Intelligence Organisation
AIB	Allied Intelligence Bureau
AIF	Australian Imperial Force
ALP	Australian Labor Party
AMF	Australian Military Forces
ANMEF	Australian Naval and Military Expeditionary Force
ANU	Australian National University
ANZAC	Australian and New Zealand Army Corps
ANZUS	Australia New Zealand United States (treaty)
ARDF	airborne radio direction finder
ARPANET	Advanced Research Projects Agency Network (USA)
ASD	Australian Signals Directorate
ASEAN	Association of South East Asian Nations
ASIO	Australian Security Intelligence Organisation

ASIS	Australian Secret Intelligence Service
ASTJIC	Australian Theatre Joint Intelligence Centre
ASV	air-to-surface vessel (radar aid)
ASWG	Army Special Wireless Group
ASWS	Army Special Wireless Section
ATIS	Allied Translator and Interpreter Section
AWAS	Australian Women's Army Service
AWM	Australian War Memorial
BEF	British Expeditionary Force
BRUSA	Britain–United States (Sigint-sharing agreement)
BTM	British Tabulating Machine company
CANUSA	Canada–United States (Sigint-sharing agreement)
CB	Central Bureau
CBNRC	Communications Branch of the National Research Council (Canada)
CEF	Canadian Expeditionary Force
CERT	Computer Emergency Response Team
CGS	Chief of the General Staff
CIA	Central Intelligence Agency (USA)
CID	Committee of Imperial Defence (UK)
C-in-C	commander-in-chief
COIC	Combined Operations Intelligence Centre
CPE	Central Photographic Establishment
CPU	central processing unit
CSE	Communications Security Establishment (Canada)
CSIR	Council for Scientific and Industrial Research (precursor to CSIRO)
CSOC	Cyber Security Operations Centre
CT	Communist Terrorist
DARPA	Defense Advanced Research Projects Agency (USA)
DC	direct current
DC	District of Columbia (US capital)
DF	direction finding
DIGO	Defence Imagery and Geopolitical Organisation
DIO	Defence Intelligence Organisation

DMI	Director(ate) of Military Intelligence
DMZ	Demilitarized Zone
DNI	Director(ate) of Naval Intelligence (RN/RAN)
DNO	Director(ate) of Naval Operations (RN/RAN)
DSB	Defence Signals Branch/Bureau
DSCS	Defense Satellite Communications System (USA)
DSD	Defence Signals Division/ Directorate
EATS	Empire Air Training Scheme
ECCM	electronic counter counter measures
ECM	electronic counter measures
EP	electronic protection (incl. ECM & ECCM)
ESM	electronic support measures
EW	electronic warfare
FARELF	Far East Land Forces (UK)
FBI	Federal Bureau of Investigation (USA)
FEAF	Far East Air Force (UK)
FECB	Far East Combined Bureau (UK)
FFV	Field Force Vietnam (USA)
FRUMEL	Fleet Radio Unit Melbourne
FRUPAC	Fleet Radio Unit Pacific (Hawaii)
FTP	file transfer protocol
G2	Military Intelligence Division (US Army)
G2-A6	Military Intelligence Division Code and Cipher Section
GC&CS	Government Code and Cypher School (UK)
GCHQ	Government Communications Headquarters (UK)
GCSB	Government Communications Security Bureau (NZ)
GHQ	General Headquarters
GOC	General Officer Commanding
GSO	General Staff Officer
H2S	air-ground-mapping radar
HF	high frequency (radio waves)
HFDF	high frequency direction finding
HIJMS	His Imperial Japanese Majesty's Ship
HMAS	Her/His Majesty's Australian Ship (RAN)
HMS	Her/His Majesty's Ship (RN)

HMSO	Her/His Majesty's Stationery Office
HPE	Hewlett Packard Enterprise
HVB	*Handelsschiffsverkehrsbuch* (German maritime code book)
IBM	International Business Machines
IFF	identification friend or foe
IGIS	Inspector-General of Intelligence and Security
IJA	Imperial Japanese Army
IJN	Imperial Japanese Navy
INTERFET	International Force East Timor
IO	information operations
ISR	intelligence, surveillance and reconnaissance
IT/ITOC	Intelligence/Interceptor/Intercepting Telephone
JARIC	Joint Air Reconnaissance Intelligence Centre (UK)
JEWOSU	Joint Electronic Warfare Operational Support Unit
JIB	Joint Intelligence Bureau
JIC	Joint Intelligence Committee
JIC FE	Joint Intelligence Centre Far East (UK)
JN-25	Japanese naval code
JSTARS	Joint Surveillance Target Attack Radar System
JTUM	Joint Telecommunications Unit Melbourne
KGB	Security Service (USSR)
LSIB	London Signal Intelligence Board
MI5	Security Service (UK)
MI6	Secret Intelligence Service (UK)
MI-8	Codes and Cipher Section (US Army)
MO5	War Office Special Duties Section (UK)
MO5b /MI1b	War Office cryptanalytic section (UK)
NAA	National Archives of Australia
NASA	National Aeronautics and Space Administration (USA)
NCR	National Cash Register company
NID	Naval Intelligence Department/Directorate (RAN/RN)
NID-25	Admiralty cryptanalytic section (UK)
NSA	National Security Agency (USA)
NVA	North Vietnamese Army
NZSIS	New Zealand Security Intelligence Service

OBU	OSIS Baseline Upgrade
OED	OBU Evolutionary Development
ONA	Office of National Assessments
ONI	Office of National Intelligence
OP-20-G	US Navy's Sigint office
OSIS	Oceanic Surveillance Information System
OSS	Office of Strategic Services (USA)
PJCAAD	Parliamentary Joint Committee on ASIO, ASIS and DSD
PJCIS	Parliamentary Joint Committee on Intelligence and Security
RAAF	Royal Australian Air Force
RAF	Royal Air Force (UK)
RAN	Royal Australian Navy
RANTEWSS	Royal Australian Navy Tactical Electronic Warfare Support Squadron
RAR	Royal Australian (Infantry) Regiment
RCA	Radio Corporation of America
RCASIA	Royal Commission on Australia's Security and Intelligence Agencies
RCIS	Royal Commission on Intelligence and Security
RCM	radio counter-measures
RCMP	Royal Canadian Mounted Police
RDF	radio direction finding (radar)
RMIT	Royal Melbourne Institute of Technology
RN	Royal Navy (UK)
RNZN	Royal New Zealand Navy
SAS	Special Air Service
SBRS	Shoal Bay Receiving Station
SEAC	South East Asia Command (UK)
SIB	Special Intelligence Bureau (Melbourne)
SIS	Secret Intelligence Service (or MI6, UK) / Signal Intelligence Service (US)
SKM	*Signalbuch der Kaiserlinchen Marine* (German naval code book)

SLO	Special Liaison Officer
SLU	Special Liaison Unit
SMS	Seiner Majestät Schiff (German 'HMS')
SOA	Special Operations Australia
SOE	Special Operations Executive (UK)
SOIC	Signal Officer-in-Chief
SRD	Services Reconnaissance Department
STC	Standard Telegraphic Code
SWPA	South West Pacific Area
TA	traffic analysis
TCP/IP	transmission control protocol/internet protocol
TIC	Target Intelligence Centre
TNA	The National Archives, Kew (UK)
TNI	Tentara Nasional Indonesia (Indonesia's national military)
UAV	unattended/unmanned aerial vehicle
UKUSA	United Kingdom–United States (Sigint sharing agreement)
UN	United Nations
UNAMET	United Nations Assistance Mission East Timor
USAAF	US Army Air Force
USN	US Navy
USS	United States (Navy) Ship
VB	*Verkerhrsbuch* (German navy code book)
VC	Viet Cong
WAAAF	Women's Australian Auxiliary Air Force
WAC	Women's Army Corps (USA)
WEC	Wireless Experimental Centre (UK)
WMI	wireless message interception
WRANS	Women's Royal Australian Naval Service
WRENS	Women's Royal Naval Service (UK)
W/T	wireless telegraphy

BIBLIOGRAPHY

Archival collections

Australian War Memorial – AWM
C1241 'Military Forces of the Commonwealth General Orders'.

National Archives of Australia – NAA
MP 729/6, 'Secret correspondence files, multiple number series with "401" infix'.
MP 729/8, 'Secret correspondence files, multiple number series'.
MP 1049/1, 'Secret and confidential correspondence files, annual single number series with "O" infix'.
MP 1049/9, 'Correspondence files (general)'.
MP 1185/8, 'Secret and confidential correspondence files, multiple number series'.
A35, 'Special bundles of correspondence relating to the New Hebrides'.
A461, 'Correspondence Files'.
A816, 'Correspondence files, multiple number series' (Department of Defence).
A1194, 'Defence Library Material'.
A1608, 'Correspondence Files'.
A2585, 'Naval Board minute books'.
A3269, 'Collection of Special Operations Australia'.
A3978, 'Confidential Reports – Navy Officers'.
A5954, 'The Shedden Collection'.
A6661, 'Special Portfolio'.
A6923, 'Army Headquarters/Army Office record sets'.
A8908, 'Reports of the Royal Commission on Intelligence and Security, numerical series'.
A10908, 'Correspondence files, single number series' (Central Bureau).
A10909, 'Correspondence and photographs, single number series' (Fleet Radio Unit, Melbourne).
A11401, 'Correspondence files, multiple number series' (Australian Signals Directorate).
A12392, 'Material relating to history of Australian Security and Intelligence Services, compiled by Department of Foreign Affairs for the Royal Commission on Intelligence and Security'.
B883, 'Second Australian Imperial Force Personnel Dossiers, 1939–1947'.
B5436, 'Folder of Central Bureau Records, "Bibliography"'.
B5554, 'Volume of technical records relating to naval codes ad cyphers' (Fleet Radio Unit, Melbourne).
B5555, 'Translations of cypher messages' (Fleet Radio Unit, Melbourne).
2DRL/0469 'Bridges, William Throsby, Sir, KCB, CMG (Major General, b.1861–d.1915)'.
3DRL/6643, 'Blamey, Sir Thomas GBE, KCB, CMG, DSO (Field Marshal, b.1884–d.1951)'.

The National Archives, United Kingdom – TNA (UK)
HW 4/24, 'HMS Anderson and special intelligence in Far East'.
HW 40/207, 'Australia: Leakage of Sigint (arising from weakness in Chinese codes)'.
HW 43/10, 'GC&CS Naval Sigint Volume I: The Organisation and Evolution of British Naval Sigint' by Frank Birch. Part 1 The Making of Sigint'.
HW 43/13, 'GC&CS Naval Sigint Volume V(a): The Organisation and Evolution of British Sigint by Franch Birch'.
HW 43/49, 'GC&CS Army and Air Force Sigint Volume VI: The Organisation and Evolution of British Sigint II' by Lt. Cmdr L.A. Griffiths RNVR.
HW 43/78, 'Allied Sigint Policy and Organisation: Chapter VII, India and the Far East' by Nigel De Grey.
HW 43/58, 'GC&CS Army and Air Force Sigint Volume XV: Intelligence – Army – Japanese' by Lt. Cmdr L.A. Griffiths, RNVR.
HW 52/93, 'Notes on the organisation of signals intelligence in Australia and New Zealand'.
HW 52/97, 'Signals exchanged between GC&CS, Central Bureau Brisbane (CBB) and other signal intelligence authorities in Australia'.
HW 57/15, 'British-Canadian-American radio intelligence discussion in Washington DC'.
ADM 223/463, 'History of Naval Intelligence and Naval Intelligence Department 1940–1945'.

Imperial War Museum – UK
Imperial War Museum, Oral History SQN LDR Sidney F. Burley, Sound Recording, at <www.iwm.org.uk/collections/item/object/80009138>, accessed 14 October 2022.

Government reports, bills, speeches & media releases

Australian Army, The Unit Guide, *The Australian Army 1939–1945*, volume 6, 6.010.
Australian Government, 'Media release: Australian Heron mission ends in Afghanistan', Defence News, 3 December 2014, at <www.news.defence.gov.au/media/media-releases/australias-heron-mission-ends-afghanistan>, accessed 14 October 2022.
Australian Government, *Reform of Australia's Electronic Surveillance Framework: Discussion Paper* (Canberra: Department of Home Affairs, 2021), available at <www.homeaffairs.gov.au/reports-and-publications/submissions-and-discussion-papers/reform-of-australias-electronic-surveillance-framework-discussion-paper>, accessed 14 October 2022.
Australian Government, 'Naval Communications Station Harold E Holt (AREA A), Exmouth, WA, Australia', Department of Agriculture, Water and the Environment, at <www.environment.gov.au/cgi-bin/ahdb/search.pl?mode=place_detail;place_id=103552>, accessed 14 October 2022.
Australian Signals Directorate, 'Narrative on the Deployment of 547 Signal Troop to Vietnam, 1966 to 1971,' declassified by ASD 15/12/2020, at <www.asd.gov.au/sites/default/files/2022-03/asd_narrative_on_the_deployment_of_547_signal_troop_to_vietnam_1966_-_1971.pdf>, accessed 14 October 2022.
Mike Burgess, 'Director-General ASD speech to the Lowy Institute,' 27 March 2019, at <www.asd.gov.au/publications/director-general-asd-speech-lowy-institute>, accessed 15 November 2022.
Chief Signal Officer, *Japanese Signal Intelligence Service*, 3rd edition (Washington, DC: US Army Signal Security Agency, 1 November 1944).
Defence Honours and Awards Tribunal, Inquiry into Recognition for Service with 547 Signal Troops in Vietnam from 1966 to 1971, at <https://defence-honours-tribunal.gov.au/

wp-content/uploads/2019/11/547-Signal-Troop-Inquiry-Report.pdf>, accessed 4 January 2023.

Department of Defence, *Annual Report 2000–01* (Canberra: Australian Government Publishing Service, 2001).

Department of Foreign Affairs and Trade (Australian Government), 'Agreement with the Government of the United States of America relating to the Establishment of the Joint Defence Space Research Facility', Australian Treaty Series Number (1966) ATS 17, at <www.info.dfat.gov.au/info/Treaties/Treaties.nsf/AllDocIDs/675B157B757190FDCA256B59007D0CF3>, accessed 14 October 2022.

Department of Foreign Affairs and Trade, Australia's International Cyber and Critical Tech Engagement Strategy 2021 (Canberra: Australian Government Publishing Service, 2001), at <www.internationalcybertech.gov.au/sites/default/files/2021-04/21045%20DFAT%20Cyber%20Affairs%20Strategy%20Internals_Acc_update_1_0.pdf>, accessed 15 November 2022.

Philip Flood, *Report of the Inquiry into Australian Intelligence Agencies July 2004* (Canberra: Commonwealth of Australia, 2004), at <www.fas.org/irp/world/australia/flood.pdf>, accessed 14 October 2022.

Government of Canada, 'CANUSA Agreement,' Communications Security Establishment – History, website of the Government of Canada, <www.cse-cst.gc.ca/en/culture-and-community/history>, accessed 25 August 2021.

Government of Canada, 'History: Communications Security Establishment', website of the Government of Canada, <www.cse-cst.gc.ca/en/culture-and-community/history>, accessed 25 August 2021.

Robert Hope, *Royal Commission on Intelligence and Security* (Canberra: Australian Government Publishing Service, 1977).

Robert Hope, *Royal Commission into Australia's Security and Intelligence Agencies* (Canberra: Australian Government Publishing Service, 1984).

Michael L'Estrange and Stephen Merchant, *Independent Intelligence Review*, June 2017 (Canberra: Commonwealth of Australia, 2017), <www.pmc.gov.au/national-security/2017-independent-intelligence-review>, accessed 14 October 2022.

National Museum of Australia, 'Defining Moments: Overland Telegraph', National Museum of Australia website, <www.nma.gov.au/defining-moments/resources/overland-telegraph>, accessed 19 November 2019.

National Security Agency/Central Security Service, 'The Rare Book Collection', <www.nsa.gov/about/cryptologic-heritage/historical-figures-publications/publications/misc/rare-books>, accessed 19 September 2019.

National Security Agency/Central Security Service, 'Venona,' NSA/CCS, at <www.nsa.gov/Helpful-Links/NSA-FOIA/Declassification-Transparency-Initiatives/Historical-Releases/Venona/smdpage14707/66/smdsort14707/title/>, accessed 15 November 2022.

Parliament of Australia, Intelligence Services Bill 2001, House of Representatives Explanatory Memorandum, <www.aph.gov.au/Parliamentary_Business/Bills_Legislation/Bills_Search_Results/Result?bId=r1350>, accessed 14 October 2022.

Linda Reynolds, 'Media release: New satellite ground station in WA fully operational', Department of Defence, Australian Government, 12 October 2020, <www.minister.defence.gov.au/minister/lreynolds/media-releases/new-satellite-ground-station-wa-fully-operational>, accessed 14 October 2022.

Linda Reynolds, 'Media release: E7-A Wedgetail joins US-led Global Coalition against Daesh', Department of Defence, Australian Government, 14 September 2019, at <www.minister.defence.gov.au/minister/lreynolds/media-releases/e-7a-wedgetail-joins-us-led-global-coalition-against-daesh>, accessed 14 October 2022.

Dennis Richardson, Anna Harmer & Tara Inverarity, 'Comprehensive Review of the Legal Framework of the National Intelligence Community, December 2019', website of the Australian Government Attorney-General's Department, <www.ag.gov.au/national-security/publications/report-comprehensive-review-legal-framework-national-intelligence-community>, accessed 14 October 2022.

Gordon J. Samuels and Michael H. Codd, *Commission of Inquiry into the Australian Secret Intelligence Service: Report on the Australian Secret Intelligence Service,* Public Edition, March 1995 (Canberra: Australian Government, 1995), at <www.apo.org.au/sites/default/files/resource-files/1995-03/apo-nid37293.pdf>, accessed 15 November 2022.

Stephen Smith, Ministerial statements, Australian Parliament House, 26 June 2013, <www.parlinfo.aph.gov.au/parlInfo/search/display/display.w3p;query=Id:%22chamber/hansardr/4d60a662-a538-4e48-b2d8-9a97b8276c77/0016%22;src1=sm1>, accessed 15 November 2022.

Robert Taschereau, *Royal Commission to Investigate the Facts Relating to and the Circumstances Surrounding the Communication by the Public Officials and Other Persons in Positions of Trust of Secret and Confidential Information to Agents of Foreign Power,* Ottawa, 1946, at <https://publications.gc.ca/site/eng/472640/publication.html>, accessed 15 November 2022.

Malcolm Turnbull, 'Launch of Australia's Cyber Security Strategy Sydney,' transcript of speech delivered on 21 April 2016, at <www.malcolmturnbull.com.au/media/launch-of-australias-cyber-security-strategy>, accessed 15 November 2022

Patrick D. Weadon, 'Sigsaly Story', National Security Agency/Central Security Service, <www.nsa.gov.au/about/cryptologic-heritage/historical-figures-publications/publications/wwii/sigsaly-story>, accessed 9 September 2019.

Alex Younger, 'Intelligence Services Speech, "C", Head of MI6,' St Andrews, Scotland, 3 December 2018, at <www.speakola.com/ideas/alex-younger-mi6-sis-st-andrews-2018>, accessed 10 November 2022.

Published monographs

Matthew M. Aid, *The Secret Sentry: The Untold History of the National Security Agency* (New York: Bloomsbury Press, 2009).

Richard J. Aldrich, *Intelligence and the War against Japan: Britain, America and the Politics of Secret Service* (Cambridge: Cambridge University Press, 2000).

Richard J. Aldrich, *GCHQ: The Uncensored Story of Britain's Most Secret Intelligence Agency* (London: Harper Press, 2010).

Christopher Andrew, *The Secret World: A History of Intelligence* (Milton Keynes: Allen Lane, Penguin, 2018).

Christopher Andrew and Keith Neilson, 'Tsarist Codebreakers and British Codes,' in Christopher Andrew (ed.), *Codebreaking and Signals Intelligence* (London: Frank Cass, 1986), pp. 6–12.

Daniel Baldino and Rhys Crawley (eds), *Intelligence and the Function of Government* (Melbourne: Melbourne University Press, 2018).

Desmond Ball, *A Suitable Piece of Real Estate: American Installations in Australia* (Sydney: Hale & Iremonger, 1980).

Desmond Ball, *A Base for Debate: The US Satellite Station at Nurrungar* (Sydney: Allen & Unwin, 1987).

Desmond Ball, *Pine Gap: Australia and the US Geostationary Signals Intelligence Satellite Program* (Sydney: Allen & Unwin, 1988).

Desmond Ball, *Australia's Secret Space Programs,* Canberra Papers on Strategy and Defence (Canberra: Strategic and Defence Studies Centre, 1988).
Desmond Ball, *Australia and the Global Strategic Balance,* Canberra Papers on Strategy and Defence, No. 49 (Canberra: Strategic and Defence Studies Centre, 1989).
Desmond Ball, *Code 777: Australia and the US Defense Satellite Communications System (DSCS),* Canberra Papers on Strategy and Defence, No. 56 (Canberra: Strategic and Defence Studies centre, 1989).
Desmond Ball, *The Intelligence War in the Gulf,* Canberra Papers on Strategy and Defence, No. 75 (Canberra: Strategic and Defence Studies Centre, 1991).
Desmond Ball and David Horner, *Breaking the Codes: Australia's KGB Network, 1944*–1950 (Sydney: Allen & Unwin, 1998).
Desmond Ball and Hamish McDonald, *Death in Balibo, Lies in Canberra* (Sydney: Allen & Unwin, 2000).
Desmond Ball and Keiko Tamura (eds), *Breaking Japanese Diplomatic Codes: David Sissons and D Special Section during the Second World War* (Canberra: Australian National University, 2013).
Geoffrey Ballard, *On ULTRA Active Service: The Story of Australia's Signals Intelligence Operations During World War II* (Melbourne: Spectrum Publications, 1991).
James Bamford, *The Puzzle Palace: A Report on America's Most Secret Agency* (Boston: Houghton Mifflin, 1982).
Craig P. Bauer, *Secret History: The Story of Cryptology* (Boca Raton, FL: CRC Press, 2013).
Jim Beach, *Haig's Intelligence GHQ and the German Army, 1916*–*1918* (Cambridge: Cambridge University Press, 2013).
Hans-Otto Behrendt, *Rommel's Intelligence in the Desert Campaign* (London: William Kimber, 1980).
Robert L. Benson, *The Venona Story* (Fort Meade, MD: Center for Cryptologic History, NSA, 2012).
Anthony Best, *British Intelligence and the Japanese Challenge in Asia, 1914*–*1941* (Basingstoke: Palgrave MacMillan, 2002).
John Blaxland, *Swift and Sure: A History of the Royal Australian Corps of Signals* (Melbourne: Royal Australian Corps of Signals Committee, 1998).
John Blaxland, *Information Era Manoeuvre: The Australian-Led Mission in East Timor* (Canberra: Land Warfare Studies Centre, 2002).
John C. Blaxland, *Strategic Cousins: Australian and Canadian Expeditionary Forces and the British and American Empires* (Montreal: McGill-Queens University Press, 2006).
John Blaxland, *The Australian Army from Whitlam to Howard* (Melbourne: Cambridge University Press, 2014).
John Blaxland, *The Protest Years: The Official History of ASIO,* vol. II (Sydney: Allen & Unwin, 2015).
John Blaxland (ed.), *East Timor Intervention: A Retrospective on INTERFET* (Melbourne: Melbourne University Press, 2015).
John Blaxland, Marcus Fielding and Thea Gellerfy (eds), *Niche Wars: Australia in Afghanistan and Iraq, 2001*–*2014* (Canberra: ANU Press, 2020).
John Blaxland, Michael Kelly and Liam Brewin Higgins (eds), *In from the Cold: Reflections on Australia's Korean War* (Canberra: ANU Press, 2020).
Jack Bleakley, *The Eavesdroppers (*Canberra: Australian Government Publishing Service, 1992).
Donald A. Borrmann, William T. Kvetkas, Charles V. Brown, Michael J. Flatley and Robert Hunt, *The History of Traffic Analysis: World War I – Vietnam* (Fort Meade, MD: Center for Cryptologic History, National Security Agency, 2013).

Jean Bou, *MacArthur's Secret Bureau: The Story of the Central Bureau, MacArthur's Signals Intelligence Organisation* (Sydney: Australian Military History Publications, 2012).
Bob Breen, *First to Fight: Australian Diggers, N.Z. Kiwis, and U.S. Paratroopers in Vietnam, 1965–66* (Sydney: Allen & Unwin, 1988).
A. Jack Brown, *Katakana Man: I Worked Only for Generals* (Canberra: Air Power Development Centre, 2005).
Keast Burke (ed.), *With Horse and Morse in Mesopotamia* (Sydney: Arthur McQuitty and Co., 1927).
Alan D. Campen and Douglas H. Dearth, *Cyberwar 2.0: Myths, Mysteries & Reality* (Fairfax, VA: AFCEA International Press, 1998).
Mike Carlton, *The Scrap Iron Flotilla* (Melbourne: William Heinemann, 2022).
John F. Clabby, *Brigadier John Tiltman, A Giant Among Cryptanalysts* (Fort Meade, MD: National Security Agency, Center for Cryptologic History, 2007).
Richard A. Clarke and Robert K. Knake, *The Fifth Domain: Defending Our Country, Our Companies and Ourselves in the Age of Cyber Threats* (New York: Penguin, 2020).
Bernard Collaery, *Oil Under Troubled Water: Australia's Timor Sea Intrigue* (Melbourne: Melbourne University Press, 2020).
Craig Collie, *Code Breakers: Inside the Shadow World of Signals Intelligence in Australia's Two Bletchley Parks* (Sydney: Allen & Unwin, 2017).
C.D. Coulthard-Clark, *The Citizen General Staff: The Australian Intelligence Corps 1907–1914* (Canberra: Military Historical Society of Australia, 1976).
F.M. Cutlack, *Official History of Australia in the Great War of 1914–18*, vol. 8: The Australian Flying Corps (Sydney: Angus & Robertson, 1923).
Charles Darby, *Australia's Liberators: B-24 Operations from Australia* (Melbourne: Red Roo Models, 2009).
Peter Dean (ed.), *Australia, 1944–45: Victory in the Pacific* (Melbourne: Cambridge University Press, 2015).
Peter Dennis and Jeffrey Grey, *Emergency and Confrontation: Australian Military Operations in Malaya and Borneo 1950–1966*, The Official History of Australia's Involvement in Southeast Asian Conflicts 1948–1975 (Sydney: Allen & Unwin, 1996).
Diodorus Siculus, *Library*, at <www.perseus.tufts.edu/hopper/text?doc=Perseus:text:1999.01.0084>, accessed 29 November 2019.
Peter Donovan and John Mack, *Code Breaking in the Pacific* (Cham: Springer International Publishing, 2014).
John F. Dooley, *Codes, Ciphers and Spies: Tales of Military Intelligence in World War I* (New York: Copernicus Books, 2016).
Edward J. Drea, *MacArthur's ULTRA: Codebreaking and the War against Japan, 1942–1945* (Wichita, KS: University of Kansas Press, 1992).
David Dufty, *The Secret Code Breakers of Central Bureau: How Australia's Signals Intelligence Helped Win the Pacific War* (Melbourne: Scribe, 2017).
Lawrence Durrant, *Seawatchers: The Story of Australia's Coast Radio Service* (Sydney: Angus and Robertson, 1986).
Peter G. Edwards, *Crises and Commitments: The Politics and Diplomacy of Australia's Involvement in South East Asia, 1948–1965* (Sydney: Allen & Unwin, 1992).
S.R. Elliott, *Scarlet to Green: A History of Intelligence in the Canadian Army, 1903–1963* (Ottawa: Canadian Intelligence and Security Association, 1981).
John Fahey, *Australia's First Spies: The Remarkable Story of Australia's Intelligence Operations, 1901–1945* (Sydney: Allen & Unwin, 2018).
John Fahey, *Traitors and Spies: Espionage and Corruption in High Places in Australia, 1901–1950* (Sydney: Allen & Unwin, 2020).

Thomas G. Fergusson, *British Military Intelligence, 1870–1914: The Development of a Modern Intelligence Organization* (Frederick, MD: University Publications of America, 1984).

John Ferris (ed.), *The British Army and Signals Intelligence During the First World War* (Stroud: Alan Sutton for the Army Records Society, 1992).

John Ferris, *Behind the Enigma: The Authorised History of GCHQ* (London: Bloomsbury, 2020).

Thomas Fleming (ed.), *The Founding Fathers: Benjamin Franklin, A Biography in His Own Words* (New York: Newsweek, Harper & Row, 1972).

William F. Flicke, *War Secrets in the Ether,* vols 1 & 2 (Laguna Hills, CA: Aegean Park Press, 1977).

LCDR Gregory J. Florence, USN, *Courting a Reluctant Ally: An Evaluation of U.S./UK Naval Intelligence Cooperation, 1935–1941* (Washington, DC: Joint Military Intelligence College, Center for Strategic Intelligence Research, 2004).

G. Hermon Gill, *Royal Australian Navy 1939–1942,* Australia in the War of 1939–1945, Series Two, Navy, vol. I (Canberra: Collins and Australian War Memorial, 1957–1985 reprint).

G. Hermon Gill, *Royal Australian Navy 1942–1945,* Australia in the War of 1939–1945, Series Two, Navy, vol. II (Canberra: Collins and Australian War Memorial, 1968–1985 reprint).

Tom Gilling, *Project Rainfall: The Secret History of Pine Gap* (Sydney: Allen & Unwin, 2019).

Douglas Gillison, *Royal Australian Air Force 1939–1942,* Australia in the War of 1939–1945, Series Three, vol. I (Canberra: Australian War Memorial, 1962).

Yves Glyden, *The Contribution of the Cryptographic Bureaus in the World War* (Laguna Hills, CA: Aegean Park Press, 1978).

Jeffrey Grey, *Up Top: The Royal Australian Navy and Southeast Asian Conflicts, 1955–1972* (Sydney: Allen & Unwin, 1998).

Allan Gyngell and Michael Wesley, *Making Australian Foreign Policy*, 2nd edition (Melbourne: Cambridge University Press, 2007).

Allan Gyngell, *Fear of Abandonment: Australia in the World since 1942* (Melbourne: Black Inc., 2017).

Cargill Hall (ed.), *Lightning Over Bougainville* (Washington, DC: Smithsonian, 1991).

Robert J. Hanyok and David P. Mowry, *West Wind Clear: Cryptology and the Winds Message Controversy – A Documentary History*, United States Cryptologic History Series IV: World War II, vol. 10 (Fort Meade, MD: Center for Cryptologic History, NSA, 2008).

David A. Hatch, *The Dawn of American Cryptology, 1900–1917* (Fort Meade, MD: Center for Cryptologic History, National Security Agency, 2019).

David A. Hatch with Robert Louis Benson, *The Korean War: The SIGINT Background*, United States Cryptologic History Series V, The Early Postwar Period, 1945–1952, vol. 3 (Fort Meade, MD, Center for Cryptologic History, National Security Agency, 2000).

Daniel R. Headrick, *The Invisible Weapon: Telecommunications and International Politics 1851–1945* (New York and Oxford: Oxford University Press, 1991).

John Herington, *Air War Against Germany & Italy 1939–1943,* Australia in the War of 1939–1945, Series III, Air, vol. 3 (Canberra: Australian War Memorial, 1954).

John Herington, *Air Power Over Europe 1944–1945,* Australia in the War of 1939–1945, Series III, Air, vol. 4 (Canberra: Australian War Memorial, 1963).

F.H. Hinsley, *Official History of British Intelligence in the Second World War: Its Influence on Strategy and Operations* (London: HMSO, 1979).

F.H. Hinsley and Alan Stripp (eds), *Code Breakers: The Inside Story of Bletchley Park* (Oxford: Oxford University Press, 1993).

F.H. Hinsley, E.E. Thomas, C.F.G. Ransom and R.C. Knight, *British Intelligence in the Second World War*, vol. I (London: HMSO, 1979).

F.H. Hinsley, E.E. Thomas, C.F.G. Ransom and R.C. Knight, *British Intelligence in the Second World War*, vol. 2 (London: HMSO, 1981).
D.M. Horner, *Crisis of Command: Australian Generalship and the Japanese Threat, 1941–1943* (Canberra: ANU Press, 1978).
David Horner, *High Command: Australia's Struggle for an Independent War Strategy, 1939–1945* (Sydney: Allen & Unwin, 1982).
David Horner, *The Gulf Commitment: The Australian Defence Force's First War* (Carlton, Vic.: Melbourne University Press, 1992).
David Horner, *Defence Supremo: Sir Frederick Shedden and the Making of Australian Defence Policy* (Sydney: Allen & Unwin, 2000).
David Horner, *Australia and the 'New World Order': From Peacekeeping to Peace Enforcement: 1988–1991*, The Official History of Australian Peacekeeping, Humanitarian and Post-Cold War Operations (Melbourne: Cambridge University Press & The Australian War Memorial, 2011).
David Horner, *The Spy Catchers: The Official History of ASIO, 1949–1963*, vol. I (Sydney: Allen & Unwin, 2014).
Ann Howard, *You'll Be Sorry! How World War II Changed Women's Lives* (Sydney: Big Sky Publishing, 1990).
Shirley Fenton Huie, *Ships Belles: The Story of the Women's Royal Naval Service in War and Peace, 1941–1985* (Sydney: Watermark Press, 2000).
John Johnson, *The Evolution of British Sigint: 1653–1939* (Cheltenham: GCHQ, 1997).
Thomas R. Johnson, *American Cryptology During the Cold War, 1945–1989*. Book I The Struggle for Centralization 1945–1960 (Fort Meade, MD: Center for Cryptologic History, National Security Agency 1995).
Arthur W. Jose, *The Official History of Australia in the War of 1914–1918*, vol. 9: The Royal Australian Navy (St Lucia: University of Queensland Press and Australian War Memorial, 1993 reprint).
David Kahn, *The Code-Breakers: The Comprehensive History of Secret Communication from Ancient Times to the Internet* (New York: Scribner, 1996).
David Kahn, *The Reader of Gentlemen's Mail: Herbert O. Yardley and the Birth of American Codebreaking* (New Haven, CT: Yale University Press, 2004).
John Keegan, *Intelligence in War: Knowledge of the Enemy from Napoleon to Al-Qaeda* (London: Hutchinson, 2003).
Sylvia Kleinart and Margo Neale (eds), *The Oxford Companion to Aboriginal Art and Culture* (South Melbourne: Oxford University Press, 2000).
Peter Kornicki, *Eavesdropping on the Emperor: Interrogators and Codebreakers in Britain's War with Japan* (London: Hurst and Company, 2021).
Władysław Kozaczuk, *Bitwa o tajemnice: Służby wywiadowcze Polski I Rzeszy Niemieckiej, 1922–1939* (Warsaw: Książka I Wiedza, 1967).
Mark Lax, *Malayan Emergency and Indonesian Confrontation 1950 to 1966*, The Australian Air Campaign Series – 2 (Newport, Sydney: RAAF History and Heritage and Big Sky Publishing, 2021).
John le Carré, *Tinker, Tailor, Soldier, Spy* (London and Sydney: Pan Books, 1975).
Kai-Fu Lee, *AI Superpower: China, Silicon Valley, and the New World Order* (Boston, MA: Mariner Books, 2018).
Christian Leuprecht and Hayley McNorton, *Intelligence as Democratic Statecraft Accountability and Governance of Civil-Intelligence Relations Across the Five Eyes Security Community – the United States, United Kingdom, Canada, Australia, and New Zealand* (Oxford: Oxford University Press, 2021).

Hong Liu, *Chinese Business: Landscapes and Strategies* (London and New York: Routledge, 2009).
Gavin Long, *The Final Campaigns* (Canberra: Australian War Memorial, 1963).
Robert Macklin, *Warrior Elite: Australia's Special Forces Z Force to the SAS Intelligence Operations to Cyber Warfare* (Sydney: Hachette, 2015).
John McCarthy, *A Last Call of Empire: Australian Aircrew, Britain and the Empire Air Training Scheme* (Canberra: Australian War Memorial, 1988).
Ian McNeill, *The Team: Australian Army Advisers in Vietnam, 1962–1972* (Canberra: Australian War Memorial, 1984).
Ian McNeill, *To Long Tan: The Australian Army and the Vietnam War, 1950–1966* (Sydney: Allen & Unwin, 1993).
Steven E. Maffeo, *Most Secret and Confidential: Intelligence in the Age of Nelson* (Annapolis, MD: Naval Institute Press, 2000).
Sharon A. Maneki, *The Quiet Heroes of the Southwest Pacific Theater: An Oral History of the Men and Women of CBB and FRUMEL* (Fort Meade, MD: Center for Cryptological History, NSA, 1996).
Dr A. Ray Miller, *The Cryptographic Mathematics of Enigma*, revised edition (Fort Meade, MD: Center for Cryptologic History, NSA, 2019).
David P. Mowry, *Listening to the Rumrunners: Radio Intelligence during Prohibition*, 2nd edition (Fort Meade, MD: Center for Cryptologic History, NSA, 2014).
David P. Mowry, *German Cipher Machines of World War II*, revised edition (Fort Meade, MD: Center for Cryptologic History, 2014).
Timothy Mucklow, *The SIGABA/ECM II Cipher Machine: 'A Beautiful Idea'* (Fort Meade, MD: Center for Cryptologic History, 2015).
Thomas S. Mullaney, *The Chinese Typewriter: A History* (Cambridge, MA: Massachusetts Institute of Technology Press, 2017).
Liza Mundy, *Code Girls: The Untold Story of the American Women Code Breakers of World War II* (New York: Hachette Books, 2017).
Michael V. Nelmes, *Tocumwal to Tarakan: Australians and the Consolidated B-24 Liberator* (Canberra: Banner Books, 1994).
Don Oberdorfer, *The Two Koreas: A Contemporary History* (Reading, MA: Addison-Wesley, 1997).
George Odgers, *Air War Against Japan 1939–1945,* Australia in the War of 1939–1945, Series III, vol. 2 (Canberra: Australian War Memorial, 1957).
Robert J. O'Neill, *Australia in the Korean War, 1950–53* (Canberra: Australian War Memorial and Australian Government Publishing Service, 1985).
Bojan Pajic, *Our Forgotten Volunteers: Australians and New Zealanders with Serbs in World War I* (Melbourne: Arcadia, Australian Scholarly Publishing, 2019).
Ron Palenski, *Men of Valour: New Zealand and the Battle for Crete* (Auckland: Hachette, 2013).
Frederick D. Parker, *A Priceless Advantage: US Navy Communications Intelligence and the Battles of the Coral Sea, Midway and Aleutians* (Fort Meade, MD: National Security Agency, 1993, 2017 reprint).
Ian Pfennigwerth, *A Man of Intelligence: The Life of Captain Theodore Eric Nave, Australian Codebreaker Extraordinary* (Sydney: Pfennigwerth, 2006).
Ian Pfennigwerth, *Missing Pieces: The Intelligence Jigsaw and RAN Operations 1939–1971*, Papers in Australian Maritime Affairs No. 25 (Canberra: Sea Power Centre-Australia, Department of Defence, 2008), at <www.navy.gov.au/w/images/PIAMA25.pdf>.
James Phelps, *Australian Codebreakers: Our Top-secret War with the Kaiser's Reich* (Sydney: Harper Collins, 2020).

Franco Piodi and Iolanda Mombelli, *The ECHELON Affair: The European Parliament and the Global Interception System* (Luxembourg: European Parliamentary Research Service, 2014).
Alan Powell, *War by Stealth: Australians and the Allied Intelligence Bureau, 1942–1945* (Melbourne: Melbourne University Press, 1996).
RAAF, *Aircraft of the Royal Australian Air Force* (Canberra: Air Force History Branch, 2021).
Jeffrey Richelson and Desmond Ball, *The Ties That Bind: Intelligence Cooperation Between the UKUSA Countries* (Sydney: Allen & Unwin, 1985).
Anthea Roberts and Nicolas Lamp, *Six Faces of Globalisation* (Cambridge, MA: Harvard University Press, 2019).
Jerry Roberts, *Lorenz: Breaking Hitler's Top Secret Code at Bletchley Park* (Stroud: The History Press, 2017).
David Rosenberg, *Inside Pine Gap: The Spy Who Came in From The Desert* (Melbourne: Hardie Grant Books, 2011).
Andrew Ross, *Armed and Ready: The Industrial Development and Defence of Australia 1900–1945* (Sydney: Department of Industry, Science and Technology, 1995).
Felicity Ruby and Peter Cronau (eds), *A Secret Australia* (Melbourne: Monash University Publishing, 2020).
James Rusbridger and Eric Nave, *Betrayal at Pearl Harbor: How Churchill Lured Roosevelt into War* (New York: Summit Books, 1991).
David Sherman, *The First Americans: The 1941 US Codebreaking Mission to Bletchley Park* (Fort Meade, MD: Center for Cryptologic History, NSA, 2016).
Emma Shortis, *Our Exceptional Friend: Australia's Fatal Alliance with the United States* (Melbourne: Hardie Grant Books, 2021).
Harold A. Skaarup, *Out of Darkness – Light: A History of Canadian Military Intelligence*, vol. 1, Pre-Confederation to 1982 (Lincoln, NE: iUniverse, 2005).
Alan Stephens, *The Royal Australian Air Force: The Australian Centenary History of Defence*, vol. 2 (Melbourne: Oxford University Press, 2001).
Craig Stockings, *Born of Fire and Ash: Australia's operations in response to the East Timor crisis 1999–2000* (Sydney: UNSW Press, 2022).
R. Tanter, D. Ball and G. van Klinken (eds), *Masters of Terror: Indonesia's Military and Violence in East Timor* (Lanham, MD: Rowman & Littlefield, 2006).
Judy Thomson, *Winning with Intelligence: A Biography of Brigadier John David Rogers, CBE, MC 1895–1978* (Sydney: Australian Military History Publications, 2000).
Blair Tidey, *Forewarned Forearmed: Australian Specialist Intelligence Support in South Vietnam, 1966–1971* (Canberra: Strategic and Defence Studies Centre, The Australian National University, 2007).
Mark Urban, *The Man Who Broke Napoleon's Codes* (London: Faber & Faber, 2001).
Martin Van Creveld, *Command in War* (Cambridge, MA: Harvard University Press, 1985).
Michael Veitch, *The Battle of the Bismarck Sea: The Forgotten Battle that Saved the Pacific* (Sydney: Hachette Press, 2021).
Patrick Walters (ed.), *ANZUS at 70: The Past, Present and Future of the Alliance* (Canberra: Australian Strategic Policy Institute, 2021).
Gary Waters, Desmond Ball and Ian Dudgeon, *Australia and Cyber-warfare* (Canberra: ANU E Press, 2008).
Ralph E. Weber, *Masked Dispatches: Cryptograms and Cryptology in American History, 1775–1900*, 3rd edition (Washington, DC: National Security Agency, Center for Cryptologic History, 2013).
Anthony R. Wells, *Between Five Eyes: 50 Years of Intelligence Sharing* (Havertown, PA: Casemate, 2020).

Nigel West, *Historical Dictionary of Signals Intelligence* (Plymouth: Scarecrow Press, 2012).
David Whitehead (translator), *Aineias the Tactician: How to Survive Under a Siege* (Oxford: Clarendon Press, 1990).
Craig Wilcox, *Australia's Boer War: The War in South Africa, 1899–1902* (Melbourne: Oxford University Press, 2002).
Jennifer Wilcox, *Solving the Enigma: History of the Cryptanalytic Bombe* (Fort Meade, MD: Center for Cryptologic History, NSA, Revised, 2006).
Jennifer Wilcox, *Revolutionary Secrets: Cryptology in the American Revolution* (Fort Meade, MD: Center for Cryptologic History, National Security Agency, 2012).
Mark Willacy, *Rogue Forces: An Explosive Insiders' Account of Australian SAS War crimes in Afghanistan* (Sydney: Simon & Schuster, 2021).
Stewart Wilson, *Catalina, Neptune and Orion in Australian Service* (Canberra: Aerospace Publications, 1991).
Barbara Winter, *The Intrigue Master: Commander Long and Naval Intelligence in Australia, 1913–1945* (Brisbane: Boolarong Press, 1995).
F.W. Winterbotham, *The Ultra Secret: The Inside Story of Operation Ultra, Bletchley Park and Enigma* (London: Orion Books, 1974).
Herbert D. Yardley, *The American Black Chamber* (New York: Ishi Press International, reprint, 2016).

Journal articles

Desmond Ball, 'Silent Witness: Australian intelligence and East Timor,' *Pacific Review*, 14:1 (2001): 35–62.
Desmond Ball, 'The Strategic Essence,' *Australian Journal of International Affairs*, 55:2 (2001): 235–48.
Jim Beach and James Bruce, 'British Signals Intelligence in the Trenches, 1915–1918: Part 1, "Listening Sets",' *Journal of Intelligence History*, 19:1 (2019): 1–23.
John Blaxland, 'On Operations in East Timor: The Experiences of the Intelligence Officer, 3rd Brigade,' *Australian Army Journal* (2000): 1–13.
John Blaxland, 'The Role of Signals Intelligence in Australian Military Operations, 1939–1972,' *Australian Army Journal*, II:2 (2008): 203–16.
William Cahill, 'Far East Air Forces RCM Operating in the Final Push on Japan,' *Air Power History*, 65:1 (2018): 37–47.
Kevin Davies, 'Field Unit 12 Takes New Technology to War in the Southwest Pacific,' *Studies in Intelligence*, 58:3 (2014): 11–20.
Edward J. Drea and Joseph E. Richard, 'New Evidence on Breaking the Japanese Army Codes,' *Intelligence and National Security*, 14:1 (1999): 62–83.
R. Erskine, 'Churchill and the Start of the Ultra-Magic Deals,' *International Journal of Intelligence and Counter Intelligence*, 10:1 (1997): 57–74.
John Ferris, 'Before "Room 40": The British Empire and Signals Intelligence, 1898–1914,' *Journal of Strategic Studies*, 12:4 (1989): 431–57.
John Ferris, 'The Road to Bletchley Park: The British Experience with Signals Intelligence, 1892–1945,' *Intelligence and National Security*, 17:1 (2002): 53–84.
John Ferris, '"Consistent with an Intention": The Far East Combined Bureau and the Outbreak of the Pacific War, 1940–41,' *Intelligence and National Security*, 27:1 (2012): 5–26.
Tim Gellel, 'The Capture of Unit 621: Lessons in Information Security from the North Africa Campaign,' *The Australian Army Journal*, XII:1 (2015): 77–89.

Frances Hackney, 'A Tribute to a Gentle Genius', *Annals: Journal of Catholic Culture* (November/December 1996): 14.
Jason Healey, 'Research Paper: The Implications of Persistent (and Permanent) Engagement in Cyberspace', *Journal of Cybersecurity*, 5:1 (2019): 1–15.
Robert Jervis, 'The Impact of the Korean War on the Cold War', *Journal of Conflict Resolution*, 24:4 (1980): 563–92.
Richard Johnson, 'Trendall, Arthur Dale (1909–1995)', *ANU Reporter*, 13 December 1995, 11. <https://oa.anu.edu.au/obituary/trendall-arthur-dale-976>.
Liam Kane, 'Allied Air Intelligence in the South West Pacific Area, 1942–1945', *Journal of Intelligence History* (2021): 1–21.
Ulug Kuzuoglu, 'Chinese Cryptography: The Chinese Nationalist Party and Intelligence Management 1927–1949', *Cryptologia*, 42:6 (2018): 514–39.
Daniel Larsen, 'British Codebreaking and American Diplomatic Telegrams, 1914–1915', *Intelligence and National Security*, 32:2 (2017): 252–63.
Fei-Wen Liu, 'Being to Becoming: Nushu and Sentiments in a Chinese Rural Community', *American Ethnologist*, 31:3 (2004): 422–39.
Ian Pfennigwerth, 'Nave, Theodore Eric (1899–1993)', *Australian Dictionary of Biography*, vol. 19, 2021, at <www.adb.anu.edu.au/biography/nave-theodore-eric-17834>, accessed 15 November 2022.
Matthew Purbrick, 'Patriotic Chinese Triads and Secret Societies: From the Imperial Dynasties, to Nationalism and Communism', *Journal of Asian Affairs*, 50: 3 (2019): 305–22.
Jon Robb-Webb, 'Anglo-American Naval Intelligence Co-operation in the Pacific, 1944–45', *Intelligence and National Security*, 22:5 (2007): 767–86.
Geoffrey Serle, 'McCay, Sir James Whiteside (1864–1930)', *Australian Dictionary of Biography*, vol. 10, 1986, at <www.adb.anu.edu.au/biography/mccay-sir-james-whiteside-7312>, accessed 15 November 2022.
Wesley K. Wark, 'Cryptographic Innocence: The Origins of Signals Intelligence in Canada in the Second World War', *Journal of Contemporary History*, 22:4 (1987): 639–65.
Wesley Wark, 'The Road to CANUSA: How Canadian Signals Intelligence Won its Independence and Helped Create the Five Eyes', *Intelligence and National Security*, 35:1 (2020): 20–34.

Journalism and commentary

Kim Beazley, 'Pine Gap at 50: The Paradox of a Joint Facility', *The Strategist*, Australian Strategic Policy Institute, 30 August 2017, at <www.aspistrategist.org.au/pine-gap-50-paradox-joint-facility/>, accessed 15 November 2022.
Danielle Cave, 'Australia and the Great Huawei Debate: Risks, Transparency and Trust', *The Strategist*, Canberra, ASPI, 11 September 2019.
Rishi Iyengar, 'Why Ukraine is Stuck with Elon for Now', *Foreign Policy*, 22 November 2022, at <www.foreignpolicy.com/2022/11/22/ukraine-internet-starlink-elon-musk-russia-war>, accessed 29 November 2022.
Wang Kaihao, 'The "Secret" Language of Women', *China Daily Europe*, 1 September 2017, at <https://europe.chinadaily.com.cn/epaper/2017-09/01/content_31577726.htm>, accessed 15 November 2022.
Andrew McLaughlin, 'RAAF's 10SQN AP3C (EW) Orions Transferred to 42WG', *Australian Defence Business Review*, 26 June 2019, at <www.adbr.com.au/raafs-10sqn-ap-3cew-orions-transferred-to-42wg/>, accessed 14 October 2022.
Dennis Moore, 'CB: Central Bureau Intelligence Corps Association Inc.', June 2000, at <www.ozatwar.com/sigint/2000_jun_cbic.pdf>, accessed 15 November 2022.

Joseph S. Nye, Jr, 'The End of Cyber-Anarchy? How to Build a New Digital Order', *Foreign Affairs*, January/February 2022, at <www.foreignaffairs.com/articles/russian-federation/2021-12-14/end-cyber-anarchy>, accessed 15 November 2022.

John Sullivan, 'Sun Tzu on Espionage or: How I Learned to Stop Worrying and Love the Double Agent', *Small Wars Journal*, 10 January 2018, at <www.smallwarsjournal.com/jrnl/art/sun-tzu-espionage-or-how-i-learned-stop-worrying-and-love-double-agent>, accessed 29 November 2019.

Richard Tanter, 'Fifty Years on, Pine Gap Should Reform to Better Serve Australia', *The Conversation*, 9 December 2016, at <www.theconversation.com/fifty-years-on-pine-gap-should-reform-to-better-serve-australia-65650>, accessed 15 November 2022.

Richard Tanter, 'Hiding From The Light: The Establishment of the Joint Australia-United States Relay Ground Station at Pine Gap,' NAPSNet Policy Forum, 2 November 2019, at <www.nautilus.org/napsnet/napsnet-policy-forum/hiding-from-the-light-the-establishment-of-the-joint-australia-united-states-relay-ground-station-at-pine-gap/>, accessed 14 October 2022.

Print and online media

ABC News (Australia)
The Age (Melbourne)
The Australian Financial Review
The Canberra Times
The Guardian
The London Gazette
The Mandarin
The Saturday Paper
The Sydney Morning Herald
Daily Telegraph (Sydney)
The West Australian

Television

Graeme Dobell, Interview with Gough Whitlam, 'Whitlam Reveals His East Timor Policy', *The World Today Archive*, 6 December 1999, <www.abc.net.au/worldtoday/stories/s71200.htm>, accessed 14 October 2022.

Dylan Welch, 'Top Intelligence Analyst Slams Pine Gap's Role in American Drone Strikes', *730 Report*, ABC Television, Wednesday 13 August 2014, <www.abc.net.au/7.30/top-intelligence-analyst-slams-pine-gaps-role-in/5669322>, accessed 14 October 2022.

Unpublished works

Craig Arthur Bellamy, 'The Beginning of the Secret Australian Radar Countermeasures Unit During the Pacific War', PhD thesis, Charles Darwin University, 2020.

John J. Bird, 'Analysis of Intelligence Support to the 1991 Persian Gulf War: Enduring Lessons,' USAWC Strategic Research Project, US Army War College, Carlisle, Pennsylvania, March 2004, available at <www.apps.dtic.mil/sti/pdfs/ADA423282.pdf>, accessed 14 October 2022.

Lewis Buckingham, 'Eulogy, Athanasius Pryor Treweek, 29 December 1911 – 20 January 1995,' (supplied by family).

Paul Kirk, 'Navy Information Warfare: Developing an Information Warfare Strategy for the Royal Australian Navy to Support Contemporary Maritime Operations', MPhil thesis, UNSW, Canberra, 2015, available at <http://hdl.handle.net/1959.4/56982>, accessed 14 October 2022.

Kieran Laurence Miles, 'A History of the Allied Interpreter Section: Southwest Pacific Area WWII', PhD thesis, The University of Queensland, 1993.

T.E. Nave, 'History', in his autobiography (copy in possession of authors).

Joseph H. Straczek, 'The Origins and Development of Royal Australian Naval Signals Intelligence in an Era of Imperial Defence 1914–1945', PhD thesis, UNSW Canberra, 2008.

R.C. Thompson, 'Australian Imperialism and the New Hebrides, 1862–1922', PhD thesis, ANU, 1970.

Correspondence and interviews

Ian Dudgeon, correspondence with authors, 24 August 2022.
Major General (ret'd) Steve Meekin, correspondence with authors, 10 October 2021.
Helen Robertson (Treweek's daughter), interview, 29 July 2022.
Brendan Sargeant, correspondence with authors, 10 September 2021.

Websites and blogs

Matthew M. Aid, 'Commentary: The National Security Agency During the Cold War,' National Security Agency Releases History of Cold War Intelligence Activities, 14 November 2008, The National Security Archive website, at <www.nsarchive2.gwu.edu/NSAEBB/NSAEBB260/index.htm>, accessed 14 October 2022.

Anon., '$20 Million JEWOSU Redevelopment Opened,' *Australian Defence Magazine*, 10 January 2008, at <www.australiandefence.com.au/D719AFE0-F806-11DD-8DFE0050568C22C9>, accessed 14 October 2022.

Anon., 'Australia Extends Heron Aircraft Deployment in Afghanistan', *Airforce Technology*, 11 December 2013, at <www.airforce-technology.com/news/newsaustralia-extends-heron-aircraft-deployment-in-afghanistan-4144870>, accessed 14 October 2022.

Anon., 'New Top Secret Sydney Cyber Security Centre Operational', Cyber Security Connect website, 30 August 2022, at <www.cybersecurityconnect.com.au/critical-infrastructure/8202-new-top-secret-sydney-cyber-security-centre-operational>, accessed 14 October 2022.

Anon., 'Ransomware Attacks Have Increased 24% in the Last Quarter', Cyber Security Connect website, 29 August 2022, at <www.cybersecurityconnect.com.au/commercial/8194-ransomware-attacks-have-increased-24-in-the-last-quarter>, accessed 14 October 2022.

Anon., 'OSIS Oceanic Surveillance Information System', GlobalSecurity website, at <www.globalsecurity.org/intell/systems/obu.htm>, accessed 14 October 2022.

Anon., 'Heron Tactical Unmanned Aerial System (UAS)', IAI website, at <www.iai.co.il/p/tactical-heron>, accessed 14 October 2022.

Australian Cyber Security Centre, 'About the ACSC', Australian Signals Directorate, Department of Defence, October 2020, Australian Cyber Security Centre website, at <www.cyber.gov.au/acsc>, accessed 14 October 2022.

Australian Information Security Association, 'Security Perspectives of a Former Spy Chief', AISA Melbourne Branch meeting, 13 February 2019, AISA website, at <www.aisa.org.au/

Public/Events/Event_Display.aspx?EventKey=d6acbf83-ed03-4590-abe0-a21b722b4cb6>, accessed 14 October 2022.

Australian Signals Directorate, 'No 3 Telecommunication Unit (3TU)', website of the Australian Signals Directorate, at <www.asd.gov.au/about/history/declassified/2022-03-15-no-3-telecommunication-unit-3tu>, accessed 14 October 2022.

Australian War Memorial, 'Prime Minister Robert G. Menzies: Wartime Broadcast', *Australian War Memorial Encyclopaedia,* at <www.awm.gov.au/articles/encyclopedia/prime_ministers/menzies>, accessed 14 October 2022.

Australian War Memorial, 'Empire Air Training Scheme', *Australian War Memorial Encyclopaedia,* at <www.awm.gov.au/articles/encyclopedia/raaf/eats>, accessed 14 October 2022.

Desmond Ball, Bill Robinson and Richard Tranter, 'Australia's Participation in the Pine Gap Enterprise,' NAPSNet Special Report, 9 June 2016, Nautilus Institute website, at <nautilus.org/napsnet/napsnet-special-reports/australias-participation-in-the-pine-gap-enterprise/>, accessed 14 October 2022.

Henry George Barnard, 'James Gerald Bridges Armstrong', Geni website, 26 May 2009, at <www.geni.com/people/James-Gerald-Bridges-Armstrong/6000000003982737032>, accessed 14 October 2022.

Jennifer Bussell, 'Cyberspace', Britannica, at <www.britannica.com/topic/cyberspace>, accessed 14 October 2022.

CBR Staff Writer, 'Australia Admits to Monitoring Electronic Signals', TechMonitor website, 27 May 1999, at <https://techmonitor.ai/technology/australia_admits_to_monitoring_electronic_signals>, accessed 14 October 2022.

Center for Cryptologic History, 'Radio Intelligence on the Mexican border, World War I: A Personal View', National Security Agency website, at <www.nsa.gov./about/cryptologic-heitage/historical-figures-publications/publications/pre-wwi/radio-intel-mexican-border/>, accessed 9 September 2019.

Creative Spirits, 'Understanding Aboriginal Painting,' Creative Spirits website, at <www.creativespirits.info/aboriginalculture/arts/understanding-aboriginal-paintings>, accessed 14 October 2022.

Defence Connect, 'Spy Base Looking for Contractors, Upgrades', *Defence Connect*, 30 January 2017, at <www.defenceconnect.com.au/intel-cyber/233-spy-base-looking-for-contractors-upgrades>, accessed 14 October 2022.

Department of Defence (Australian Government), 'Electronic Warfare Operations', Defence Science and Technology Group website, at <www.dst.defence.gov.au/capability/electronic-warfare-operations>, accessed 14 October 2022.

Department of Defence (Australian Government), 'Information Warfare Division', 3 March 2021, Information Warfare Division, Joint Capabilities Group website, Department of Defence, at <www.defence.gov.au/jcg/iwd.asp>, accessed 15 November 2022.

Department of Prime Minister and Cabinet (Australian Government), '2017 Independent Intelligence Review', at <www.pmc.gov.au/national-security/2017-independent-intelligence-review>, accessed 14 October 2022.

J. Fenton, H. O'Flynn, S. Hart, P. Murray and M. J. Davies (eds), 'The Unclassified History of 547 Signal Troop in South Vietnam,' at <www.angelfire.com/empire/547sigs/History of 547 Signal Troop.pdf >, accessed 14 October 2022.

Peter Few, 'JTUM Unit History – Joint Telecommunications Unit, Melbourne, 1981–1991', 3 Telecommunication Unit Association website, at <www.3teluunitassn.com/files/ex-P-Few-JTUM-history-article.pdf>, accessed 14 October 2022.

Peter Few, 'Some SIGSOP History', 3 Telecommunication Unit Association website, at <www.3teluunitassn.com/files/1_Wireless_Unit_history_article.pdf>, accessed 14 October 2022.

Steve Hart with Ernie Chamberlain, 'Story 3 – A Tactical SIGINT Success Story', Pronto in South Vietnam: Royal Australian Corps of Signals website, at <www.pronto.au104.org/547Sigs/547story3.html>, accessed 14 October 2022.

David A. Hatch with Robert Louis Benson, 'The Korean War: The SIGINT Background,' National Security Agency/Central Security Service website, at <www.nsa.gov/about/cryptologic-heritage/historical-figures-publications/publications/korean-war/koreanwar-sigint-bkg/>, accessed 14 October 2022.

The Historical Archive, 'The History of Electricity – A Timeline', The Historical Archive website, at <www.thehistoricalarchive.com/happenings/57/the-history-of-electricity-a-timeline/>, accessed 12 November 2019.

Hewlett Packard Enterprise, 'From the Cray-1 to HPE Today,' Hewlett Packard Enterprise website, at <www.hpe.com/us/en/compute/hpc/cray.html>, accessed 14 October 2022.

David Jones, 'Banks Outpace Other Industries in Cyber Investments, Defense Strategies: Report', Cybersecurity Dive website, 15 November 2021, at <www.cybersecuritydive.com/news/banks-cyber-security-investments/610045/>, accessed 14 October 2022.

Jeff Malone, 'A Tale of the Kangaroo and the Crow: Electromagnetic Spectrum Operations Challenges for the Australian Defence Force', PowerPoint presentation, 26–28 May 2015, Stockholm, Sweden, at <http://repository.jeffmalone.org/files/personal/A%20Tale%20of%20the%20Kangaroo%20and%20the%20Crow.pdf>, accessed 14 October 2022.

'Morse Code & The Telegraph', History website, at <www.history.com/topics/inventions/telegraph>, accessed 12 November 2019.

Naval Historical Society of Australia, 'The Sydney Cenotaph and its Guardians', Royal Australian Navy website, <https://navyhistory.org.au/the-sydney-cenotaph-and-its-guardians/>, accessed 14 October 2022.

Joris Nieuwint, 'When the Allies Killed Over 20,000 of Their Own Countrymen as They Sank Japanese Hell Ships that Transported Them,' War History Online website, 17 January 2016, at <www.warhistoryonline.com/featured/japanese-hellships.html>, accessed 27 February 2020.

Jeremy Norman, 'The Invention of Movable Type in China Circa 1041–1048', Jeremy Norman's History of Information website, at <www.historyofinformation.com/detail.php?entryid=1642>, accessed 20 November 2019.

Kelly Pang, 'Chinese Ancient Currency', China Highlights website, at <www.chinahighlights.com/travelguide/culture/chinese-ancient-currency.htm>, accessed 14 October 2022.

RAAF, 'History of the Air Force', Royal Australian Air Force website, at <www.airforce.gov.au/about-us/history>, accessed 14 October 2022.

Rohit Ranjan, 'Quad Senior Cyber Group Meets in Sydney to Strengthen Cybersecurity Cooperation says WH', RepublicWorld website, 26 March 2022, at <www.republicworld.com/world-news/rest-of-the-world-news/quad-senior-cyber-group-meets-in-sydney-to-strengthen-cybersecurity-cooperation-says-wh-articleshow.html>, accessed 14 October 2022.

Anthea Roberts, Henrique Choer Moraes and Victor Ferguson, 'The Geoeconomic World Order', Lawfare blog, 19 November 2018, at <www.lawfareblog.com/geoeconomic-world-order>, accessed 14 October 2022.

Anthea Roberts, Henrique Choer Moraes and Victor Ferguson, 'The U.S.-China Trade War is a Competition for Technological Leadership', Lawfare blog, 21 May 2019, at <www.lawfareblog.com/us-china-trade-war-competition-technological-leadership>, accessed 14 October 2022.

Tim Sherratt, 'Historic Hansard: Commonwealth of Australia parliamentary debates presented in an easy-to-read format for historians and other lovers of political speech,' Honest History website, at <www.honesthistory.net.au/wp/sherratt-tim-historic-hansard-commonwealth-of-australia-parliamentary-debates-presented-in-an-easy-to-read-format-for-historians-and-other-lovers-of-political-speech/>, accessed 14 October 2022.

Siegphyl, 'Mission of Secret Australian Submarines During the Cold War Revealed', War History Online website, 14 November 2013, at <www.warhistoryonline.com/war-articles/missions-secret-australian-submarines-cold-war-revealed.html>, accessed 14 October 2022.

The Soufan Center, 'IntelBrief: Could Confrontation in Cyberspace Escalate the War in Ukraine?', The Soufan Center website, 10 June 2022, at <https://thesoufancenter.org/intelbrief-2022-june-10/>, accessed 14 October 2022.

David Stevens, 'The RAN – A Brief History', Royal Australian Navy website, at <www.navy.gov.au/history/feature-histories/ran-brief-history>, accessed 21 January 2020.

Casey Tonkin, '$1.3b. Drone Scrapped for REDSPICE', Information Age, Australian Computer Society website, 5 April 2022, at <http://ia.acs.org.au/article/2022/1-3b-drone-scrapped-for-redspice.html>, accessed 14 October 2022.

Brad Williams, 'Nakasone: Cold War-style deterrence "does not comport to cyberspace", *Breaking Defense*, 4 November 2021, <www.breakingdefense.com/2021/11/nakasone-cold-war-style-deterrence-does-not-comport-to-cyberspace/>, accessed 14 October 2022>.

Edward Zheng, Kevin Stewart and Michael Caplan, 'A Spice Up to Australia's Cyber Capabilities?' Gilbert+Tobin, Digital Hub, 13 April 2022, <www.gtlaw.com.au/knowledge/spice-australias-cyber-capabilities-what-we-know-so-far-about-project-redspice>, accessed 14 October 2022.

NOTES

Introduction
1. For the United Kingdom, most prominent are works written by Richard Aldrich and John Ferris. The United States has had several works written on its national Sigint arrangements including by James Bamford, Jeffrey Richelson, David Kahn, Sharon Maneki, Liza Mundy and others as listed in the bibliography.
2. The relevant works of these authors are listed in the bibliography.
3. See Shannon Jenkins, 'ASD's decision to cancel official history contract to be probed at Senate estimates,' *The Mandarin,* 21 September 2020, <www.themandarin.com.au/140230-asds-decision-to-cancel-official-history-contract-to-be-probed-at-senate-estimates/>, accessed 15 November 2022; Anthony Galloway, 'Cyber spy agency dumps military historian from writing its official history', *The Sydney Morning Herald,* 18 September 2020; Paul Daley, 'Australia's spy agency has dismissed its official historian. But why?,' *The Guardian,* Monday 28 September 2020, <www.theguardian.com/commentisfree/2020/sep/28/australias-spy-agency-has-dismissed-its-official-historian-but-why>, accessed 15 November 2022; and Ben Packham, 'Spy agency and Australian National University fight over axed book contract,' *The Australian,* 22 September 2020, <www.theaustralian.com.au/higher-education/spy-agency-and-australian-national-university-fight-over-axed-book-contract/news-story/0f6e502d994666db522e49b85b40bdfc>, accessed 15 November 2022.
4. Extract from Recommendations of the Defence Committee on the Australian contribution to the British Commonwealth Signal Intelligence Organisation: NAA: A5954 2363/3, 137.
5. John le Carré, *Tinker, Tailor, Soldier, Spy* (London/Sydney: Pan Books, 1975), p. 306.
6. Alex Younger, 'MI6 "C" speech on fourth generation espionage,' transcript of speech delivered at St Andrews University, UK government website, 3 December 2018, <www.gov.uk/government/speeches/mi6-c-speech-on-fourth-generation-espionage>, accessed 25 October 2022.
7. Nigel de Grey, 'Allied Sigint-Policy and Organisation', Chap. VIII, 'Australia', 17 TNA (UK): HW 43/78.
8. Peter Donovan and John Mack, *Code Breaking in the Pacific* (Cham: Springer International Publishing, 2014), p. 75.
9. Sharon A. Maneki, *The Quiet Heroes of the Southwest Pacific Theater: An Oral History of the Men and Women of CBB and FRUMEL* (Fort Meade, MD: Center for Cryptological History, NSA, 1996), p. 81.
10. F.H. Hinsley, *Official History of British Intelligence in the Second World War: Its Influence on Strategy and Operations* (London: HMSO, 1979), p. x.

1 Early cryptology, secrets & intelligence
1. Dennis Richardson, Anna Harmer and Tara Inverarity, *Comprehensive Review of the Legal Framework of the National Intelligence Community,* December 2019, vol. 1, 10. Recommendation 1.

2   Steven E. Maffeo, *Most Secret and Confidential: Intelligence in the Age of Nelson* (Annapolis: Naval Institute Press, 2000), p. xvii.
3   Christopher Andrew, *The Secret World: A History of Intelligence* (Milton Keynes: Allen Lane, Penguin, 2018), pp. 1–2, 13–26.
4   Craig P. Bauer, *Secret History: The Story of Cryptology* (Boca Raton, FL: CRC Press, 2013), p. 4; David Kahn, *The Code-Breakers: The Comprehensive History of Secret Communication from Ancient Times to the Internet* (New York: Scribner, 1996), pp. 82–83; and Diodorus Siculus, *Library*, at <www.perseus.tufts.edu/hopper/text?doc=Perseus:text:1990.01.0084:book=13:chapter=106&highlight=skytale>, accessed 29 November 2019.
5   Bauer, *Secret History*, pp. 4–5.
6   Folder of Central Bureau Records, 'Bibliography', NAA: B5436, Part L. The bibliography lists 124 publications between the 4th century BC and 1920. *Le tacticien* was published in 1757 in Amsterdam and translated into French by Marshal the Count of Beausobre. A later version of Aenias' work in Latin was published as 'Commentarius de Toleranda Obsidione' (or 'How to survive under a siege') in Leipzig in 1818. English translations have emerged recently. See David Whitehead (translator), *Aineias the Tactician: How to Survive under a Siege* (Oxford: Clarendon Press, 1990), pp. 84–90.
7   Bauer, *Secret History*, pp. 11–12; and Kahn, *The Code-Breakers*, pp. 83–84.
8   Bauer, *Secret History*, p. 26.
9   Bauer, *Secret History*, pp. 3, 54–69.
10   See John Sullivan, 'Sun Tzu on Espionage or: How I Learned to Stop Worrying and Love the Double Agent', *Small Wars Journal*, at <www.smallwarsjournal.com/jrnl/art/sun-tzu-espionage-or-how-i-learned-stop-worrying-and-love-double-agent>, accessed 29 November 2019.
11   Andrew, *The Secret World*, 3, pp. 98–99.
12   Ulug Kuzuoglu, 'Chinese Cryptography: The Chinese Nationalist Party and Intelligence Management, 1927–1949', *Cryptologia* (2018), pp. 42–46, 514–39.
13   Jeremy Norman, 'The Invention of Movable Type in China Circa 1041–1048,' at <www.historyofinformation.com/detail.php?entryid=1642>, accessed 20 November 2019.
14   'Chinese Ancient Currency', at <www.chinahighlights.com/travelguide/culture/Chinese-ancient-currency.htm>, accessed 20 November 2019.
15   Hong Liu, *Chinese Business: Landscapes and Strategies* (London and New York: Routledge, 2009).
16   Matthew Purbrick, 'Patriotic Chinese Triads and Secret Societies: From the Imperial Dynasties, to Nationalism and Communism', *Journal of Asian Affairs*, 50:3 (2019): 305–22.
17   Wang Kaihao, 'The "Secret" Language of Women', *China Daily Europe*, 1 September 2017, at <www.europe.chinadaily.com.cn/epaper/2017-09/01/content_31577726.htm>, accessed 25 October 2022.
18   Fei-Wen Liu, 'Being to Becoming: Nushu and Sentiments in a Chinese Rural Community,' *American Ethnologist*, 31:3 (2004): 422–39.
19   Andrew, *The Secret World*, p. 127; Bauer, *Secret History*, pp. 14–15; and Kahn, *The Code-Breakers*, p. 125.
20   Kahn, *The Code-Breakers*, p. 130.
21   Kahn, *The Code-Breakers*, pp. 130, 135–36.
22   Bauer, *Secret History*, p. 75.
23   Kahn, *The Code-Breakers*, pp. 130, 135–36.
24   Bauer, *Secret History*, p. 74.

25 John F. Dooley, *Codes, Ciphers and Spies: Tales of Military Intelligence in World War I* (New York: Copernicus Books, 2016), pp. 49–50
26 NSA/CSS, 'The Rare Book Collection', <www.nsa.gov/about/cryptologic-heritage/historical-figures-publications/publications/misc/rare-books/>, accessed 19 September 2019.
27 Andrew, *The Secret World*, p. 5.
28 Thomas G. Ferguson, *British Military Intelligence, 1870–1914: The Development of a Modern Intelligence Organization* (Frederick, MD: University Publications of America, 1984), pp. 8–9, 88.
29 Andrew, *The Secret World*, p. 175.
30 Andrew, *The Secret World*, pp. 176–77.
31 Ferguson, *British Military Intelligence, 1870–1914*, pp. 8–9, 88.
32 John Johnson, *The Evolution of British Sigint, 1653–1939* (Cheltenham: GCHQ, 1997), pp. 14–19.
33 Kahn, *The Code-Breakers*, pp. 163–64.
34 Andrew, *The Secret World*, pp. 270–72.
35 Andrew, *The Secret World*, pp. 288–90.
36 Mark Urban, *The Man Who Broke Napoleon's Codes* (London: Faber & Faber, 2001), pp. 103–104.
37 Cited in Urban, *The Man Who Broke Napoleon's Codes*, pp. 103–104.
38 Bauer, *Secret History*, p. 104.
39 Maffeo, *Most Secret and Confidential*, pp. 73–75.
40 Maffeo, *Most Secret and Confidential*, pp. 84–85; and Daniel R. Headrick, *The Invisible Weapon: Telecommunications and International Politics 1851–1945* (New York and Oxford: Oxford University Press, 1991), pp. 11–12.
41 Andrew, *The Secret World*, pp. 5–6.
42 Jennifer Wilcox, *Revolutionary Secrets: Cryptology in the American Revolution* (Fort Meade, MD: Center for Cryptologic History, National Security Agency, 2012), pp. 3, 24.
43 Ralph E. Weber, *'Masked Dispatches': Cryptograms and Cryptology in American History, 1775–1900*, 3rd edition (Fort Meade, MD: National Security Agency, Center for Cryptologic History, 2013), p. 1.
44 Weber, *'Masked Dispatches'*, p. xi.
45 Wilcox, *Revolutionary Secrets*, p. 46.
46 Andrew, *The Secret World*, pp. 5–6.
47 Andrew, *The Secret World*, pp. 339, 341.
48 Andrew, *The Secret World*, p. 354.
49 Martin Van Creveld, *Command in War* (Cambridge, MA: Harvard University Press, 1985), p. 82, cited in Andrew, *The Secret World*, p. 339.
50 Urban, *The Man Who Broke Napoleon's Codes*, p. 69.
51 Andrew, *The Secret World*, p. 345.
52 Andrew, *The Secret World*, p. 347.
53 Fergusson, *British Military Intelligence, 1870–1914*, pp. 47, 133, 134, 136.
54 Andrew, *The Secret World*, pp. 347–48.
55 Urban, *The Man Who Broke Napoleon's Codes*, p. 242.
56 John Keegan, *Intelligence in War: Knowledge of the Enemy from Napoleon to Al-Qaeda* (London: Hutchinson, 2003), p. 99; and Maffeo, *Most Secret and Confidential*, pp. 68–69.
57 Andrew, *The Secret World*, p. 354.
58 Andrew, *The Secret World*, pp. 357–58.
59 Andrew, *The Secret World*, pp. 360–62.

60 See Sullivan, 'Sun Tzu on Espionage or: How I Learned to Stop Worrying and Love the Double Agent'.
61 Andrew, *The Secret World*, pp. 309, 363.
62 Thomas Fleming (ed.), *The Founding Fathers: Benjamin Franklin: A Biography in His Own Words* (New York: Newsweek, Harper & Row, 1972), p. 92.
63 Anon, 'The History of Electricity – A Timeline', The Historical Archive, at <www.thehistoricalarchive.com/happenings/57/the-history-of-electricity-a-timeline/>, accessed 12 November 2019.
64 'Morse Code & The Telegraph', History.com editors, at <www.history.com/topics/inventions/telegraph>, accessed 12 November 2019.
65 Johnson, *The Evolution of British Sigint*, p. 19.
66 Johnson, *The Evolution of British Sigint*, pp. 21–23.
67 Headrick, *The Invisible Weapon*, pp. 14–15.
68 Keegan, *Intelligence in War*, p. 99.
69 Andrew, *The Secret World*, pp. 402–404; and Fergusson, *British Military Intelligence, 1870–1914*, p. 137.
70 Keegan, *Intelligence in War*, p. 100.
71 Andrew, *The Secret World*, pp. 402–404.
72 Headrick, *The Invisible Weapon*, pp. 4–5.
73 Kahn, *The Code-Breakers*, p. 204.
74 Dooley, *Codes, Ciphers and Spies*, p. 35.
75 Dooley, *Codes, Ciphers and Spies*, p. 50.
76 Fergusson, *British Military Intelligence, 1870–1914*, p. 29.
77 Headrick, *The Invisible Weapon*, pp. 19–21.
78 Jim Beach, *Haig's Intelligence GHQ and the German Army, 1916–1918* (Cambridge: Cambridge University Press, 2013), p. 23; and Fergusson, *British Military Intelligence, 1870–1914*, p. 30.
79 John Ferris, 'Before "Room 40": The British Empire and Signals Intelligence, 1898–1914,' *Journal of Strategic Studies*, 12:4 (1989), p. 433.
80 Christopher Andrew and Keith Neilson, 'Tsarist Codebreakers and British Codes,' in Christopher Andrew, *Codebreaking and Signals Intelligence* (London: Frank Cass, 1986), p. 2; and Ferris, 'Before "Room 40"', p. 432.
81 Andrew, *The Secret World*, pp. 410–11.
82 Weber, *Masked Dispatches*, p. xi.
83 Kahn, *The Code-Breakers*, p. 220.
84 Andrew, *The Secret World*, pp. 410–11.
85 Keegan, *Intelligence in War*, pp. 100–101.
86 Andrew, *The Secret World*, pp. 410–11.
87 Andrew, *The Secret World*, pp. 412–14.
88 Weber, *Masked Dispatches*, p. 188.
89 Fergusson, *British Military Intelligence, 1870–1914*, p. 29.
90 Keegan, *Intelligence in War*, p. 103.
91 Thomas S. Mullaney, *The Chinese Typewriter: A History* (Cambridge, MA: Massachusetts Institute of Technology Press, 2017), p. 81.
92 Mullaney, *The Chinese Typewriter: A History*, pp. 115–16.
93 John F. Clabby, *Brigadier John Tiltman, a Giant Among Cryptanalysts* (Fort Meade, MD: National Security Agency, Center for Cryptologic History, 2007), p. 17.
94 Mullaney, *The Chinese Typewriter: A History*, pp. 115–16.
95 Anon, 'Defining Moments: Overland Telegraph', at <www.nma.gov.au/defining-moments/resources/overland-telegraph>, accessed 19 November 2019.

96 Ferris, 'Before "Room 40"', p. 435.
97 Keegan, *Intelligence in War,* p. 101.
98 Keegan, *Intelligence in War*, p. 101; and Headrick, *The Invisible Weapon*, pp. 117–18.
99 John Ferris, 'The Road to Bletchley Park: The British Experience with Signals Intelligence, 1892–1945,' *Intelligence and National Security*, 17:1 (2002), p. 59.
100 Ferris, 'Before "Room 40"', p. 432.
101 Donovan and Mack, *Code Breaking in the Pacific*, p. 20.
102 Donovan and Mack, *Code Breaking in the Pacific*, p. 20.
103 Fergusson, *British Military Intelligence, 1870–1914,* pp. 15–17, 22, 43, 46; and C.D. Coulthard-Clark, *The Citizen General Staff: The Australian Intelligence Corps, 1907–1914* (Canberra: Military Historical Society of Australia, 1976), pp. 4–5.
104 Ferris, 'Before "Room 40"', p. 442.

2   From Federation & the Anglo–Boer War to the First World War

1 See Craig Wilcox, *Australia's Boer War: The War in South Africa, 1899–1902* (Melbourne: Oxford University Press, 2002).
2 Thomas G. Fergusson, *British Military Intelligence, 1870–1914: The Development of a Modern Intelligence Organization* (Frederick, MD: University Publications of America, 1984), pp. 103–104 and 114–15.
3 John Ferris, 'Before "Room 40": The British Empire and Signals Intelligence, 1898–1914,' *Journal of Strategic Studies*, 12:4 (1989), p. 435.
4 Fergusson, *British Military Intelligence, 1870–1914*, p. 189.
5 Fergusson, *British Military Intelligence, 1870–1914*, p. 160; and Ferris, 'Before "Room 40"', p. 436.
6 Fergusson, *British Military Intelligence, 1870–1914*, pp. 219–20.
7 Francis Maxwell diary, entries 28.6 and 19.7 1901, cited in Ferris, 'Before "Room 40"', p. 436.
8 Ferris, 'Before "Room 40"', p. 436.
9 Fergusson, *British Military Intelligence, 1870–1914*, p. 160.
10 C.D. Coulthard-Clark, *The Citizen General Staff: The Australian Intelligence Corps 1907–1914* (Canberra: Military Historical Society of Australia, 1976), pp. 2–3.
11 John Ferris, 'The Road to Bletchley Park: The British Experience with Signals Intelligence, 1892–1945,' *Intelligence and National Security*, 17:1 (2002), p. 60.
12 Fergusson, *British Military Intelligence, 1870–1914*, pp. 166–67.
13 Fergusson, *British Military Intelligence, 1870–1914*, p. 189.
14 John Keegan, *Intelligence in War: Knowledge of the Enemy from Napoleon to Al-Qaeda* (London: Hutchinson, 2003), pp. 102–103.
15 Peter Donovan and John Mack, *Code Breaking in the Pacific* (Cham: Springer International Publishing, 2014), p. 5.
16 Daniel R. Headrick, T*he Invisible Weapon: Telecommunications and International Politics 1851–1945* (New York and Oxford: Oxford University Press, 1991), pp. 125–26.
17 Keegan, *Intelligence in War*, p. 105.
18 Ferris, 'Before "Room 40"', p. 448.
19 Christopher Andrew and Keith Neilson, 'Tsarist Codebreakers and British Codes,' in Christopher Andrew (ed.), *Codebreaking and Signals Intelligence* (London: Frank Cass, 1986), p. 2.
20 Andrew and Neilson, 'Tsarist Codebreakers and British Codes,' p. 10.
21 Military Forces of the Commonwealth General Orders 1/1902. Printed and Published for the Government of the Commonwealth of Australia by Robert S. Bain, Government Printer for the State of Victoria. AWM: C1241.

22 Letter of recommendation from James Burns for Wilson Le Couteur to Prime Minister Edmund Barton, July 2nd, 1901. NAA: A35, Bundle 2/7, BC31404093.
23 Wilson Le Couteur sgd letter to Prime Minister Edmund Barton, 8 February 1901, NAA: A35, Bundle 2/1, BC31404093.
24 R.C. Thompson, 'Australian Imperialism and the New Hebrides, 1862–1922', PhD thesis, ANU, 1970, p. 374.
25 Thompson, 'Australian Imperialism and the New Hebrides,' p. 375.
26 Thompson, 'Australian Imperialism and the New Hebrides,' p. 403.
27 Atlee Hunt, Secretary of the Department of External Affairs sgd letter to Wilson Le Couteur, Melbourne, 1 August 1901. NAA: A35, Bundle 2/12, BC 31404104.
28 Coulthard-Clark, *The Citizen General Staff*, pp. 7–9; and John Fahey, *Australia's First Spies: The Remarkable Story of Australia's Intelligence Operations, 1901–1945* (Sydney: Allen & Unwin, 2018), pp. 15–16.
29 Military Forces of the Commonwealth, General Order No. 3 of 6 January 1905. Government Printer, Melbourne.
30 Military Forces of the Commonwealth General Order No. 4 of 7 January 1905. Government Printer, Melbourne.
31 Military Order 414 dated 1 January 1909, Commonwealth Forces Military Orders 1908. Department of Defence Melbourne 7th January 1908. By authority J. Kemp, Government Printer, Melbourne.
32 Coulthard-Clark, *The Citizen General Staff*, pp. 7–9, 12.
33 Ferguson, *British Military Intelligence, 1870–1914*, p. 13.
34 The connections between Australia and Canada are explored in John C. Blaxland, *Strategic Cousins: Australian and Canadian Expeditionary Forces and the British and American Empires* (Montreal: McGill-Queens University Press, 2006).
35 Military Order 305 of 1907. Commonwealth Forces Military Orders, 1-313-1907. Government Printer, Melbourne.
36 Geoffrey Serle, 'McCay, Sir James Whiteside (1864–1930)', *Australian Dictionary of Biography*, vol. 10, 1986, at <www.adb.anu.edu.au/biography/mccay-sir-james-whiteside-7312>, accessed 25 October 2022.
37 Copy of Original Diary of Major General W.T. Bridges, 13 March 1915, Australian War Memorial 2DRL/0469, <www.awm.gov.au/collection/C89052>, accessed 25 October 2022.
38 Letter from Field Marshal Lord William Birdwood, 12 October 1915, from Australian War Memorial Collection of Letters and cablegrams written by Field Marshal Lord William Birdwood to Sir Ronald Crawfurd Munro Ferguson, 1915. AWM: RCDIG0000039.
39 Standing Orders 1911 for the Australian Intelligence Corps, NAA: A1194/1, 20.41/6751.
40 Military Order 41, 1908. Commonwealth Forces Military Orders, Department of Defence, 1908. By authority: J. Kemp, Government Printer, Melbourne.
41 Military Order 444 of 1912. Commonwealth Forces Military Orders 1-719-1912. Government Printer, Melbourne.
42 Report on an Intelligence Staff Tour held in New South Wales under the Direction of Colonel the Honourable J.W. McCay, V.D. Director of Intelligence From 24th to 30th January 1912. By Authority: J. Kemp, Government Printer, Melbourne, NAA: A1194 A1194/1, Control Symbol 12.42/4778.
43 Military Order 390 of 14 July 1914. Commonwealth Forces Military Orders 1-709 1914. By Authority Albert J. Mullett, Government Printer Melbourne.

44 Tributes in Parliament on the death of Sir James McCay, 30.10.30, NAA: 461 A461/10, Control Symbol 700/1/134.
45 Arthur W. Jose, *The Official History of Australia in the War of 1914–1918*, vol. 9: The Royal Australian Navy (Brisbane: University of Queensland Press and Australian War Memorial, 1993 reprint), pp. 436–38.
46 NAA: A1608, Item No. B. 15/1/1, cited in Coulthard-Clark, *The Citizen General Staff*, p. 53.
47 Ian Pfennigwerth, *Missing Pieces: The Intelligence Jigsaw and RAN Operations 1939–1971*, Papers in Australian Maritime Affairs No. 25 (Canberra: Sea Power Centre-Australia, Department of Defence, 2008), p. 5, available at <www.navy.gov.au/sites/default/files/documents/PIAMA25.pdf>, accessed 25 October 2022.
48 O. Murray sgd letter from Admiralty to Under Secretary of State Colonial Office, 5 September 1913. NAA: A6661, 1357 BC 422481, p. 56.
49 Pfennigwerth, *Missing Pieces*, p. 2.
50 W. Graham Greene sgd Letter from Admiralty to Under Secretary of State Colonial Office, 19 May 1914. NAA: A6661, 1357 BC 422481, pp. 34–36.
51 Fahey, *Australia's First Spies*, pp. 39–41.
52 Headrick, *The Invisible Weapon*, pp. 98–100.
53 Ferris, 'Before "Room 40"', pp. 443–44, 447.
54 Ferris, 'The Road to Bletchley Park', p. 60.
55 Fergusson, *British Military Intelligence, 1870–1914*, p. 220.
56 Fergusson, *British Military Intelligence, 1870–1914*, pp. 218–19.
57 Ferris, 'Before "Room 40"', p. 449.
58 Ferris, 'The Road to Bletchley Park', p. 60.

3 First World War Allied Sigint

1 Thomas G. Ferguson, *British Military Intelligence, 1870–1914: The Development of a Modern Intelligence Organization* (Frederick, MD: University Publications of America, 1984), pp. 7–9.
2 S.R. Elliott, *Scarlet to Green: A History of Intelligence in the Canadian Army, 1903–1963* (Ottawa: Canadian Intelligence and Security Association, 1981), p. 23.
3 John Keegan, *Intelligence in War: Knowledge of the Enemy from Napoleon to Al-Qaeda* (London: Hutchinson, 2003), p. 106.
4 John Ferris, 'The Road to Bletchley Park: The British Experience with Signals Intelligence, 1892–1945,' *Intelligence and National Security*, 17:1 (2002), p. 59.
5 John Ferris, 'Before "Room 40": The British Empire and Signals Intelligence, 1898–1914,' *Journal of Strategic Studies*, 12:4 (1989): 431–57; and John Johnson, *The Evolution of British Sigint: 1653–1939* (Cheltenham: GCHQ, 1997), pp. 27–29.
6 David Kahn, *The Code-Breakers: The Comprehensive History of Secret Communication from Ancient Times to the Internet* (New York: Scribner, 1996), pp. 272–73.
7 John Johnson, *The Evolution of British Sigint: 1653–1939* (Cheltenham: GCHQ, 1997), p. 27.
8 Bojan Pajic, *Our Forgotten Volunteers: Australians and New Zealanders with Serbs in World War I* (Melbourne: Australian Scholarly Publishing, 2019), p. 235.
9 Pajic, *Our Forgotten Volunteers*, p. 1.
10 Naval Historical Society of Australia, 'The Sydney Cenotaph and its Guardians', at <www.navyhistory.org.au/tag/memorial/page/2/>, accessed 18 December 2019.
11 Daniel R. Headrick, *The Invisible Weapon: Telecommunications and International Politics 1851–1945* (New York and Oxford: Oxford University Press, 1991), pp. 140–42.

12 Ferris, 'The Road to Bletchley Park,' pp. 62–63.
13 Decoded telegram sent to District Naval Officers (DNO's) all States and SDNO's Thursday Island and Newcastle 13 August 1914, 3.10 pm. NAA: MP1049/1, 1914/0351 BC413224, p. 216.
14 Navy Office Melbourne telephoned telegrams in code to DNO Port Melbourne, 9/8/14, 1300H and 9/8/14 1456. NAA: MP4019/1, 1914/0351 BC413224, pp. 309, 311.
15 DNO Victoria letter to Officer Commanding, Naval Intelligence Office, 12 August 1914. NAA: MP1049/1, 1914/0351 BC413224, p. 225.
16 Admiralty decoded telegram to Navy Office, Melbourne 7/9/1914 8.50pm NAA: MP1049/1, 1914/0351 BC413224, p. 173.
17 James Phelps, *Australian Code Breakers: Our Top Secret War with the Kaiser's Reich* (Sydney: Harper Collins, 2020).
18 John Fahey, *Australia's First Spies: The Remarkable Story of Australia's Intelligence Operations, 1901–1945* (Sydney: Allen & Unwin, 2018), pp. 47, 49, 50; and Joseph H. Straczek, 'The Origins and Development of Royal Australian Naval Signals Intelligence in an Era of Imperial Defence 1914–1945', PhD thesis, UNSW Canberra, 2008, p. 29.
19 Johnson, *The Evolution of British Sigint*, pp. 30–31.
20 Arthur W. Jose, *The Official History of Australia in the War of 1914–1918*, vol. 9: The Royal Australian Navy (Brisbane: University of Queensland Press and Australian War Memorial, 1993 reprint), pp. 9–11; Fahey, *Australia's First Spies*, p. 45; and Keegan, *Intelligence in War*, pp. 124–25.
21 For a detailed account of the capture, see Jose, *The Royal Australian Navy*, pp. 74–99.
22 Jose, *The Royal Australian Navy*, p. 11.
23 Jose, *The Royal Australian Navy*, p. 30; Keegan, *Intelligence in War*, p. 142; and Headrick, *The Invisible Weapon*, pp. 160–61.
24 See Jose, *The Royal Australian Navy*, pp. 179–207.
25 Keegan, *Intelligence in War*, pp. 126–32.
26 Jose, *The Royal Australian Navy*, pp. 179–93; Keegan, *Intelligence in War*, pp. 131–32; and William F. Flicke, *War Secrets in the Ether* (Laguna Hills, CA: Aegean Park Press, 1977), p. 61.
27 Geoffrey Ballard, *On ULTRA Active Service: The Story of Australia's Signals Intelligence Operations During World War II* (Richmond: Spectrum Publications, 1991), p. 44.
28 Keegan, *Intelligence in War*, p. 142.
29 Jose, *The Royal Australian Navy*, p. 439.
30 Jose, *The Royal Australian Navy*, p. 439.
31 Johnson, *The Evolution of British Sigint*, p. 31.
32 Kahn, *The Code-Breakers*, pp. 272–73.
33 Johnson, *The Evolution of British Sigint*, p. 31.
34 Donald A. Borrmann, William T. Kvetkas, Charles V. Brown, Michael J. Flatley and Robert Hunt, *The History of Traffic Analysis: World War I–Vietnam* (Fort Meade, MD: Center for Cryptologic History, National Security Agency, 2013), pp. 14–16; and Kahn, *The Code-Breakers*, pp. 272–73.
35 Kahn, *The Code-Breakers*, p. 272.
36 Borrmann et al., *The History of Traffic Analysis*, pp. 14–16; and Kahn, *The Code-Breakers*, pp. 272–73.
37 Ferris, 'The Road to Bletchley Park', p. 62.
38 C.D. Coulthard-Clark, *The Citizen General Staff: The Australian Intelligence Corps 1907–1914* (Canberra: Military Historical Society of Australia, 1976), pp. 54–55.
39 Blamey would end the war as Monash's Chief of Staff and subsequently, in the Second World War, he was Commander-in-Chief of Australian Military Forces.

40 John Ferris (ed.), *The British Army and Signals Intelligence during the First World War* (Stroud: Alan Sutton for the Army Records Society, 1992), p. 3.
41 Headrick, *The Invisible Weapon*, p. 155.
42 Jim Beach, *Haig's Intelligence GHQ and the German Army, 1916–1918* (Cambridge: Cambridge University Press, 2013), p. 23.
43 Johnson, *The Evolution of British Sigint*, pp. 34–35.
44 Thomas G. Fergusson, *British Military Intelligence, 1870–1914: The Development of a Modern Intelligence Organization* (Frederick, MD: University Publications of America, 1984), p. 239.
45 Ferris, *The British Army and Signals Intelligence*, pp. 3–4; and Johnson, *The Evolution of British Sigint*, p. 35.
46 Fergusson, *British Military Intelligence, 1870–1914*, p. 239.
47 Flicke, *War Secrets in the Ether*, pp. 28–29.
48 Jim Beach and James Bruce, 'British Signals Intelligence in the Trenches, 1915–1918: Part 1, Listening Sets,' *Journal of Intelligence History*, 19:1 (2020), p. 1.
49 Beach and Bruce, 'British Signals Intelligence in the Trenches, 1915–1918: Part 1,' p. 1.
50 Ferris, *The British Army and Signals Intelligence*, pp. 321–22.
51 Beach and Bruce, 'British Signals Intelligence in the Trenches, 1915–1918: Part 1,' pp. 5, 10.
52 Flicke, *War Secrets in the* Ether, p. 26.
53 Beach and Bruce, 'British Signals Intelligence in the Trenches, 1915–1918: Part 1,' pp. 5, 10.
54 Ferris, *The British Army and Signals Intelligence*, pp. 13–14.
55 Ferris, *The British Army and Signals Intelligence*, p. 14.
56 Jim Beach and James Bruce, 'British Signals Intelligence in the Trenches, 1915–1918: Part 1,' p. 20.
57 Jim Beach and James Bruce, 'British Signals Intelligence in the Trenches, 1915–1918: Part 2, Interpreter Operators,' *Journal of Intelligence History*, 19:1 (2020), p. 2.
58 Beach and Bruce identify correspondence showing Canadians were recruited to serve as interpreter operators. Textual references suggest Australians were involved as well, although no documentary evidence confirming that has been uncovered. See Beach and Bruce, 'British Signals Intelligence in the Trenches: Part 2,' p. 14.
59 Beach, *Haig's Intelligence*, pp. 38–40, cited in Beach and Bruce, 'British Signals Intelligence in the Trenches: Part 2,' p. 18.
60 Ferris, *The British Army and Signals Intelligence*, p. 22.
61 Beach and Bruce, 'British Signals Intelligence in the Trenches: Part 1,' p. 12.
62 Beach and Bruce, 'British Signals Intelligence in the Trenches: Part 1,' p. 16.
63 Beach and Bruce, 'British Signals Intelligence in the Trenches: Part 1,' p. 22.
64 Beach, *Haig's Intelligence*, p. 160; Ferris, *The British Army and Signals Intelligence*, pp. 15, 36; and Priestly, *Signal Service*, p. 165 (cited by Beach).
65 Beach and Bruce, 'British Signals Intelligence in the Trenches,' p. 155.
66 Peter Donovan and John Mack, *Code Breaking in the Pacific* (Cham: Springer International Publishing, 2014), pp. 5–6; Yves Glyden, *The Contribution of the Cryptographic Bureaus in the World War* (Laguna Hills, CA: Aegean Park Press, 1978), pp. 55–60; and Flicke, *War Secrets in the Ether*, pp. 3–7.
67 Glyden, *The Contribution of the Cryptographic Bureaus in the World War*, p. 56.
68 Glyden, *The Contribution of the Cryptographic Bureaus in the World War*, p. 64.
69 Donovan and Mack, *Code Breaking in the Pacific*, pp. 5–6; and Flicke, *War Secrets in the Ether*, pp. 3–7.

70 Flicke, *War Secrets in the Ether*, pp. 21, 23, 24; and Glyden, *The Contribution of the Cryptographic Bureaus in the World War*, p. 38.
71 Flicke, *War Secrets in the Ether*, p. 33.
72 GCHQ, 'Defending Our Skies: How Signals Intelligence helped combat air raids in World War I,' at <www.gchq.gov.uk/information/defending-our-skies>, accessed 25 October 2022.
73 Ferris, *The British Army and Signals Intelligence*, pp. 14–15.
74 See F.M. Cutlack, *Official History of Australia in the Great War of 1914–18*, vol. VIII: The Australian Flying Corps (Sydney: Angus & Robertson, 1923), pp. 427–29.
75 Cutlack, *Official History of Australia in the Great War*, vol. VIII, pp. 280–81.
76 Beach, *Haig's Intelligence GHQ and the German Army, 1916–1918*, pp. 162–63.
77 Ferris, *The British Army and Signals Intelligence*, pp. 8–9.
78 Ferris, 'The Road to Bletchley Park', p. 61.
79 Ferris, *The British Army and Signals Intelligence*, pp. 18–19.
80 Glyden, *The Contribution of the Cryptographic Bureaus in the World War*, pp. 20–21.
81 Glyden, *The Contribution of the Cryptographic Bureaus in the World War*, p. 62.
82 Johnson, *The Evolution of British Sigint*, p. 37.
83 Christopher Andrew, *The Secret World: A History of Intelligence* (Milton Keynes: Allen Lane, Penguin, 2018), p. 536; and Headrick, *The Invisible Weapon*, p. 167.
84 Daniel Larsen, 'British Codebreaking and American Diplomatic Telegrams, 1914–1915', *Intelligence and National Security*, 32:2 (2017), pp. 252–63.
85 Andrew, *The Secret World*, p. 536; and Flicke, *War Secrets in the Ether*, p. 41.
86 Andrew, *The Secret World*, p. 536.
87 David A. Hatch, *The Dawn of American Cryptology, 1900–1917* (Fort Meade, MD: Center for Cryptologic History, National Security Agency, 2019), pp. 31–32; Headrick, *The Invisible Weapon*, p. 169; and Johnson, *The Evolution of British Sigint*, pp. 38–39.
88 Flicke, *War Secrets in the Ether*, p. 55.
89 Flicke, *War Secrets in the Ether*, p. 52; and Beach, *Haig's Intelligence GHQ and the German Army*, p. 163.
90 Beach, *Haig's Intelligence GHQ and the German Army*, p. 163.
91 Beach, *Haig's Intelligence GHQ and the German Army*, p. 163.
92 Center for Cryptologic History, 'Radio Intelligence on the Mexican border, World War I: A Personal View', National Security Agency at <www.nsa.gov./about/cryptologic-heitage/historical-figures-publications/publications/pre-wwi/radio-intel-mexican-border/>, accessed 9 September 2019; and Hatch, *The Dawn of American Cryptology*, pp. 4–9, 47, 48.
93 Hatch, *The Dawn of American Cryptology*, p. 54.
94 John F. Dooley, *Codes, Ciphers and Spies: Tales of Military Intelligence in World War I* (New York: Copernicus Books, 2016), p. 5.
95 By 1917, the MID would include MI-1 Administration (Personnel & Office Management), MI-2 Collection & Dissemination of Foreign Intelligence, MI-3 Counterespionage (Military), MI-4 Counterespionage (Civilian), and MI-8 Cable & Telegraph (Code and Cipher section). By 1918, this also included MI-5 Military Attaches, MI-6 Translation, MI-7 Graphics (Maps), MI-9 Field Intelligence, MI-10 Censorship, MI-11 Passport & Port Control, and MI-12 Graft & Fraud. See Dooley, *Codes, Ciphers and Spies*, p. 26.
96 Hatch, *The Dawn of American Cryptology*, p. 11; and Herbert D. Yardley, *The American Black Chamber* (New York: Ishi Press International, reprint, 2016), p. 7.
97 Yardley, *The American Black Chamber*, p. x; and Dooley, *Codes, Ciphers and Spies*, p. 25.

98   Dooley, *Codes, Ciphers and Spies*, pp. ix, 37.
99   Dooley, *Codes, Ciphers and Spies*, p. 33.
100   Dooley, *Codes, Ciphers and Spies*, pp. 27–28.
101   Dooley, *Codes, Ciphers and Spies*, p. 28.
102   Dooley, *Codes, Ciphers and Spies*, p. 26.
103   Glyden, *The Contribution of the Cryptographic Bureaus in the World War*, p. 35.
104   Glyden, *The Contribution of the Cryptographic Bureaus in the World War*, p. 36.
105   Beach and Bruce, 'British Signals Intelligence in the Trenches, 1915–1918: Part 2,' p. 26.
106   Coulthard-Clark, *The Citizen General Staff*, p. 51.
107   Ferris, 'The Road to Bletchley Park,' p. 61.
108   Beach and Bruce, 'British Signals Intelligence in the Trenches, 1915–1918: Part 2,' p. 26.
109   Ferris, 'The Road to Bletchley Park,' p. 61.
110   Ferris, *The British Army and Signals Intelligence*, pp. 334–35.
111   Jean Bou, *MacArthur's Secret Bureau: The Story of the Central Bureau, MacArthur's Signals Intelligence Organisation* (Sydney: Australian Military History Publications, 2012), p. 12.
112   Bou, *MacArthur's Secret Bureau*, p. 12.
113   Keast Burke (ed.), *With Horse and Morse in Mesopotamia* (Sydney: Arthur McQuitty and Co., 1927), p. 37.
114   Bou, *MacArthur's Secret Bureau*, p. 13.
115   Burke, *With Horse and Morse in Mesopotamia*, p. 37; and Bou, *MacArthur's Secret Bureau*, p. 13.
116   Bou, *MacArthur's Secret Bureau*, pp. 12–13.
117   Extract of letter from Gerard Clauson to Lieutenant G.G. Crocker, 9 October 1917, cited in Ferris, *The British Army and Signals Intelligence*, p. 212.
118   Ferris, *The British Army and Signals Intelligence*, p. 335.
119   Flicke, *War Secrets in the Ether*, p. 78.
120   Ferris, 'The Road to Bletchley Park,' pp. 64–65.
121   Ferris, 'The Road to Bletchley Park,' p. 65.
122   Ferris, 'The Road to Bletchley Park,' p. 66.
123   Ferris, 'The Road to Bletchley Park,' p. 66.
124   Christopher Andrew and Keith Neilson, 'Tsarist Codebreakers and British Codes,' in Christopher Andrew (ed.), *Codebreaking and Signals Intelligence* (London: Frank Cass, 1986), p. 10.
125   Andrew and Neilson, 'Tsarist Codebreakers and British Codes,' pp. 2, 10.
126   Andrew and Neilson, 'Tsarist Codebreakers and British Codes,' p. 11.
127   Joseph H. Straczek, 'The Origins and Development of Royal Australian Naval Signals Intelligence in an Era of Imperial Defence 1914–1945', PhD thesis, UNSW Canberra, 2008, p. 47.

4   Australian Sigint & the interwar years

1   Wesley K. Wark, 'Cryptographic Innocence: The Origins of Signals Intelligence in Canada in the Second World War', *Journal of Contemporary History*, 22:4 (1987), p. 639.
2   John Fahey, *Australia's First Spies: The Remarkable Story of Australia's Intelligence Operations, 1901–1945* (Sydney: Allen & Unwin, 2018), pp. 85–86; Joseph H. Straczek, 'The Origins and Development of Royal Australian Naval Signals Intelligence in an Era of Imperial Defence 1914–1945', PhD thesis, UNSW Canberra, 2008. pp. 49–50; and John Johnson, *The Evolution of British Sigint: 1653–1939* (Cheltenham: GCHQ, 1997), p. 43.
3   Johnson, *The Evolution of British Sigint*, pp. 43–46, 52–54.
4   Johnson, *The Evolution of British Sigint*, p. 45.

5   John Ferris, 'The Road to Bletchley Park: The British Experience with Signals Intelligence, 1892–1945,' *Intelligence and National Security*, 17:1 (2002), pp. 53–54.
6   Ferris, 'The Road to Bletchley Park', p. 67.
7   Johnson, *The Evolution of British Sigint*, pp. 54–55.
8   F.H. Hinsley, E.E. Thomas, C.F.G. Ransom and R.C. Knight, *British Intelligence in the Second World War*, vol. I (London: HMSO, 1979), p. 23.
9   Hinsley et al., *British Intelligence in the Second World War*, pp. 23–24.
10 Christopher Andrew, *The Secret World: A History of Intelligence* (Milton Keynes: Allen Lane/Penguin, 2018), pp. 576–78.
11 Johnson, *The Evolution of British Sigint*, pp. 48–49.
12 Andrew, *The Secret World*, pp. 577–78.
13 Andrew, *The Secret World*, pp. 578–79.
14 Andrew, *The Secret World*, pp. 579–80.
15 Richard J. Aldrich, *GCHQ: The Uncensored Story of Britain's Most Secret Intelligence Agency* (London: Harper Press, 2010), p. 18.
16 Andrew, *The Secret World*, p. 583; and Johnson, *The Evolution of British Sigint*, p. 49.
17 Andrew, *The Secret World*, p. 584.
18 Andrew, *The Secret World*, p. 584.
19 Johnson, *The Evolution of British Sigint*, p. 48.
20 David A. Hatch with Robert Louis Benson, *The Korean War: The SIGINT Background*, United States Cryptologic History Series V, The Early Postwar Period, 1945–1952, vol. 3 (Fort Meade, MD, Center for Cryptologic History, National Security Agency, 2000), p. 4.
21 Johnson, *The Evolution of British Sigint*, p. 50.
22 Daniel R. Headrick, *The Invisible Weapon: Telecommunications and International Politics 1851–1945* (New York and Oxford: Oxford University Press, 1991), pp. 221–22.
23 Johnson, *The Evolution of British Sigint*, p. 50.
24 Headrick, *The Invisible Weapon*, pp. 221–22.
25 Johnson, *The Evolution of British Sigint*, pp. 55–56; and Headrick, *The Invisible Weapon*, pp. 221–22.
26 Headrick, *The Invisible Weapon*, pp. 181–83, 202–206, 212, 218.
27 Herbert D. Yardley, *The American Black Chamber* (New York: Ishi Press International, reprint, 2016), p. xi.
28 Yardley, *The American Black Chamber*, p. xi; and David Kahn, *The Reader of Gentlemen's Mail: Herbert O. Yardley and the Birth of American Codebreaking* (New Haven, CT: Yale University Press, 2004), pp. 63–80.
29 Yardley, *The American Black Chamber*, p. xi, 5; Kahn, *The Reader of Gentlemen's Mail*, pp. 97, 102; and Johnson, *The Evolution of British Sigint*, p. 50.
30 Kahn, *The Reader of Gentlemen's Mail*, pp. 104–20; and Yardley, *The American Black Chamber*, p. 8.
31 Yardley, *The American Black Chamber*, pp. xii–xiii; and Kahn, *The Reader of Gentlemen's Mail*, pp. 121–36.
32 Chief Signal Officer, *Japanese Signal Intelligence Service*, 3rd edition (Washington, DC: US Army Signal Security Agency, 1 November 1944), p. 58.
33 David P. Mowry, *Listening to the Rumrunners: Radio Intelligence during Prohibition*, 2nd edition (Fort Meade, MD: Center for Cryptologic History, 2014), pp. 3–4.
34 Mowry, *Listening to the Rumrunners*, pp. 5, 11.
35 Mowry, *Listening to the Rumrunners*, pp. 30, 32, 34.
36 Edward J. Drea, *MacArthur's ULTRA: Codebreaking and the War against Japan, 1942–1945* (Wichita, KS: University Press of Kansas, 1992), pp. 10–11.

37 Peter Donovan and John Mack, *Code Breaking in the Pacific* (Cham: Springer International Publishing, 2014), p. x.
38 Donovan and Mack, *Code Breaking in the Pacific*, p. 22.
39 Donovan and Mack, *Code Breaking in the Pacific*, pp. 10–11.
40 Donovan and Mack, *Code Breaking in the Pacific*, pp. 10–11.
41 Donovan and Mack, *Code Breaking in the Pacific*, p. 12.
42 Donovan and Mack, *Code Breaking in the Pacific*, p. x.
43 Drea, *MacArthur's ULTRA*, p. 16.
44 Fahey, *Australia's First Spies*, p. 88.
45 Nigel de Grey, 'Allied Sigint-Policy and Organisation', Chap VIII, 'Australia', 17 TNA (UK): HW 43/78.
46 Anon, 'Lord Jellicoe's Mission', *The Age*, 14 August 1919, p. 7.
47 Donovan and Mack, *Code Breaking in the Pacific*, p. 68.
48 Jozef Straczek, 'The Empire is Listening: Naval Signals Intelligence in the Far East to 1942,' *Journal of the Australian War Memorial*, 35, 1 Dec 2001, para 10; and Fahey, *Australia's First Spies*, p. 48.
49 Royal Australian Navy Summary of Constitution and Policy 1926, Chapter VII, Intelligence Organisation pp. 68, 69. Naval Office Melbourne, June, 1926. NAA: A5954 2378/1 BC 735531.
50 Secretary, Department of the Navy (sgd) Letter to Honorary Lieutenant A.C. Gregory, 29 March 1920, NAA: BC406713, pp. 15, 16.
51 Commodore Henry Cochrane, Second Naval Member, Minute to Rear-Admiral Percy Grant, First Naval Member, 'Naval Intelligence Reports from Broome', 1 July 1920, NAA: MP1049/1, 1920/0476 BC396875, p. 17.
52 Alex Flint (sgd) for Secretary of the Admiralty, letter to Under Secretary of State, Colonial Office, 20 January 1920, NAA: MP1185/8 1846/4/25, BC398924, p. 290.
53 Minute, F.G. Cresswell, Director Signal Section to Head, N Branch of 8 June 1921. NAA: MP 1049/9, 1997/5/196, Item no 505956, p. 311.
54 Minute, Rear Admiral Commanding H.M. Australian Fleet, to Secretary, Naval Office, 31 March 1922, NAA: MP1049/9, 1997/5/196, Item no 505956, p. 301.
55 Flying Officer E.A. Mustard DFC, Islands and Harbours of Mandated Territories visited by him while attached to HMAS *Adelaide* during Royal Australian Navy Winter Cruise 29 June to 3 August 1923, NAA: A9376, 90 BC1101463.
56 Anthony Best, *British Intelligence and the Japanese Challenge in Asia, 1914–1941* (Basingstoke: Palgrave MacMillan, 2002), p. 93.
57 Straczek, 'The Empire is Listening,' para 5.
58 RTB, 'HMS Anderson and Special Intelligence in the Far East,' 1 NID Vol. 10A, TNA (UK) HW 4/24.
59 See Ian Pfennigwerth, *A Man of Intelligence: The Life of Captain Theodore Eric Nave, Australian Codebreaker Extraordinary* (Sydney: Pfennigwerth, 2006).
60 The date, in fact, was 1925.
61 Frank Birch, *The Official History of British Sigint 1914–1945*, vol. I, Part I (Milton Keynes: The Military Press, 2004), p. 12.
62 Extract from letter dated 1 September 1925 from Paymaster Lieutenant Nave, Hong Kong, NAA: MP1049/9, 1997/5/196, Item ID 505956.
63 According to Captain Shaw in 'History of HMS Anderson' 21, cited in Frank Birch GC&CS Volume V (a) 'The Organisation and Evolution of British Naval Sigint', 88, TNA (UK) HW 43/13
64 Ian Pfennigwerth, 'Nave, Theodore Eric (1899–1993)', *Australian Dictionary of Biography*, vol. 19, 2021, at <www.adb.anu.edu.au/biography/nave-theodore-eric-17834>, accessed 2 November 2022; and Pfennigwerth, *A Man of Intelligence*.

65 T.E. Nave, 'History' (Autobiography, copy in possession of authors), p. 4.
66 Report (By R.T.B.), 'HMS Anderson and Special Intelligence in the Far East', TNA (UK) HW 4/24.
67 Officers (RAN) Personal Record Theodore Eric Nave, NAA: A3978, NAVE T E BC8360031, p. 11.
68 Letter, Charles Walker on behalf of Lords of the Admiralty to The Secretary, Navy Office, Melbourne, 9 November 1927, NAA: MP1049/9, 1997/5/196, p. 182.
69 Officers (RAN) Personal Record – Theodore Eric Nave, NAA: A3978 NAVE T E BC8360031, p. 24.
70 Pfennigwerth, *A Man of Intelligence*, p. 114.
71 Minute from the Assistant Chief of Naval Staff to the Chief of Naval Staff, 17/1/24 NAA: MP1049/9, 1997/5/196, p. 295.
72 Fahey, *Australia's First Spies*, p. 96.
73 Letter, F.G. Cresswell to Chief of Naval Staff, 7 July 1926, NAA: MP1049/9, 1997/5/196, 505956, pp. 220–21.
74 Minute, F.G. Cresswell sgd., 7 July 1926, MP1049/9, 1997/5/196, p. 219.
75 Minute, F.G. Cresswell sgd., 2/8/27, NAA: MP1049/9, 1997/5/19, 50596, p. 321.
76 Minute, Second Naval Member to First Naval Member, 12 September 1927, NAA: MP 1049/9, 1997/5/196, p. 191.
77 Fahey, *Australia's First Spies*, p. 107.
78 Fahey, *Australia's First Spies*, pp. 111–13.
79 Fahey, *Australia's First Spies*, pp. 108–109.
80 Straczek, 'The Origins and Development of Royal Australian Naval Signals Intelligence', p. 119.
81 Fahey, *Australia's First Spies*, pp. 108–109; and letter from Secretary Naval Board 'Procedure Y – Reception in HMA Squadron' at NAA: MP1049, 1997/5/196, cited in Straczek, 'The Origins and Development of Royal Australian Naval Signals Intelligence', p. 129.
82 Extract from Committee of Imperial Defence Paper 418-C, NAA: A5954, 1961/3 BC655420, p. 21.
83 Fahey, *Australia's First Spies*, pp. 116–21.
84 Straczek, 'The Origins and Development of Royal Australian Naval Signals Intelligence', p. 113.
85 Johnson, *The Evolution of British Sigint*, p. 51.
86 Frank Birch, 'The Making of Sigint' 18, TNA (UK) HW 43/10.
87 Admiral Sir William Milbourne James, Memorandum to DNI, 4.12.36 (NID, vol. 26), cited in Birch, *The Official History of British Sigint 1914–1945*, vol. I, part I, p. 22.
88 Birch, *The Official History of British Sigint 1914–1945*, vol. I, part I, p. 23.
89 Charles Morgan, 'Early Days of Special Intelligence', 12 NID vol. 1, pp. 9 & 17, TNA (UK) ADM 223/463.
90 Birch, *The Official History of British Sigint 1914–1945*, p. 23.
91 Fahey, *Australia's First Spies*, pp. 116–21.
92 Johnson, *The Evolution of British Sigint*, p. 51.
93 Memorandum, J.W.A. Waller sgd. 14 January 1936. NAA: MP1049/9, 1997/5/196, 505956, p. 3.
94 Straczek, 'The Origins and Development of Royal Australian Naval Signals Intelligence', pp. 148–51, 165.
95 Straczek, 'The Origins and Development of Royal Australian Naval Signals Intelligence', pp. 105–106.

96 'Royal New Zealand Navy and Naval Facilities in New Zealand,' dated 30 April 1944, cited in Straczek, 'The Origins and Development of Royal Australian Naval Signals Intelligence,' pp. 107–108.
97 Hinsley et al., *British Intelligence in the Second World War*, p. 53.
98 Lawrence Durrant, *Seawatchers: The Story of Australia's Coast Radio Service* (Sydney: Angus & Robertson, 1986), p. 84.
99 Shirley Fenton Huie, *Ships Belles: The Story of the Women's Royal Naval Service in War and Peace 1941–1985* (Sydney: Watermark Press, 2000), pp. 197–200.
100 Huie, *Ships Belles*, pp. 200, 203.
101 Huie, *Ships Belles*, p. 202.
102 G. Hermon Gill, *Royal Australian Navy 1942–1945*, Australia in the War of 1939–1945, Series Two, Navy, vol. II (Canberra: Collins and Australian War Memorial, 1968–1985 reprint), pp. 122–24, 334–36, 396.
103 Fahey, *Australia's First Spies*, p. 84.
104 Donovan and Mack, *Code Breaking in the Pacific*, p. 75.
105 Nigel de Grey, 'Allied Sigint-Policy and Organisation', Chap VIII, 'Australia', p. 1, TNA (UK) HW 43/78.
106 Donovan and Mack, *Code Breaking in the Pacific*, p. 70.
107 Fahey, *Australia's First Spies*, p. 110. Fahey is quoting CBCS *Year Book Australia*, No. 26 (Canberra: Commonwealth of Australia, 1933), p. 329.

5 Sigint in the Second World War, 1939–41

1 F.W. Winterbotham, *The Ultra Secret: The Inside Story of Operation Ultra, Bletchley Park and Enigma* (London: Orion Books, 1974). Notably, however, in 1967 a Polish historian, Władysław Kozaczuk, published an account of the Polish breakthrough against Enigma: *The Secret Battle*. See Władysław Kozaczuk, *Bitwa o tajemnice: Sluzby wywiadowcze Polski I Rzeszy Niemieckiej 1922–1939* (Warsaw: Ksiazka I Wiedza, 1967).
2 The list is extensive (several noted in the bibliography) and includes works by David Kahn, Edward Drea, Stephen Budiansky, Harry Hinsley, Richard Aldrich, John Ferris and Christopher Andrew.
3 This includes the work of Geoffrey Ballard (on the Australian Army) and Jack Bleakley (on the RAAF).
4 This includes works by David Horner, Desmond Ball, John Fahey, Ian Pfennigwerth, David Dufty, Peter Donovan and John Mack, among others.
5 F.H. Hinsley, *Official History of British Intelligence in the Second World War: Its Influence on Strategy and Operations* (London: HMSO, 1979), p. x.
6 Australian War Memorial, 'Prime Minister Robert G. Menzies: Wartime broadcast', *Encyclopaedia*, <www.awm.gov.au/articles/encyclopaedia/prime_ministers/menzies>, accessed 21 January 2020.
7 Desmond Ball and David Horner, *Breaking the Codes: Australia's KGB Network, 1944–1950* (Sydney: Allen & Unwin, 1998), p. 19.
8 Barbara Winter, *The Intrigue Master: Commander Long and Naval Intelligence in Australia, 1913–1945* (Brisbane: Boolarong Press, 1995), p. 49. See also John Ferris, '"Consistent with an Intention": The Far East Combined Bureau and the Outbreak of the Pacific War, 1940–41,' *Intelligence and National Security*, 27:1 (2012): pp. 5–26.
9 *Cryptographic Organisation in Australia*, Minute from DNI to Chief of Naval Staff, 28 November 1939, NAA: A6923, 37/401/425.
10 *Cryptographic Organisation in Australia*, Minute from Chief of Naval Staff to the Chief of the General Staff, 12.12.1939, 315, NAA: A6923, 37/401/425.

11 *Cryptographic Organisation in Australia*, Minute from Air Vice-Marshall to Chief of Naval Staff, 29. 12. 1939, NAA: A7942 Z146.
12 Frank Birch (edited by John Jackson) *The Official History of British Sigint 1914–1945*, vol. I, part I (Milton Keynes: The Military Press, 2004), p. 11.
13 Birch, *The Official History of British Sigint 1914–1945*, p. 11.
14 Birch, *The Official History of British Sigint 1914–1945*, p. 47.
15 *Cryptographic Organisation in Australia*, Minute from Chief of the General Staff to Chief of Naval Staff of 16 December 1939, 297, NAA: A6923, 37/401/425.
16 Birch, *The Official History of British Sigint 1914–1945*, p. 11.
17 Memorandum, *Cryptographic Organization in Australia*, F.G. Shedden sgd, 27 April 1940, NAA: A6923 37/401/425; and Letter, Prime Minister Menzies to the Dominions Office, 11 April 1940, NAA: A6923, 37/401.425.
18 Interservice Special Intelligence School 'ISIS' cited in Frank Birch, 'GC&CS Volume V (a) The Organisation and Evolution of British Naval Sigint', 90, TNA (UK) HW 43/13; and 'GC&CS Miscellaneous Papers Q/2022 8.10.40,' cited in Frank Birch, 'GC&CS Volume V (a) The Organisation and Evolution of British Naval Sigint', 90, TNA (UK) HW 43/13.
19 David Dufty, *The Secret Code Breakers of Central Bureau: How Australia's Signals Intelligence Helped Win the Pacific War* (Melbourne: Scribe, 2017), p. 23.
20 Frances Hackney, 'A Tribute to a Gentle Genius', *Annals: Journal of Catholic Culture*, November/December 1996, p. 14.
21 Interview with Mrs Helen Roberts (Treweek's daughter), 29 July 2022; and Lewis Buckingham, 'Eulogy, Athanasius Pryor Treweek 29/12/1911 – 20/1/1995'.
22 Jean Bou, *MacArthur's Secret Bureau: The Story of the Central Bureau, MacArthur's Signals Intelligence Organisation* (Sydney: Australian Military History Publications, 2012), pp. 20–21; and Desmond Ball and Keiko Tamura (eds), *Breaking Japanese Diplomatic Codes: David Sissons and D Special Section During the Second World War* (Canberra: Australian National University, 2013), pp. 16, 23.
23 Nigel de Grey, 'Allied Sigint Policy and Organisation', Chapter VIII, 'Australia', p. 3, fn 2, TNA (UK) HW 43/78.
24 Nigel de Grey, 'Allied Sigint Policy and Organisation', Chapter VIII, 'Australia', p. 2 TNA (UK) HW 43/78.
25 Pfennigwerth, *A Man of Intelligence*, p. 129.
26 Bleakley, *The Eavesdroppers*, p. 6; Pfennigwerth, *Missing Pieces*, p. 24.
27 Ball and Tamura, *Breaking Japanese Diplomatic Codes*, p. 17.
28 Ian Pfennigwerth, *Missing Pieces: The Intelligence Jigsaw and RAN Operations 1939–1971*, Papers in Australian Maritime Affairs No. 25 (Canberra: Sea Power Centre – Australia, 2008), p. 28, available at <www.navy.gov.au/sites/default/files/documents/PIAMA25.pdf>, accessed 25 October 2022.
29 Mike Carlton, *The Scrap Iron Flotilla* (Sydney: William Heinemann, 2022), pp. 122–23, 125.
30 Pfennigwerth, *Missing Pieces*, pp. 29–36.
31 Pfennigwerth, *Missing Pieces*, pp. 37–47.
32 G. Hermon Gill, *Royal Australian Navy 1939–1942*, Australia in the War of 1939–1945, Series Two, Navy, vol. I (Canberra: Collins and Australian War Memorial, 1957–1985 reprint), pp. 420–21.
33 John Blaxland, 'Intelligence and Special Operations in the Southwest Pacific, 1942–45,' in Peter Dean (ed.), *Australia, 1944–45: Victory in the Pacific* (Melbourne: Cambridge University Press, 2015), p. 149; Judy Thomson, *Winning With Intelligence: A Biography of Brigadier John David Rogers, CBE, MC 1895–1978* (Sydney: Australian Military History

Publications, 2000), p. 138; and Desmond Ball and David Horner, *Breaking the Codes: Australia's KGB Network, 1944*–1950 (Sydney: Allen & Unwin, 1998), p. 37.
34 Nigel de Grey, 'Allied Sigint-Policy and Organisation', Chap VIII, 'Australia', p. 8, TNA (UK) HW 43/78.
35 Report, Lt. Colonel J.W. Ryan AIF to the Director of Military Intelligence and Director of Intelligence RAAF, 'Report on Special Wireless Units (Signals) 1940–45', 19 December 1945, NAA: A10908, p. 2.
36 Geoffrey Ballard, *On ULTRA Active Service: The Story of Australia's Signals Intelligence Operations During World War II* (Melbourne: Spectrum Publications, 1991), pp. 43, 44.
37 Nigel de Grey, 'Allied Sigint-Policy and Organisation', Chap VIII, 'Australia', p. 8, TNA (UK) HW 43/78.
38 Report, Lt. Colonel J. W. Ryan AIF to the Director of Military Intelligence and Director of Intelligence RAAF, 'Report on Special Wireless Units (Signals) 1940–45', 19 December 1945, NAA: A10908, p. 2.
39 Nigel de Grey, 'Allied Sigint-Policy and Organisation', Chap VIII, 'Australia', p. 8, TNA (UK) HW 43/78.
40 Sandford, Alastair Wallace, 'Attestation Form for Special Forces Raised for Service in Australia or Abroad', NAA: B883, Control Symbol SX11231.
41 Ballard, *On ULTRA Active Service*, p. 43; and Collie, *Code Breakers*, p. 53.
42 Pfennigwerth, *A Man of Intelligence*, p. 208.
43 'Commander T.E. Nave R.N.', Minute Sandford to DDMI of 31 October 1942, NAA: A6923, 37/401/425.
44 Sandford, Alastair Wallace, 'Attestation Form for Special Forces Raised for Service in Australia or Abroad'.
45 Ballard, *On ULTRA Active Service*, p. 71.
46 Alastair Wallace Sandford, Officer's Record of Service, 23.3.43. NAA: B883, Control Symbol SX11231, p. 8.
47 Stuart Milner-Barry, 'Hut 6, Early Days,' in F.H. Hinsley and Alan Stripp (eds), *Code Breakers: The Inside Story of Bletchley Park* (Oxford: Oxford University Press, 1993), p. 98.
48 Ron Palenski, *Men of Valour: New Zealand and the Battle for Crete* (Auckland: Hachette, 2013), p. 44.
49 Milner-Barry, 'Hut 6, Early Days', p. 98.
50 Hans-Otto Behrendt, *Rommel's Intelligence in the Desert Campaign* (London: William Kimber, 1980), p. 16.
51 F.H. Hinsley, E.E. Thomas, C.F.G. Ransom and R.C. Knight, *British Intelligence in the Second World War*, vol. 2 (London: HMSO, 1981), p. 298.
52 William F. Flicke, *War Secrets in the Ether* (Laguna Hills, CA: Aegean Park Press, 1977), pp. 192–93, 195; and Daniel R. Headrick, *The Invisible Weapon: Telecommunications and International Politics 1851–1945* (New York and Oxford: Oxford University Press, 1991), pp. 229–30.
53 Behrendt, *Rommel's Intelligence in the Desert Campaign*, p. 15.
54 Flicke, *War Secrets in the Ether*, pp. 192–93, 196–97; and Behrendt, *Rommel's Intelligence in the Desert Campaign*, p. 17.
55 See Tim Gellel, 'The Capture of Unit 621: Lessons in Information Security from the North Africa Campaign', *The Australian Army Journal*, XII:1 (2015), pp. 77–89; and Hinsley, et al., *British Intelligence in the Second World War*, vol. 2, p. 404.
56 Behrendt, *Rommel's Intelligence in the Desert Campaign*, p. 225.
57 Behrendt, *Rommel's Intelligence in the Desert Campaign*, p. 17; and Headrick, *The Invisible Weapon*, p. 230.

58 Flicke, *War Secrets in the Ether*, pp. 192–93, 197–98.
59 David Stevens, 'The RAN – A Brief History', Royal Australian Navy website, at <www.navy.gov.au/history/feature-histories/ran-brief-history>, accessed 21 January 2020. For a detailed account, see Gill, *Royal Australian Navy 1939–1942,* Australia in the War of 1939–1945, Series Two, Navy, vol. I.
60 RAAF, 'History of the Air Force', Royal Australian Air Force website, at <www.airforce.gov.au/about-us/history>, accessed 21 January 2020. For a detailed account of RAAF developments during this period and an overview of the formation of the RAAF, see Douglas Gillison, *Royal Australian Air Force 1939–1942,* Australia in the War of 1939–1945, Series Three, vol. I (Canberra: Australian War Memorial, 1962).
61 See Andrew Ross, *Armed and Ready: The Industrial Development and Defence of Australia 1900–1945* (Sydney: Department of Industry, Science and Technology, 1995).
62 See Australian War Memorial, 'Empire Air Training Scheme', *Encyclopaedia,* <www.awm.gov.au/articles/encyclopaedia/raaf/eats>, accessed 21 January 2020; Alan Stephens, *The Royal Australian Air Force: The Australian Centenary History of Defence*, vol. 2 (Melbourne: Oxford University Press, 2001); and John McCarthy, *A Last Call of Empire: Australian Aircrew, Britain and the Empire Air Training Scheme* (Canberra: Australian War Memorial, 1988).
63 See John Herington, *Air War Against Germany & Italy 1939–1943, Australia in the War of 1939–1945,* Series III, Air, vol. III (Canberra: Australian War Memorial, 1954), pp. 38, 154, 160, 274, 286.
64 See Herington, *Air War Against Germany & Italy 1939–1943,* pp. 410, 416, 436, 585.
65 Herington, *Air War Against Germany & Italy 1939–1943,* pp. 448, 471, 593; and John Herington, *Air Power Over Europe 1944–1945, Australia in the War of 1939–1945,* Series III, Air, vol. IV (Canberra: Australian War Memorial, 1963), pp. 197, 296, 298.
66 Herington, *Air Power Over Europe 1944–1945,* p. 219.
67 Herington, *Air War Against Germany & Italy 1939–1943,* p. 585.
68 See Australian War Memorial, 'Empire Air Training Scheme'; Alan Stephens, *The Royal Australian Air Force: The Australian Centenary History of Defence,* vol. 2 (Melbourne: Oxford University Press, 2001); and McCarthy, *A Last Call of Empire.*
69 Craig Arthur Bellamy, 'The Beginning of the Secret Australian Radar Countermeasures Unit During the Pacific War,' PhD thesis, Charles Darwin University, 2020.
70 Ian Pfennigwerth, 'Nave, Theodore Eric (1899–1993)', *Australian Dictionary of Biography,* vol. 19, 2021, at <www.adb.anu.edu.au/biography/nave-theodore-eric-17834>, accessed 21 January 2020; and Pfennigwerth, *A Man of Intelligence.*
71 Douglas Menzies sgd Minute, 'Defence Committee meeting held Friday 28 November 1941', NAA: A6923, 37/401/425 BC3023506, pp. 197–98.
72 David Jenkins, 'Our War of Words,' *Sydney Morning Herald*, 19 September 1992, p. 37.
73 Richard Johnson, 'Trendall, Arthur Dale (1909–1995)', *ANU Reporter,* 13 December 1995, p. 11, at <www.oa.anu.edu.au/obituary/trendall-arthur-dale-976>, accessed 31 October 2022.
74 Ball and Tamura, *Breaking Japanese Diplomatic Codes,* p. 6.
75 Ball and Tamura, *Breaking Japanese Diplomatic Codes,* p. 131.
76 Ball and Tamura, *Breaking Japanese Diplomatic Codes,* p. 131.
77 Peter Kornicki, *Eavesdropping on the Emperor: Interrogators and Codebreakers in Britain's War with Japan* (London: Hurst and Company, 2021), p. 210.
78 Marjorie Jacobs, 'Oriental Studies in the University of Sydney', *The Australian Quarterly,* 25:2 (June 1953), p. 84.
79 Kornicki, *Eavesdropping on the Emperor,* p. 211.

80 David Kahn, *The Reader of Gentlemen's Mail: Herbert O. Yardley and the Birth of American Codebreaking* (New Haven, CT: Yale University Press, 2004), pp. 202–204.
81 Kahn, *The Reader of Gentlemen's Mail*, pp. 211–13.
82 David Dufty, *The Secret Code Breakers of Central Bureau: How Australia's Signals Intelligence Helped Win the Pacific War* (Melbourne: Scribe, 2017), p. 56.
83 Dufty, *The Secret Code Breakers of Central Bureau*, p. 56.
84 Dufty, *The Secret Code Breakers of Central Bureau*, p. 57.
85 Ragnar Colvin, Minutes of Meeting of the Naval Board Held at Navy Office, Melbourne, on Wednesday 18 January 1939, Department of the Navy (II) Minute Book [Index and Minutes] 1939–1941, NAA: A2585/XR.
86 Ragnar Colvin, Minutes of Meeting of the Naval Board Held at Navy Office Melbourne Thursday 22 March 1939. Department of the Navy (II) Minute Book [Index and Minutes] 1939-1941, NAA: A2585/XR.
87 Ragnar Colvin, Minutes of Meeting of the Naval Board Held at Navy Office Melbourne 18 October 1939, Department of the Navy (II) Minute Book [Index and Minutes] 1939-1941, NAA: A2585/XR.
88 Naval Board Minutes, 30 January 1941, NAA: A2585, 1939/41, 270.
89 Dufty, *The Secret Code Breakers of Central Bureau*, p. 57.
90 Dufty, *The Secret Code Breakers of Central Bureau*, p. 57.
91 Shirley Fenton Huie, *Ships Belles: The Story of the Women's Royal Naval Service in War and Peace, 1941–1985* (Sydney: Watermark Press, 2000), p. 27.
92 Ragnar Colvin, 'Minutes of Meeting of the Naval Board Held at Navy Office Melbourne Thursday 13 February 1941', Reference Copy, 276, NAA: A2585, 1939/1941.
93 Ragnar Colvin, 'Minutes of Meeting of the Naval Board Held at Navy Office Melbourne Thursday 20 March 1941, 284, NAA: A2585, 1939/1941.
94 J.W. Durnford, 'Minutes of Meeting of the Naval Board Held at Navy Office Melbourne Friday 18th April'. Department of Navy (II) Naval Board Minute Book: [Index and Minutes] 1939–1941, NAA: CRS 2585/XR.
95 Huie, *Ships Belles*, p. 24.
96 Dufty, *The Secret Code Breakers of Central Bureau*, p. 61.
97 Douglas Gillison, *Royal Australian Air Force 1939–1942, Australia in the War of 1939–1945,* Series Three, vol. I (Canberra: Australian War Memorial, 1962), p. 99.
98 Gillison, *Royal Australian Air Force 1939–1942,* p. 100.
99 The Unit Guide, *The Australian Army 1939–1945*, vol. 6, 6.010.
100 Ann Howard, *You'll Be Sorry! How World War II Changed Women's Lives* (Sydney: Big Sky Publishing, 1990), p. 24.
101 Ann Howard, *You'll Be Sorry!*, p. 25.
102 Liza Mundy, *Code Girls: The Untold Story of the American Women Code Breakers of World War II* (New York: Hachette Books, 2017), p. 247.
103 Mundy, *Code Girls*, pp. 12–13.
104 Mundy, *Code Girls*, p. 17.
105 Captain H.R. Sandwith, from a paper he presented Joint U.S.-British-Canadian Discussions on Radio Intelligence, Naval Department, Washington DC on April 6, 1942. HW 57/15, Appendix iii, p. 7.
106 Birch, *The Official History of British Sigint 1914–1945,* vol. I, Part I, p. 171; and Huie, *Ships Belles,* p. 71.
107 Ball and Tamura, *Breaking Japanese Diplomatic Codes*, p. 89.
108 Mundy, *Code Girls*, p. 46.
109 Robert Hope, *Royal Commission on Intelligence and Security* (Canberra: Australian Government Publishing Service, 1977), 6th Report, p. 835.

110 Obituary by Don Wormald, 'Berenice Wormald 1922–2021, Coder, war widow had choice words for Menzies,' *The Sydney Morning Herald*, 4 January 2022, p. 37.
111 LCDR Gregory J. Florence, USN, *Courting a Reluctant Ally: An Evaluation of U.S./UK Naval Intelligence Cooperation, 1935–1941* (Washington, DC: Joint Military Intelligence College, Center for Strategic Intelligence Research, 2004), pp. 20–26.
112 Florence, *Courting a Reluctant Ally*, p. 31.
113 Florence, *Courting a Reluctant Ally*, pp. 32–33.
114 Florence, *Courting a Reluctant Ally*, p. 38.
115 Thomas A. Brooks, 'Foreword,' in Florence, *Courting a Reluctant Ally*, p. vii.
116 Thomas A. Brooks, 'Foreword,' in Florence, *Courting a Reluctant Ally*, p. vii.
117 Bou, *MacArthur's Secret Bureau*, p. 25.
118 Florence, *Courting a Reluctant Ally*, p. 60; and David Sherman, *The First Americans: The 1941 US Codebreaking Mission to Bletchley Park* (Fort Meade, MD: Center for Cryptologic History, NSA, 2016), p. 6.
119 Florence, *Courting a Reluctant Ally*, pp. 71, 82.
120 Florence, *Courting a Reluctant Ally*, p. 71.
121 Bou, *MacArthur's Secret Bureau*, p. 25; and Florence, *Courting a Reluctant Ally*, p. 82.
122 Florence, *Courting a Reluctant Ally*, p. 82.
123 Bou, *MacArthur's Secret Bureau*, p. 25.
124 Stuart Milner-Barry, 'Hut 6: Early Days,' in F.H. Hinsley and Alan Stripp (eds), *Codebreakers: The Inside Story of Bletchley Park* (Oxford: Oxford University Press, 1993), p. 92.
125 David P. Mowry, *German Cipher Machines of World War II*, revised edition (Fort Meade, MD: Center for Cryptologic History, NSA, 2014), pp. 1, 15, 16.
126 Miller, *The Cryptographic Mathematics of Enigma*, revised edition (Fort Meade, MD: Centre for Cryptologic History, NSA, 2019), p. 4; and Mowry, *German Cipher Machines of World War II*, p. 1.
127 Mowry, *German Cipher Machines of World War II*, pp. 6, 8.
128 Mowry, *German Cipher Machines of World War II*, pp. 7, 8.
129 Mowry, *German Cipher Machines of World War II*, pp. 8–10.
130 Mowry, *German Cipher Machines of World War II*, p. 11.
131 Mowry, *German Cipher Machines of World War II*, pp. 13–17.
132 Mowry, *German Cipher Machines of World War II*, pp. 18–19, 24, 32, 35.
133 Craig P. Bauer, *Secret History: The Story of Cryptology* (Boca Raton, FL: CRC Press, 2013); and Captain Jerry Roberts, *Lorenz: Breaking Hitler's Top Secret Code at Bletchley Park* (Stroud: History Press, 2017), pp. 65, 72.
134 Michael Kerrigan, *Enigma: How Breaking The Code Helped Win World War II* (London: Amber Books Ltd, 2018), pp. 177–79; and Roberts, *Lorenz*, pp. 79–84, 107.
135 Kerrigan, *Enigma*, pp. 178–79; and Roberts, *Lorenz*, p. 80.
136 Kerrigan, *Enigma*, pp. 180–82; and Roberts, *Lorenz*, pp. 141–46.
137 Kerrigan, *Enigma*, pp. 184–86.
138 Roberts, *Lorenz*, pp. 150–52.
139 'Report on United States-British Staff Conversations', 27 March 1941, *Stark papers*, US Department of the Navy, cited in Florence, *Courting a Reluctant Ally*, p. 78.
140 Florence, *Courting a Reluctant Ally*, pp. 79, 85.
141 Florence, *Courting a Reluctant Ally*, pp. 83–84; Richard J. Aldrich, *Intelligence and the War against Japan: Britain, America and the Politics of Secret Service* (Cambridge: Cambridge University Press, 2000), p. 78; and Sherman, *The First Americans*, pp. 28–31.
142 R. Erskine, 'Churchill and the Start of the Ultra-Magic Deals,' *Cryptologia*, 10:1 (1997), p. 63, as cited in Aldrich, *Intelligence and the War against Japan*, p. 78.

143 Florence, *Courting a Reluctant Ally*, p. 84.
144 Friedman, 'Certain Aspects of "Magic" in the Cryptological background of the Various Investigations into the Attack on Pearl Harbor', pp. 36–37, cited in Sherman, *The First Americans*, p. 42.
145 Hinsley et al., *British Intelligence in the Second World War*, vols I–III (London: HMSO, 1979–83); and Florence, *Courting a Reluctant Ally*, p. 80.
146 Bou, *MacArthur's Secret Bureau*, p. 25.
147 Florence, *Courting a Reluctant Ally*, p. 86.
148 Ferris, '"Consistent with an Intention", The Far East Combined Bureau and the Outbreak of the Pacific War, 1940–41', *Intelligence and National Security*, 27:1 (2012), p. 23.
149 Bou, *MacArthur's Secret Bureau*, p. 25; and Florence, *Courting a Reluctant Ally*, p. 7.
150 Ian Pfennigwerth, *A Man of Intelligence: The Life of Captain Theodore Eric Nave, Australian Codebreaker Extraordinary* (Sydney: Rosenberg Publishing, 2006), p. 175.
151 Robert J. Hanyok and David P. Mowry, *West Wind Clear: Cryptology and the Winds Message Controversy – A Documentary History,* United States Cryptologic History Series IV: World War II, vol. X (Fort Meade, MD: Center for Cryptologic History, NSA, 2008), p. viii; and Pfennigwerth, *A Man of Intelligence*, p. 175.
152 Dufty, *The Secret Code Breakers of Central Bureau*, p. 89; and Hanyok and Mowry, *West Wind Clear*, pp. 26, 38.
153 Hanyok and Mowry, *West Wind Clear*, p. viii.
154 Letter, Second Navy Office to Defence Secretary (Most Secret), 28 November 1941 – copy in A. Jack Brown, *Katakana Man: I worked only for Generals* (Canberra: Air Power Development Centre, 2005), p. 24.
155 Hanyok and Mowry, *West Wind Clear*, pp. xi, 38, 41, 46, 48.
156 Dufty, *The Secret Code Breakers of Central Bureau*, p. 93.
157 Pfennigwerth, *A Man of Intelligence*, p. 175.
158 Hanyok and Mowry, *West Wind Clear*, p. vii.
159 See James Rusbridger and Eric Nave, *Betrayal at Pearl Harbor: How Churchill Lured Roosevelt into War* (New York: Summit Books, 1991).
160 Dufty, *The Secret Code Breakers of Central Bureau*, p. 94.
161 Dufty, *The Secret Code Breakers of Central Bureau*, p. 96.
162 Hanyok and Mowry, *West Wind Clear*, p. 97.
163 Hanyok and Mowry, *West Wind Clear*, p. 98.
164 Pfennigwerth, *A Man of Intelligence*, p. 184.
165 See D.M. Horner, *Crisis of Command: Australian Generalship and the Japanese Threat, 1941–1943* (Canberra: ANU Press, 1978), pp. 37, 39.
166 G. Hermon Gill, *Royal Australian Navy 1939–1942,* Australia in the War of 1939–1945, Series Two, Navy, vol. I (Canberra: Collins and Australian War Memorial, 1957–1985 reprint), pp. 516–19.
167 See D.M. Horner, *Crisis of Command*, pp. 39, 43.

6 Sigint arrangements & the War in the Pacific

1 Letter, Dennison to Travis, (Notes on the organisation of signals intelligence in Australia and New Zealand) 10 December 1942, TNA (UK) HW 52/93.
2 Ian Pfennigwerth, *Missing Pieces: The Intelligence Jigsaw and RAN Operations 1939–1971,* Papers in Australian Maritime Affairs, No. 25 (Canberra: Sea Power Centre–Australia, 2008), p. 5, at <www.navy.gov.au/sites/default/files/documents/PIAMA25.pdf>, p. 9.
3 Signal from C. in C. Eastern Fleet 9,00am Singapore time 22 December 1941 to Australian Commonwealth Navy Board (ACNB) Melbourne. GC&CS Miscellaneous

4   Signal from ACNB OF 23 December 1941 to C. in C. E. F., GC&CS Miscellaneous Papers Q/2022, 0129Z/23.12.41, cited in Frank Birch, 'GC&CS Volume V (a) The Organisation and Evolution of British Naval Sigint', p. 112, TNA (UK) HW 43/13.
5   Signal from C. in C. East Indies of 25 December 1941 to C. in C. E. F., GC&CS Miscellaneous Papers Q/2022, 0318Z/25.12.41, cited in Frank Birch, 'GC&CS Volume V (a) The Organisation and Evolution of British Naval Sigint, p. 112, TNA (UK) HW 43/13.
6   T.E. Nave, 'History', in his autobiography, p. 351.
7   Frank Birch, 'GC&CS Volume V (a) The Organisation and Evolution of British Naval Sigint', p. 131, TNA (UK) HW43/13.
8   Nave, 'History,' p. 357.
9   Nigel de Grey, 'Allied Sigint-Policy and Organisation', Chap VIII, 'Australia', p. 3, TNA (UK) HW 43/78.
10 Minute, 'Hollerith machines', Dir DSB to Defence Controller Intelligence, 21 August 1947, NAA: 42/301/1035; and Sharon A. Maneki, *The Quiet Heroes of the Southwest Pacific Theater: An Oral History of the Men and Women of CBB and FRUMEL* (Fort Meade, MD: Center for Cryptological History, NSA, 1996), p. 88.
11 Ian Pfennigwerth, *A Man of Intelligence: The Life of Captain Theodore Eric Nave, Australian Codebreaker Extraordinary* (Sydney: Self-published, 2006), p. 186.
12 Nigel de Grey, 'Allied Sigint-Policy and Organisation', Chap VIII, 'Australia', p. 4, TNA (UK) HW 43/78.
13 Edward J. Drea, *MacArthur's ULTRA: Codebreaking and the War against Japan, 1942–1945* (Wichita, KS: University Press of Kansas, 1992), pp. 24–25; and Jean Bou, *MacArthur's Secret Bureau: The Story of the Central Bureau, MacArthur's Signals Intelligence Organisation* (Sydney: Australian Military History Publications, 2012), p. 7.
14 Maneki, *The Quiet Heroes of the Southwest Pacific Theater*, p. 63.
15 Maneki, *The Quiet Heroes of the Southwest Pacific Theater*, pp. 78, 79.
16 Maneki, *The Quiet Heroes of the Southwest Pacific Theater*, pp. 66, 79.
17 Maneki, *The Quiet Heroes of the Southwest Pacific Theater*, p. 70.
18 Frank Birch, GC&CS Volume V (a), 'The Organisation and Evolution of British Naval Sigint', p. 180, TNA (UK) HW 43/13.
19 The Holden Agreement is actually a memorandum for Commander E.W. Travis RN signed by C.F. Holden, USN, Director of Naval Communications, 2 October 1942. C-in-C signified his concurrence on 20 October 1942. Cited in Frank Birch, GC&CS Volume V (a), 'The Organisation and Evolution of British Naval Sigint', p. 181, TNA (UK) HW 43/13.
20 Frank Birch, GC&CS Volume V (a), 'The Organisation and Evolution of British Naval Sigint', p. 173, TNA (UK) HW 43/13.
21 Frank Birch, GC&CS Volume V (a), 'The Organisation and Evolution of British Naval Sigint', p. 174, TNA (UK) HW 43/13.
22 'Organisation and Evolution of British Naval Sigint', pp. 211–12, quoted in Nigel de Grey, 'Allied Sigint Policy and Organisation', pp. 211–12, Chapter IV, p. 298 TNA (UK) HW 43/76.
23 Frank Birch, GC&CS Volume V (a), 'The Organisation and Evolution of British Naval Sigint', p. 175, TNA (UK) HW 43/13.
24 Frank Birch, GC&CS Volume V (a), 'The Organisation and Evolution of British Naval Sigint', p. 175, TNA (UK) HW 43/13.

25 'The Holden Agreement, 2nd October 1942,' Appendix B, p. 556, TNA (UK) HW 43/14.
26 Nigel de Grey, 'Allied Sigint-Policy and Organisation', Chap VIII, 'Australia', p. 4, TNA (UK) HW 43/78.
27 Report, 'General Notes on Special and W/T intelligence in Australia and New Zealand', Colegrave, (with cover note from Humphrey Sandwith of 4 December 1942), TNA (UK) HW 52/93.
28 Pfennigwerth, *A Man of Intelligence*, pp. 196–97.
29 Eric Nave, messages to GC&CS OF 14.10.1942 and 17.10.1942, Melbourne Organisation and Personnel File 40/2/12, TNA (UK) HW 52/93.
30 Copy of message 692 from Admiralty to A.C.N.B. OF 24/12/1942, NAA: A6923, 37/401/425, BC3023506, p. 120.
31 Minute Commander T.E. Nave RN to D.D.M.I., A.W. Sandford sgd. 31 October 1942. NAA: A6923, 37/401/425; and Signal to London from Landforces Melbourne, sgd. R.A. Little, General Staff, Military Intelligence, NAA: A6923, 37/401/425.
32 Signal to Signal Officer in Chief, Australian Army Melbourne from Signal Officer GHQ Akin sgd. January 11, 1943, NAA: A6923, 37/401/425.
33 Land Forces Minute from A/DDMI to DMI of 25 January 1943, NAA: A6923, 37/401/425.
34 Christopher Andrew, *The Secret World: A History of Intelligence* (Milton Keynes: Allen Lane/Penguin, 2018), p. 642.
35 See Anthony R. Wells, *Between Five Eyes: 50 Years of Intelligence Sharing* (Havertown, PA: Casemate, 2020).
36 Desmond Ball and Keiko Tamura (eds), *Breaking Japanese Diplomatic Codes: David Sissons and D Special Section during the Second World War* (Canberra: Australian National University, 2013), p. x.
37 Ball and Tamura, *Breaking Japanese Diplomatic Codes*, p. 8.
38 Special Intelligence Section Report – Japanese Diplomatic cyphers. 1946–1946 NAA: 6923, 1/REFERENCE COPY, p. 5.
39 Minister Stockholm to M.F.A. Tokyo, Internal Situation in Germany, Secret Agent's Report, July 26, 1944, FRUMEL WWII Diplomatic Intercept (Germany/Japan) 1943/1944, NAA: A10909, p. 1.
40 Ball and Tamura, *Breaking Japanese Diplomatic Codes*, p. 8.
41 Minute, Chief of Naval Staff to Chief of Staff, 5 November 1942, NAA: 6923 37/401/425, p. 164.
42 Minute MIS 172, 1 February 1943 to Director Military Intelligence, NAA: A6923 37/401/425, p. 84.
43 Ball and Tamura, *Breaking Japanese Diplomatic Codes*, pp. 89–90.
44 Ball and Tamura, *Breaking Japanese Diplomatic Codes*, pp. 139–40. Davies was later an academic at Queens Belfast and the University of New England and Laird was to be Head of English at the Royal Military College, Duntroon.
45 Minute on Diplomatic Section of 16 May 1944, from A.D.M.I to Major Webb, Central Bureau, NAA: A6923, 37/401/425, p. 17.
46 Report, 'Technical Records relating to Naval Codes and Cyphers (The Red Book)', NAA: B5554, Folio 375; and Ball and Tamura, *Breaking Japanese Diplomatic Codes*, pp. 22, 24–26.
47 Minute – Allied Land Forces in the SW Pacific MIS 2041 dated 30 July 1942 – 'Security of most secret material obtained from the interception of enemy wireless communications', NAA: A6923, S1I/10, folios 13 and 12.

48 Minute – Land Headquarters CCO 13/1/101 dated 2 January 1943 – 'Procedures for handling "Y" information', NAA: A6923, SI/10, folios 17, 16 and 15.
49 Minute – Land Headquarters MIS 291, dated 17 February 1943 – '"Y" Security', NAA: A6923, SI/10, folios 22 and 21.
50 Letter – AAS 204 dated 20 January 1943, Lieutenant-General E.K. Smart to Lieutenant-General J. Northcott, NAA: A6923, SI/10, folios 32 to 24.
51 Letters and minutes between Major-General Dewing and DDMI Lieutenant-Colonel Little, 10 and 11 March 1943, NAA: A6923, SI/10, folios 37, 36 and 35.
52 Minute – DDMI to DMI MS735 dated 3 May 1943 'Regulations for Special and 'Y' Int', NAA: A6923 SI/10, folios 47 to 41.
53 Minute – DDMI to DMI MS735 dated 3 May 1943 'Regulations for Special and "Y" Int', NAA: A6923 SI/10, folios 42 and 41.
54 Minute – Assistant Director Central Bureau Lieutenant-Colonel Sandford to DMI 7975L, dated 29 October 1943, 'Nomenclature and Security Instructions relating to Intelligence from "Y" sources', NAA: A6923, SI/10, folios 60 to 58.
55 Minute – Untitled but begins ... 'At meeting of London "Y" Board on 7th October, 1943, following nomenclature for activities relating to study of foreign signals was agreed and will come into force on 15th November, 1943', NAA: A6923, SI/10, folios 74 to 68.
56 F.W. Winterbotham, *The Ultra Secret: The Inside Story of Operation Ultra, Bletchley Park and Enigma* (London: Orion Books, 1974), p. 172.
57 Message – 'We shall shortly inaugurate at GCCS a central signals registry...' NAA: A6923, SI/10, folios 136 to 132.
58 Staff Memorandum – 'Security Classification', NAA: A6923, SI/10, folios 148 to 140.
59 John Fahey, *Australia's First Spies: The Remarkable Story of Australia's Intelligence Operations, 1901–1945* (Sydney: Allen & Unwin, 2018), p. 232.
60 Imperial War Museum, Oral History SQN LDR Sidney F. Burley, available at <www.iwm.org.uk/collections/item/object/80009138>; and document Air Ministry Special Liaison Unit No 5 RAF. Burley served as a more junior officer with SLU 5 in Algeria, Tunisia and Italy from March 1943 to May 1944, when he assumed command of SLU 5 at Bari in Italy, from May 1944 to August 1944, before moving to India and subsequently Australia.
61 Nigel De Grey, 'Allied Sigint – Policy and Organisation' Chapters VII-X and Appendices, UK Archives, HW 43/78, pages 47 and the diagram on the subsequent un-numbered page.
62 Lieutenant Commander L.A. Griffiths, G.C.&C.S., 'Army and Air Force SIGINT – III Part 3', pp. 61, 62, 'Japanese', TNA (UK) HW 43/49.
63 Report, 'Central Bureau Technical Records Part A – Organization', 'Organization and General History', (undated) NAA: B5436, Part A, p. 5.
64 Captain A.W. Sandford, 'Memorandum to GS/ 1 Australia Corps', 8 March 1942, NAA: A6923, SI/2, p. 283.
65 Note to file, untitled, illegible signature, 31 March 1942, NAA: A6923, SI/2, p. 285.
66 Colonel C.G. Roberts, 'Director of Military Intelligence Memorandum on Interception of Enemy Wireless Traffic to Deputy Chief of General Staff', 5 April 1942, NAA: A6923, SI/2, p. 279.
67 Report, 'Central Bureau Technical Records. Part A – Organization', 'Organization and General History', (undated) NAA: B5436, Part A.
68 Minute, DMI to DCGS, 'Interception of Enemy Wireless Traffic', 5 April 1942, NAA: 6923, SI/2, p. 249.
69 'Central Bureau Technical Records. PART A – Organization', (undated) NAA: B5436, Part A.

70 See also the assessment by Dr Hooper, Head of the Japanese Air Intelligence Section later in this chapter.
71 A. Sinkov, A.W. Sandford, H. Roy Booth, 'Central Bureau Technical Records'. PART K-Critique, NAA: B5436, PART K, 6.
72 Report, 'Central Bureau Technical Records. PART A – Organization', 'Organization and General History', (undated) NAA: B5436, PART A.
73 See John Blaxland, 'Intelligence and Special Operations in the Southwest Pacific, 1942–45', in Peter Dean (ed.), *Australia, 1944–45: Victory in the Pacific* (Melbourne: Cambridge University Press, 2015), pp. 145–68.
74 Maneki, *The Quiet Heroes of the Southwest Pacific Theater*, p. 1.
75 Pfennigwerth, *A Man of Intelligence*, p. 208.
76 Pfennigwerth, *A Man of Intelligence*, p. 210.
77 Land Forces message to London 06.30 16 January 1943, NAA: 6923, 37/401/425.
78 Minute, A.W. Sandford sgd F.R. Burton for Chief of the General Staff, 23 January 1943, NAA: 6923, 37/401/425.
79 'Minutes of Meeting to Discuss Liaison between Australian Army (Diplomatic Section) and G.C.& C.S.(Civil) 30th April 1943', NAA: A6923, 37/401/425.
80 Nigel de Grey, 'Allied Sigint-Policy and Organisation', Chap VIII, 'Australia', pp. 21–22, TNA (UK): HW43/78; and Report, 'Central Bureau Technical Records. Part A – Organization', 'Organization and General History', (undated) NAA: B5436, Part A.
81 Nigel de Grey, 'Allied Sigint-Policy and Organisation', Chap VIII, 'Australia', pp. 21–22, TNA (UK): HW43/78; and Report, 'Central Bureau Technical Records. Part A – Organization', 'Organization and General History', (undated) NAA: B5436, Part A, p. 22.
82 Nigel de Grey, 'Allied Sigint-Policy and Organisation', Chap VIII, 'Australia', pp. 14–15, TNA (UK): HW 43/78; and Report, 'Central Bureau Technical Records. Part A – Organization', 'Organization and General History', (undated) NAA: B5436, Part A.
83 Report, 'Central Bureau Technical Records. PART A – Organization', 'Organization and General History', (undated) NAA: B5436, PART A.
84 Nigel de Grey, 'Allied Sigint-Policy and Organisation', Chap VIII, 'Australia', pp. 15–16, TNA (UK): HW 43/78.
85 Major General (Ret'd) Steve Meekin, correspondence with authors, 10 October 2021.
86 Lieutenant Colonel A.W. Sandford, letter of 30 August 1945 to Lieutenant-Colonel R.A. Little, NAA:6923, Item 16/6/289, Australian Military Forces, Central Bureau Administration, p. 4.
87 Ball and Tamura, *Breaking Japanese Diplomatic Codes*, p. 132.
88 'No.45555' *The London Gazette* (Supplement), 31 December 1971, p. 35.
89 Report, 'Central Bureau Technical Records. PART A – Organization', 'Organization and General History', (undated) NAA: B5436, Part A.
90 Translation involved scanning to grade incoming reports in terms of potential intelligence value, translating the document in order of priority, final checking for mistakes in interpretation and accuracy, final editing and then recording ensuring a message could be traced. See Report, 'Central Bureau Technical Records. Part I – Translation Section', (undated) NAA: B5436, Part I.
91 Report, 'Central Bureau Technical Records. PART A – Organization', 'Organization and General History', (undated) NAA: B5436, PART A.
92 Report, 'Central Bureau Technical Records. PART A – Organization', 'Organization and General History', (undated) NAA: B5436, PART A.
93 Drea, *MacArthur's ULTRA*, p. 20.

94 Nigel de Grey, 'Allied Sigint-Policy and Organisation', Chap VIII, 'Australia', 17, TNA (UK) HW 43/78.
95 A. Sinkov, A.W. Sandford, H. Roy Booth, Critique, Central Bureau Technical Records. PART K, (undated) NAA: B5436, Part K.
96 Drea, *MacArthur's ULTRA*, pp. 21–22.
97 For a history of ATIS, see Kieran Laurence Miles, 'A History of the Allied Interpreter Section: Southwest Pacific Area WW II,' PhD Thesis, University of Queensland, 1993.
98 'Central Bureau Technical Records. PART A – Organization', 'Organization and General History', (undated) NAA: B5436, PART A; and Report, 'Central Bureau Technical Records. PART K – Critique', (undated) Colonel A. Sinkov, Lieut. Colonel A.W. Sandford and Wing Commander H. Roy Booth, NAA: B5436, PART K.
99 Drea, *MacArthur's ULTRA*, pp. 21–22.
100 Drea, *MacArthur's ULTRA*, pp. 21–22.
101 See (multiple reports) 'Special Intelligence – Australian Military Forces', NAA: A6923, SI/1.
102 Nigel de Grey, 'Allied Sigint-Policy and Organisation', Chap VIII, 'Australia', p. 19, TNA (UK) HW 43/78.
103 Nigel de Grey, 'Allied Sigint-Policy and Organisation', Chap VIII, 'Australia', p. 19, TNA (UK) HW 43/78, p. 49.
104 Nigel de Grey, 'Allied Sigint-Policy and Organisation', Chap VIII, 'Australia', p. 19, TNA (UK) HW 43/78, p. 50.
105 'Central Bureau Technical Records. Part A – Organization', 'Organization and General History', (undated) NAA: B5436, Part A.
106 Nigel de Grey, 'Allied Sigint-Policy and Organisation', Chap VIII, 'Australia', pp. 40–42, TNA (UK): HW 43/78.
107 Nigel de Grey, 'Allied Sigint-Policy and Organisation', Chap VIII, 'Australia', p. 43, TNA (UK): HW 43/78.
108 Timothy Mucklow, *The SIGABA/ECM II Cipher Machine: 'A Beautiful Idea'* (Fort Meade, MD: Center for Cryptologic History, 2015), pp. 2–3.
109 Mucklow, *The SIGABA/ECM II Cipher Machine*, p. 5.
110 Mucklow, *The SIGABA/ECM II Cipher Machine*, p. 19.
111 Mucklow, *The SIGABA/ECM II Cipher Machine*, p. 26.
112 Patrick D. Weadon, 'Sigsaly Story', *NSA/CSS*, at <www.nsa.gov.au/about/cryptologic-heritage/historical-figures-publications/publications/wwii/sigsaly-story>, accessed 9 September 2019.
113 Memo, 'Minutes of the Y Committee, Committee Meeting No. 10', Central Bureau, 4 June 1942, NAA: A6923, SI/2.
114 Memo, 'Minutes of the Y Committee, Committee Meeting No. 13', Central Bureau, 24 June 1942, and 'Minutes of the Y Committee, Committee Meeting No. 14', Central Bureau, 2 July 1942, NAA: A6923, SI/2.
115 Minute, 'Hollerith machines', Dir DSB to Defence Controller Intelligence, 21 August 1947, NAA: 42/301/1035; David Dufty, *The Secret Code Breakers of Central Bureau: How Australia's Signals Intelligence Helped Win the Pacific War* (Melbourne: Scribe, 2017), p. 213; and Maneki, *The Quiet Heroes of the Southwest Pacific Theater*, p. 1.
116 Maneki, *The Quiet Heroes of the Southwest Pacific Theater*, p. 46.
117 Dufty, *The Secret Code Breakers of Central Bureau*, p. 215.
118 Dufty, *The Secret Code Breakers of Central Bureau*, pp. 216, 217.
119 Minute, CGS to 'The Secretary', 'International Business Machines for Central Bureau', 13 March 1945, NAA: A6923, SI/2.

120 Minute, CGS to 'The Secretary', 'International Business Machines for Central Bureau', 13 March 1945, NAA: A6923, SI/2; and Minute, AWS/LF 5416/44, Comd Central Bureau to DMI Adv LHQ, 'IBM Machines for Central Bureau, 5 December 1944, NAA: A6923, SI/2.
121 Minute, ADMI to Director Staff Duties, 'International Business Machines for Central Bureau', 9 Mar 1945, NAA: A6923, SI/2.
122 Minute, CGS to Secretary, Army, 'IBMs for Central Bureau', 25 May 1945, NAA: A6923, SI/2.
123 Minute, ADMI to Director Staff Duties, 'MI 16.6.186', 3 Jan 1945, NAA: A6923, SI/2.
124 Minute, DMI to CGS, 'IB Machines for Central Bureau', 9 Mar 1945, NAA: A6923, SI/2.
125 Minute, Lt. Col. A.W. Sandford to ADMI, 'International Business Machines for Central Bureau', 15 Aug 1945, NAA: A6923, SI/2.
126 Memo, Defence Secretary to Secretary Department of the Treasury (Defence Division), 'Transfer of Hollerith Equipment from Navy to Defence Signals Bureau', 9 December 1947 at NAA: 42/301/1035.
127 Dufty, *The Secret Code Breakers of Central Bureau*, p. 289.
128 Dufty, *The Secret Code Breakers of Central Bureau*, p. 289.
129 Dufty, *The Secret Code Breakers of Central Bureau*, p. 289.
130 Frederick G. Shedden, Secretary War Cabinet Minute 26th May 1945 (4198) AGENDUM No.205/1945, AGENDUM No.212/1945, Allied Central Bureau, NAA: MP729/8, 41/431/118, p. 3.
131 Nigel de Grey, 'Allied Sigint-Policy and Organisation', Chap VIII, 'Australia', p. 20, TNA (UK): HW 43/78.
132 Report, 'Central Bureau Technical Records. PART J – Field Sections', (undated) NAA: B5436, PART J.
133 Report, 'Central Bureau Technical Records, Organisation', NAA: B5436.
134 Report, Lt. Colonel J.W. Ryan AIF to the Director of Military Intelligence and Director of Intelligence RAAF, 'Report on Special Wireless Units (Signals) 1940-45', 19 December 1945, NAA: A10908, p. 2.
135 Peter Few, 'SOME SIGSOP HISTORY', n.d., at <www.3teluunitassn.com/files/1_Wireless_Unit_history_article.pdf>, accessed 31 October 2022.
136 Report, 'Central Bureau Technical Records, Organisation', NAA: B5436, PART A, p. 13.
137 Craig Arthur Bellamy, 'The Beginning of the Secret Australian Radar Countermeasures Unit During the Pacific War', PhD thesis, Charles Darwin University, 2020, p. 7.
138 George Stevens, 'The Advent of Radar in the Royal Australian Navy', *Journal of Australian Naval History*, 3:2 (2006), p. 13,
139 Stevens, 'The Advent of Radar in the Royal Australian Navy', p. 9, cited in Bellamy, 'The Beginning of the Secret Australian Radar Countermeasures Unit During the Pacific War', p. 8.
140 Maj. Gen. R.N. Sutherland, U.S. Army Chief of Staff, 'Operations Instructions number 36, 5 July 1943, from General Headquarters Southwest Pacific Area', NAA: MP729, 57/404/605, p. 11.
141 William Cahill, 'Far East Air Forces RCM Operating in the Final Push on Japan', *Air Power History*, 65:1 (2018) pp. 37, 38.
142 General MacArthur, 'Operations Instruction No. 36' for an 'Organization and Operation Radar and Radio Countermeasures Division, GHQ, SWPA', signed 5 July 1943, in NAA: MP729/6, 57/404/605, p. 11.
143 Charles Darby, *Australia's Liberators: B-24 Operations from Australia* (Melbourne: Red Roo Models, 2009), p. 27.

144 Yagi-Uda arrays were invented in Japan and named after their inventors Shintaro Uda and Prof. Hidetsugu Yagi of Tohoku University, who were at the forefront of radar research in the mid-1920s. To Japan's detriment, however, their pioneering work was not supported by the Japanese Army that came to power in the 1930s and little further research was permitted. Apparently it came as a surprise for the Japanese war to discover that the British early warning radar sets captured in Singapore in 1941 used 'their' Yagi arrays. Yet more unexpected was the use of radars incorporating Yagis at Midway, where four of their six biggest fleet carriers had been sunk. See Darby, *Australia's Liberators*, pp. 143, 146.
145 Darby, *Australia's Liberators*, p. 140.
146 Darby, *Australia's Liberators*, p. 149.
147 Darby, *Australia's Liberators*, p. 152.
148 Darby, *Australia's Liberators*, p. 191.
149 See Kevin Davies, 'Field Unit 12 Takes New technology to War in the Southwest Pacific', *Studies in Intelligence,* 58:3 (2014): 11–20.
150 Darby, *Australia's Liberators*, pp. 149, 152.
151 Darby, *Australia's Liberators*, p. 157.
152 Darby, *Australia's Liberators*, p. 162.
153 Darby, *Australia's Liberators*, p. 192.
154 Available for inclusion were the AN/AP1 Dina (to jam early-warning radars), the AN/APT-2 Carpet (to jam anti-aircraft gun-laying and searchlight direction radars), the AN/APT-3 Mandrel (a 100MW high power anti-aircraft artillery radar jammer), and a range of other items (tuned to jam the Japanese searchlight/gun-laying radar) as well as the AN/APT4 Boardroom, AN/APT-5 Carpet IV, AN/APQ-2 Rug, AN/APQ-9 Carpet III, or an ARQ-8. See Darby, *Australia's Liberators,* p. 163.
155 See Michael V. Nelmes, *Tocumwal to Tarakan: Australians and the Consolidated B-24 Liberator* (Canberra: Banner Books, 1994); and 'Australia@war', <www.mail.ozatwar.com/201flight.htm>, accessed 31 October 2022.
156 Nelmes, *Tocumwal to Tarakan*; and 'Australia@war,' <www.mail.ozatwar.com/200flight.htm>, accessed 31 October 2022.
157 Darby, *Australia's Liberators*, p. 163.

7 Wartime Sigint successes, bureaucratic & other challenges

1 Nigel de Grey, 'Allied Sigint-Policy and Organisation', Chap VIII, 'Australia', p. 17, TNA (UK): HW 43/78.
2 Nigel de Grey, 'Allied Sigint-Policy and Organisation', Chap VIII, 'Australia', p. 8, TNA (UK): HW 43/78.
3 A. Sinkov, A.W. Sandford, H. Roy Booth, Central Bureau Technical Records, *Critique,* NAA: Series B5436, PART K, 4.
4 Colonel Tiltman, 'G.C.&C.S., liaison letter 2.10.42 to Major Stevens, the British representative in Washington on Japanese military Sigint affairs G.C. & C.S.', Miscellaneous Papers, Q/2022, 2.10.42, TNA (UK) HW 43/49.
5 Colonel Tiltman, 'G.C.&C.S., liaison letter to Major Stevens, the British representative in Washington on Japanese military Sigint affairs G.C. & C.S.', Miscellaneous Papers, Q/2022, 2.10.42, TNA (UK) HW 43/49.
6 Letter, Major Stevens replying to Tiltman 12.10.42 G.C.& C.S. Miscellaneous Papers, Q/2022,12.10.42, TNA (UK) HW 43/49.
7 Letter, Major Stevens replying to Tiltman 12.10.42 G.C.& C.S. Miscellaneous Papers, Q/2022,12.10.42, TNA (UK) HW 43/49.
8 Letter, Major Stevens replying to Tiltman 12.10.42 G.C.& C.S. Miscellaneous Papers, Q/2022,12.10.42, TNA (UK) HW 43/49.

9   Letter, Major Stevens replying to Tiltman 12.10.42 G.C.& C.S. Miscellaneous Papers, Q/2022,12.10.42, TNA (UK) HW 43/49.
10 Lieutenant Commander L.A. Griffiths, R.N.V.R., 'G.C.& C.S. Army and Air Force SIGINT – III ' Part 3, 'Japanese', Chapter VI, 'The Build-up in the South-West Pacific', TNA (UK) HW 43/49.
11 Lieutenant Commander L.A. Griffiths, R.N.V.R., 'G.C.& C.S. Army and Air Force SIGINT', Vol. XV, Part 1, 'The Work of the Main Centres', Chapter VII 'The Central Bureau, Brisbane', p. 133, TNA (UK) HW 43/58.
12 Griffiths, 'G.C.& C.S. Army and Air Force SIGINT', pp. 133–34.
13 Griffiths, 'G.C.& C.S. Army and Air Force SIGINT', p. 135.
14 Griffiths, 'G.C.& C.S. Army and Air Force SIGINT', p. 135.
15 Griffiths, 'G.C.&C.S. Army and Air Force SIGINT', p. 137.
16 Griffiths, 'G.C.&C.S. Army and Air Force SIGINT', p. 136.
17 Lieutenant Commander L.A. Griffiths, R.N.V.R., 'G.C.&C.S. Army and Air Force SIGINT – III ' Part 3, 'Japanese', Chapter XVIII, 'Organisational development' p. 229, TNA (UK) HW 43/49.
18 GC&CS Miscellaneous Papers, Q/2031, 4.12.44, TNA (UK) HW 43/49.
19 Letter from Lieutenant-Colonel 'Mic' Sandford to Cmdr Sir Edward Travis of 12 February 1945. G.C.C.S. Miscellaneous Papers S.W.P.A. – Liaison, TNA (UK) HW 52/97.
20 Letter from Lieutenant-Colonel 'Mic' Sandford to Cmdr Sir Edward Travis of 12 February 1945, TNA (UK) HW 52/97.
21 Letter from Lieutenant-Colonel 'Mic' Sandford to Cmdr Sir Edward Travis of 12 February 1945, TNA (UK) HW 52/97.
22 Letter from Lieutenant-Colonel 'Mic' Sandford to Cmdr Sir Edward Travis of 12 February 1945, TNA (UK) HW 52/97.
23 Frederick D. Parker, *A Priceless Advantage: US Navy Communications Intelligence and the Battles of the Coral Sea, Midway and Aleutians* (Fort Meade, MD: National Security Agency, 1993, 2017 reprint), p. 2.
24 Parker, *A Priceless Advantage*, p. 19.
25 Parker, *A Priceless Advantage*, p. 21.
26 Daniel R. Headrick, *The Invisible Weapon: Telecommunications and International Politics 1851–1945* (New York & Oxford: Oxford University Press, 1991), pp. 244–45.
27 Parker, *A Priceless Advantage*, pp. 3–4.
28 G. Hermon Gill, *Royal Australian Navy 1942–1945*, Australia in the War of 1939–1945, Series Two, Navy, vol. II (Canberra: Collins and Australian War Memorial, 1968–1985 reprint), p. 41.
29 Parker, *A Priceless Advantage*, p. 26; and Ian Pfennigwerth, *Missing Pieces: The Intelligence Jigsaw and RAN Operations 1939–1971*, Papers in Australian Maritime Affairs No. 25 (Canberra: Sea Power Centre – Australia, 2008), at <www.navy.gov.au/sites/default/files/documents/PIAMA25.pdf>, p. 77.
30 Report, 'FRUMEL records (incomplete) of Communications Intelligence relating to the Coral Sea Battle', NAA: B5555, 3, File 3, 5.
31 Pfennigwerth, *Missing Pieces*, p. 81.
32 Report, 'FRUMEL records (incomplete) of Communications Intelligence relating to the Coral Sea Battle', NAA: B5555, 3, File 3, 11.
33 Report, 'FRUMEL records (incomplete) of Communications Intelligence relating to the Coral Sea Battle', NAA: B5555, 3, File 3, 11.
34 Parker, *A Priceless Advantage*, p. 27.
35 Gill, *Royal Australian Navy 1942–1945*, vol. II, p. 53.

36 Jack B. Newman sgd, 'FRUMEL records (incomplete) of communications intelligence relating to the Midway Battle', NAA: Item ID 856346, Series B5555, Control Symbol 4, p. 5.
37 Newman, 'FRUMEL records (incomplete) of communications intelligence relating to the Midway Battle', p. 6.
38 Newman, 'FRUMEL records (incomplete) of communications intelligence relating to the Midway Battle', p. 7.
39 John Fahey, *Australia's First Spies: The Remarkable Story of Australia's Intelligence Operations, 1901–1945* (Sydney: Allen & Unwin, 2018), p. 207.
40 Newman, 'FRUMEL records (incomplete) of communications intelligence relating to the Midway Battle', p. 7.
41 Newman, 'FRUMEL records (incomplete) of communications intelligence relating to the Midway Battle', p. 7.
42 Newman, 'FRUMEL records (incomplete) of communications intelligence relating to the Midway Battle', p. 8.
43 Newman, 'FRUMEL records (incomplete) of communications intelligence relating to the Midway Battle', p. 8.
44 Newman, 'FRUMEL records (incomplete) of communications intelligence relating to the Midway Battle', p. 8.
45 Newman, 'FRUMEL records (incomplete) of communications intelligence relating to the Midway Battle', p. 10.
46 Newman, 'FRUMEL records (incomplete) of communications intelligence relating to the Midway Battle', p. 10.
47 Newman, 'FRUMEL records (incomplete) of communications intelligence relating to the Midway Battle', p. 10.
48 Newman, 'FRUMEL records (incomplete) of communications intelligence relating to the Midway Battle', p. 10.
49 Newman, 'FRUMEL records (incomplete) of communications intelligence relating to the Midway Battle', p. 13.
50 Newman, 'FRUMEL records (incomplete) of communications intelligence relating to the Midway Battle', p. 15.
51 Newman, 'FRUMEL records (incomplete) of communications intelligence relating to the Midway Battle', NAA: Item ID 856346, Series B5555, Control Symbol 4, p. 16.
52 Sharon A. Maneki, *The Quiet Heroes of the Southwest Pacific Theater: An Oral History of the Men and Women of CBB and FRUMEL* (Fort Meade, MD: Center for Cryptological History, NSA, 1996), p. 81.
53 See Joris Nieuwint, 'When the Allies Killed Over 20,000 of Their Own Countrymen as They Sank Japanese Hell Ships That Transported Them,' War History Online, January 17, 2016, at <www.warhistoryonline.com/featured/japanese-hellships.html>, accessed 27 February 2020.
54 Liam Kane, 'Allied Air Intelligence in the South West Pacific Area, 1942–1945,' *Journal of Intelligence History* (2021): pp. 1–21.
55 George Odgers, *Air War Against Japan 1939–1945*, Australia in the War of 1939–1945, Series III, Vol. II (Canberra: Australian War Memorial, 1957), pp. 41, 59, 111, 118, 138, 157, 280.
56 A. Jack Brown, *Katakana Man: I Worked Only for Generals* (Canberra: Air Power Development Centre, 2005), pp. 33, 56, 73.
57 Brown, *Katakana Man*, p. 44.
58 Michael Veitch, *The Battle of the Bismarck Sea: The Forgotten Battle that Saved the Pacific* (Sydney: Hachette Press, 2021).

59 Jean Bou, *MacArthur's Secret Bureau: The Story of the Central Bureau, MacArthur's Signals Intelligence Organisation* (Sydney: Australian Military History Publications, 2012), pp. 1–3.
60 Bou, *MacArthur's Secret Bureau*, pp. 5–6.
61 Burke Davis, *Get Yamamoto* (London: Arthur Baker, 1969), pp. 9, 14.
62 The claim about RAAF No. 1 Wireless Unit is drawn from Jack Bleakley, *The Eavesdroppers* (Canberra: Australian Government Publishing Service, 1992), p. 95; and Brown, *Katakana Man*, pp. 29–30. Bou, however, credits FRUMEL and, in doing so, draws on Cargill Hall (ed.), *Lightning over Bougainville* (Washington DC: Smithsonian, 1991), pp. 6–7. Bou, *MacArthur's Secret Bureau*, pp. 6–9.
63 'Central Bureau Technical Records. PART A – Organization', 'Organization and General History', (undated) NAA: B5436, PART A.
64 Edward J. Drea and Joseph E. Richard, 'New Evidence on Breaking the Japanese Army Codes,' *Intelligence and National Security*, 14:1 (1999), pp. 62–83.
65 Drea and Richard, 'New Evidence on Breaking the Japanese Army Codes,' p. 67.
66 Drea and Richard, 'New Evidence on Breaking the Japanese Army Codes,' pp. 69–70.
67 'Central Bureau Technical Records. PART A – Organization', 'Organization and General History', (undated) NAA: B5436, PART A.
68 The story is given a full chapter in Craig Collie, *Code Breakers: Inside the Shadow World of Signals Intelligence in Australia's Two Bletchley Parks* (Sydney: Allen & Unwin, 2017), pp. 202–20.
69 Nigel de Grey, 'Allied Sigint-Policy and Organisation', Chap VIII, 'Australia', 23TNA (UK): HW 43/78; and Maneki, *The Quiet Heroes of the Southwest Pacific Theater*, p. 40.
70 Abraham Sinkov interview cited in Maneki, *The Quiet Heroes of the Southwest Pacific Theater*, p. 41.
71 Geoffrey Ballard, *On ULTRA Active Service: The Story of Australia's Signals Intelligence Operations During World War II* (Melbourne: Spectrum Publications, 1991), p. 43.
72 Barbara Winter, *The Intrigue Master: Commander Long and Naval Intelligence in Australia, 1913–1945* (Brisbane: Boolarong Press, 1995), p. 51.
73 Winter, *The Intrigue Master*, p. 52.
74 Desmond Ball and Keiko Tamura (eds), *Breaking Japanese Diplomatic Codes: David Sissons and D Special Section during the Second World War* (Canberra: Australian National University, 2013), p. 19; David Dufty, *The Secret Code Breakers of Central Bureau: How Australia's Signals Intelligence Helped Win the Pacific War* (Melbourne: Scribe, 2017), p. 82; and Winter, *The Intrigue Master*, pp. 51–52.
75 Winter, *The Intrigue Master*, p. 51; and Ian Pfennigwerth, *A Man of Intelligence: The Life of Captain Theodore Eric Nave, Australian Codebreaker Extraordinary* (Sydney: Pfennigwerth, 2006), p. 168.
76 Dufty, *The Secret Code Breakers of Central Bureau*, pp. 240–42.
77 James Bamford, T*he Puzzle Palace: A Report on America's Most Secret Agency* (Boston, MA: Houghton Mifflin, 1982), pp. 398–99.
78 Central Bureau Technical Records – Part A Organisation. NAA: B5436, Part A, 11.
79 Dufty, *The Secret Code Breakers of Central Bureau*, pp. 268–69; and Harold A. Skaarup, *Out of Darkness – Light: A History of Canadian Military Intelligenc*e, vol. 1, Pre-Confederation to 1982 (Lincoln, NE: iUniverse, 2005), pp. 148–49.
80 Memo, Chief of the General Staff to Secretary, 'No. 1 Special Wireless and Intelligence Group – Canadian Army', 24 February 1945, NAA: A6923, 16/6/502, pp. 85–86.
81 'Order of Detail by Chief of the General Staff Respecting Military Forces of Canada proceeding to Australia to Act in Combination with AMF Serving in Pacific Theatre',

J.C. Murchie, Chief of the General Staff [Canada], 11 December 1944, NAA: A6923, 16/6/502, p. 120.

82   Memo, Chief of the General Staff to Secretary, 'No. 1 Special Wireless and Intelligence Group – Canadian Army', 24 February 1945, NAA: A6923, 16/6/502, pp. 85–86; and Minute, 'Canadian Spec W/T Sec', Lt. Col. ADMI to BGS, 7 February 1945, NAA: A6923, 16/6/502, p. 93.

83   Minute, Signal Officer-in-Chief to DSD (through DMI), 'Withdrawal of Royal Canadian Signal Group', 31 August 1945, NAA: A6923, 16/6/502, p. 25.

84   Ballard, *On ULTRA Active Service*, p. 284. Travis's assistant on this visit was F.H. Hinsley. Later, as Professor of History at Cambridge, Hinsley edited the multi-volume official history of British Intelligence in the Second World War.

85   Ballard, *On ULTRA Active Service*, pp. 298–99.

86   Letter, Dennison to Travis, ('Notes on the organisation of signals intelligence in Australia and New Zealand'), 10 December 1942, TNA (UK) HW 52/93.

87   See F.W. Winterbotham, *The Ultra Secret: The Inside Story of Operation Ultra, Bletchley Park and Enigma* (London: Orion Books, 1974), pp. 168–76; and John Fahey, *Traitors and Spies: Espionage and Corruption in High Places in Australia, 1901–1950* (Sydney: Allen & Unwin 2020), pp. 310, 354.

88   Letter, DDMI, LTCOL Little to DMI LHQ, 1 Feb 1943, NAA: A816, BC3023506.

89   Letter, Sir Reginald Cross, to John Curtin, 'Wireless stations carrying intercepted enemy traffic', 5 January 1943, NAA: A816, BC171228, 48/302/64, cited in Fahey, *Traitors and Spies*, p. 239.

90   See David Horner, *The Spy Catchers: The Official History of ASIO 1949–1963*, vol. I (Sydney: Allen & Unwin, 2014), p. 29.

91   Horner, *The Spy Catchers*, pp. 30–31.

92   CXG 584, for C.S.S. only personal from X.A., 16.12.44, File 17H, 'Australia Leakages', TNA (UK) HW 40/207.

93   Letter, General Sir Thomas Blamey to acting Minister for the Army, J.M. Fraser, 6 January 1945, AWM: 3DRL/6643, 2/59.

94   See Letter, Blamey to J.M. Fraser, 6 January 1945, AWM: 3DRL/6643, 2/59, cited in Desmond Ball and David Horner, *Breaking the Codes: Australia's KGB Network, 1944–1950* (Sydney: Allen & Unwin, 1998); Horner, *The Spy Catchers*, pp. 29–32; and Brown, *Katakana Man*, pp. 99–101.

95   Nigel de Grey, 'Allied Sigint-Policy and Organisation', Chap VIII, 'Australia', p. 37 TNA (UK): HW 43/78.

96   This is the focus of much of the work in John Fahey, *Traitors and Spies*, chapters 15–18.

97   Ball and Tamura, *Breaking Japanese Diplomatic Codes*, p. 10.

98   Desmond Ball and Keiko Tamura, *Breaking Japanese Diplomatic Codes: David Sissons and D Special Section during the Second World War* (Canberra: ANU Press, 2013), pp. 7–14.

99   F.H. Hinsley, E.E. Thomas, C.F.G. Ransom and R.C. Knight, *British Intelligence in the Second World War*, vol. I (London: HMSO, 1979), p. 199.

100 Earlier code names for the project included 'Bride', 'Eider' and 'Acorn'. See 'Venona, 1 April 1945,' Declassified document, National Security Agency/Central Security Service website, at <www.nsa.gov/portals/75/documents/news-features/declassified-documents/venona/dated/1945/1apr_kgb_ny.pdf>, accessed 2 November 2022.

101 Nigel West, *Historical Dictionary of Signals Intelligence* (Plymouth, UK: Scarecrow Press, 2012), pp. 223–27.

102 Robert L. Benson, *The Venona Story* (Fort Meade, MD: Center for Cryptologic History, NSA, 2012), p. 1.

103 Benson, *The Venona Story*, pp. 3–11.
104 See Matthew M. Aid, *The Secret Sentry: The Untold History of the National Security Agency* (New York: Bloomsbury Press, 2000), p. 19; 'Venona,' Declassified documents, National Security Agency/Central Security Service website, at <www.nsa.gov/Helpful-Links/NSA-FOIA/Declassification-Transparency-Initiatives/Historical-Releases/Venona/>, accessed 2 November 2022; and West, *Historical Dictionary of Signals Intelligence*; and Benson, *The Venona Story*.
105 West, *Historical Dictionary of Signals Intelligence*, pp. 223–27.
106 See, for instance, 'Reissue: Biographical Details relating to Francisca Burny Alias "Sister"', 25 April 1945, Ref No: S/NBF/T11 of 21 January 1955, in 'Venona,' <www.nsa.gov./News-Featrures/Declassified-Documents/Venona/Dated/1945/April/>, accessed 2 November 2022. See also 'Reissue: "Klod" to Recommend Sources From Progressive Parties For Study of Internal Australian Affairs (1945)', 1 July 1945, Ref No. 3/NBF/T320, issued 15 December 1955, in 'Venona,' <www.nsa.gov./News-Featrures/Declassified-Documents/Venona/Dated/1945/Jul/>, accessed 2 November 2022 ; and Benson, *The Venona Story*, p. 53.
107 Report by Joint Planning Committee, 'Higher Defence Organisation-Joint Intelligence Committee –Proposed Terms of Reference', C.G. Oldham, Secretary Joint Planning Committee, NAA: Box 2363, File 1, 3.
108 The Official History of the Operations and Administration of Special Operations Australia [(SOA), also known as the Inter-Allied Services Department (ISD) and Services Reconnaissance Department (SRD)] Volume 2 – Operations – copy no.1 [for Director, Military Intelligence (DMI), Headquarters (HQ), Australian Military Forces (AMF) Melbourne, NAA: A3269, 08/A, p. 90 (hereafter SOA-ISD-SRD History for DMI).
109 For more on the Allied Intelligence Bureau, see Alan Powell, *War By Stealth: Australians and the Allied Intelligence Bureau, 1942–1945* (Melbourne: Melbourne University Press, 1996); and Gavin Long, *The Final Campaigns* (Canberra: Australian War Memorial, 1963), 'The Allied Intelligence Bureau, Appendix 4, pp. 617–22.
110 SOA-ISD-SRD History for DMI, p. 80.
111 SOA-ISD-SRD History for DMI, p. 80.
112 SOA-ISD-SRD History for DMI, p. 82.
113 SOA-ISD-SRD History for DMI, p. 86.
114 Special Intelligence Section Report – Japanese Diplomatic Cyphers 1946-1946 NAA: 6923, 1/Reference Copy, p. 5.
115 SOA-ISD-SRD History for DMI, p. 94.
116 SOA-ISD-SRD History for DMI, p. 94.
117 SOA-ISD-SRD History for DMI, p. 98.
118 SOA-ISD-SRD History for DMI, p. 104.
119 Nigel de Grey, 'Allied Sigint-Policy and Organisation', Chap VIII, 'Australia', p. 28 TNA (UK): HW 43/78.
120 Jon Robb-Webb, 'Anglo-American Naval Intelligence Co-operation in the Pacific, 1944–45,' *Intelligence and National Security,* 22:5 (2007), pp. 767–86.
121 Desmond J. Ball, 'Allied Intelligence Cooperation involving Australia during World War II,' *Australian Outlook,* 32:3 (1978), pp. 299–309.
122 Ball, 'Allied Intelligence Cooperation involving Australia during World War II,' p. 309.
123 See, for instance, Brown, *Katakana Man*, p. 150.

8 Postwar Sigint to Vietnam

1 Memorandum Special Intelligence Section from DDMI to DMI of 24 October 1942, p. 178, NAA: A6923, 37/401/425.

2   Cabinet Agendum no. 1213, 'Joint Intelligence Organisation – Post-war (iii) (h)' sgd. F.M. Forde, Acting Minister for Defence, 19 July 1946, NAA: A5954, 2363/2.
3   'Joint Intelligence Organisation – Post-War, Notes on Cabinet Agendum, 3. Signals Intelligence', 23 July 1946, NAA: A5954, 2363/2, Item ID 681171.
4   Joint Intelligence Committee Report No.29/1947, 'Detailed Recommendations for Joint Intelligence Organization', NAA: A5954 2363/2, Item ID 681171.
5   Joint Intelligence Committee Report No.29/1947, 'Detailed Recommendations for Joint Intelligence Organization', NAA: A5954 2363/2, Item ID 681171, 8 (a) ii and iii.
6   Joint Intelligence Committee Report No.29/1947, 'Detailed Recommendations for Joint Intelligence Organization', NAA: A5954, 2363/2 Item ID 681171. 8 (c).
7   Letter, F.G. Shedden to Sir Edward Travis, 26 September 1947, NAA: A5954, 2363/2, Item ID 681171.
8   Blair Tidey, *Forewarned Forearmed: Australian Specialist Intelligence Support in South Vietnam, 1966–1971* (Canberra: Strategic and Defence Studies Centre, Australian National University, 2007), p. 78.
9   ASD, 'No 3 Telecommunication Unit (3TU)', at <www.asd.gov.au/asd-declassified/3tu>, accessed 22 September 2022.
10 See David Horner, *The Spy Catchers: The Official History of ASIO, 1949–1963*, vol. I (Sydney: Allen & Unwin, 2014), pp. 38–39.
11 Brigadier B. Combes, 'Report on Joint Intelligence Organisation – Post-War', NAA: A12392, 1, p. 56.
12 F.M. Forde, Acting Minister for Defence, Cabinet Agendum No.1213, 'Joint Intelligence Organisation – Post-War', 19 July 1946, History Compiled by Department of Foreign Affairs: Post-War Development of Australia's Intelligence Machinery: A Documentary Survey Part 1, NAA: A12392, 1, pp. 68–78.
13 Letter to Secretary Department of Navy, NAA: A5954, 2363/3 Item Id 652528.
14 Extract from Recommendations of the Defence Committee on the Australian Contribution to the British Commonwealth Signal Intelligence Organisation, NAA: A5954, 2363/3, p. 137.
15 Memorandum, 'Joint Intelligence Organisation – Post-War', F.G. Shedden sgd. 2 August 1946, NAA: A5954, 2363/2 Item ID 681171.
16 'Obituaries: Lt-Cdr Teddy Poulden', *Daily Telegraph* (London), 20 November 1992, cited in Horner, *The Spy Catchers*, p. 40.
17 Interview, David Horner with C.C.F. Spry and A.P. Fleming, 17 June 1993, cited in Horner, *The Spy Catchers*, pp. 39–40.
18 Thomas R. Johnson, *American Cryptology During the Cold War, 1945–1989. Book I The Struggle for Centralization 1945–1960* (Fort Meade, MD: Center for Cryptologic History, National Security Agency 1995), p. 160.
19 See Horner, *The Spy Catchers*, pp. 56–59.
20 This is the subject of Horner, *The Spy Catchers*, chapters 3 and 4; and is also covered in John Ferris, *Behind the Enigma: The Authorised History of GCHQ* (London: Bloomsbury, 2020), p. 374.
21 Letter from copies of papers from the Harry S. Truman Presidential Library in Independence, Missouri, in CCH Series XVI, cited in Johnson, *American Cryptology During the Cold War, 1945–1989*, p. 19.
22 See David Horner, *Defence Supremo: Sir Frederick Shedden and the Making of Australian Defence Policy* (Sydney, Allen & Unwin, 2000); and Johnson, *American Cryptology During the Cold War, 1945–1989*, p. 19.
23 Letter, Sir Frederick Shedden to Sir Edward Travis, 26 September 1947, NAA: A5954, 2363/2 Item Id 681171.

24 See Horner, *The Spy Catchers,* p. 120.
25 Horner, *The Spy Catchers,* pp. 134–38.
26 See John C. Blaxland, *Strategic Cousins: Australian and Canadian Expeditionary Forces and the British and American Empires* (Montreal: McGill-Queens University Press, 2006).
27 Government of Canada, 'CANUSA Agreement', at <www.cse-cst.gc.ca/en/culture-and-community/history/archives/canusa-agreement>, accessed 25 August 2021.
28 Wesley Wark, 'The Road to CANUSA: How Canadian Signals Intelligence Won its Independence and Helped Create the Five Eyes', *Intelligence and National Security,* 35:1 (2020), pp. 20–34
29 Harold A. Skaarup, *Out of Darkness – Light: A History of Canadian Military Intelligence,* vol. 1, Pre-Confederation to 1982 (Lincoln, NE: Universe, 2005), pp. 136–40.
30 Wark, 'The Road to CANUSA'.
31 Wark, 'The Road to CANUSA,' p. 22.
32 Robert Taschereau, 'Royal Commission to Investigate the Facts Relating to and the Circumstances Surrounding the Communication by the Public Officials and Other Persons in Positions of Trust of Secret and Confidential Information to Agents of Foreign Power,' Ottawa, 1946, at <www.epe.lac-bac.gc.ca/100/200/301/pco-bcp/commissions-ef/taschereau1946ii-eng/taschereau1946ii-eng.htm>, accessed 2 November 2022.
33 Wark, 'The Road to CANUSA,' p. 23.
34 Government of Canada, 'History: Communications Security Establishment', at <www.cse-cst.gc.ca/en/culture-and-community/history>, accessed 25 August 2021.
35 Wark, 'The Road to CANUSA,' p. 23.
36 Bill Crean quoted in Wark, 'The Road to CANUSA,' p. 25.
37 Wark, 'The Road to CANUSA,' p. 28.
38 Don Oberdorfer, *The Two Koreas: A Contemporary History* (Reading, MA: Addison-Wesley, 1997), p. 8.
39 For an overview of Australia's role in the war, see John Blaxland, Michael Kelly and Liam Brewin Higgins (eds), *In from the Cold: Reflections on Australia's Korean War* (Canberra: ANU Press, 2020).
40 John Blaxland, *Swift and Sure: A History of the Royal Australian Corps of Signals* (Melbourne: Royal Australian Corps of Signals Committee, 1998), p. 64.
41 Blaxland, *Swift and Sure,* p. 56.
42 Blaxland, *Swift and Sure,* p. 64.
43 Ian Pfennigwerth, *Missing Pieces: The Intelligence Jigsaw and RAN Operations 1939–1971,* Papers in Australian Maritime Affairs No. 25 (Canberra: Sea Power Centre–Australia, 2008), p. 151.
44 See Mark Lax, *Malayan Emergency and Indonesian Confrontation 1950 to 1966,* The Australian Air Campaign Series – 2 (Sydney: RAAF History and Heritage and Big Sky Publishing, 2021). This work lists in detail the RAAF force elements deployed on operations during this period and includes no mention of Sigint or Elint functions or capabilities.
45 Lax, *Malayan Emergency and Indonesian Confrontation,* p. 39.
46 Lax, *Malayan Emergency and Indonesian Confrontation,* p. 39.
47 See Robert O'Neill, 'Setting a New Paradigm in World Order: The United Nations Action in Korea', in Blaxland, Kelly and Brewin Higgins, *In from the Cold,* pp. 29–48.
48 Pfennigwerth, *Missing Pieces,* p. 157.
49 David A. Hatch with Robert Louis Benson *The Korean War: The SIGINT Background,* at <www.nsa.gov/about/cryptologic-heritage/historical-figures-publications/publications/korean-war/koreanwar-sigint-bkg/>, accessed 31 October 2022; and Johnson, *American Cryptology During the Cold War, 1945–1989,* p. 33.

50 Robert Jervis, 'The Impact of the Korean War on the Cold War', *Journal of Conflict Resolution*, 24:4 (December 1980), pp. 563–92, 568.
51 Hatch with Benson, *The Korean War: The SIGINT Background*.
52 Hatch with Benson, *The Korean War: The SIGINT Background*.
53 Johnson, *American Cryptology During the Cold War, 1945–1989*, Book I, p. 43.
54 Johnson, *American Cryptology During the Cold War, 1945–1989*, Book I, p. 43.
55 Johnson, *American Cryptology During the Cold War, 1945–1989*, Book I, p. 43.
56 Johnson, *American Cryptology During the Cold War, 1945–1989*, Book I, pp. 45, 46.
57 Matthew M. Aid, Commentary, *The National Security Agency during the Cold War*, at <www.nsarchive2.gwu.edu/NSAEBB/NSAEBB260/index.htm>, accessed 31 October 2022.
58 RASCM Rogers, letter to Bruton, 25 August 1952 in Blaxland *Swift and Sure*, p. 67.
59 Oberdorfer, *The Two Koreas*, p. 9.
60 Pfennigwerth, *Missing Pieces*, p. 150.
61 Robert J. O'Neill, *Australia in the Korean War 1950–53*, vol. 2 (Canberra: Australian War Memorial and Australian Government Publishing Service, 1985), p. 409.
62 Allan Gyngell, *Fear of Abandonment: Australia in the World since 1942* (Melbourne: Black Inc., 2017).
63 Desmond Ball, 'The Strategic Essence', *Australian Journal of International Affairs*, 55:2 (2001), p. 237.
64 Ball, 'The Strategic Essence', pp. 235–48.
65 Ball, 'The Strategic Essence'.
66 The Defence Signals Bureau was renamed the Defence Signals Branch in October 1949, and again as the Defence Signals Division in January 1964.
67 Joint Intelligence Committee (S) Report No. 5/1955, 'Principles of Collaboration with Commonwealth Countries Other than the United Kingdom', Meeting held at Victoria Barracks Melbourne on Tuesday, 8 November 1955, NAA: A11401, S112, p. 76.
68 Appendix J Annexure J1, 'UKUSA Arrangements Affecting Australia and New Zealand', NAA: A11401, 1/1/6 PART 2, p. 81.
69 Appendix J Annexure J1, 'UKUSA Arrangements Affecting Australia and New Zealand', NAA: A11401, 1/1/6 PART 2, p. 81.
70 NSA Center for Cryptologic History, 'Cryptologic Almanac 50th Anniversary Series: Six Decades of Second Party Relations', 24 February 1998, DOCID 3559613, <www.warwick.ac.uk/fac/soc/pais/people/aldrich/vigilant/lectures/gchq/brusa1946/six_decades_of_second_party_relations.pdf>, accessed 31 October 2022.
71 Robert Macklin, *Warrior Elite: Australia's Special Forces* (Sydney: Hachette, 2015), chapter 7.
72 Horner, *The Spy Catchers*, p. 378.
73 John Ferris, *Behind the Enigma: The Authorised History of GCHQ* (London: Bloomsbury, 2020), p. 375.
74 Peter Few, 'Some SIGSOP History,' at <www.3teluunitassn.com/files/1_Wireless_Unit_history_article.pdf>, accessed 31 October 2022; and 'TELEG/SIGSOP POSTINGS TO HONG KONG', at <www.3teluunitassn.com>, accessed 31 October 2022.
75 Ferris, *Behind the Enigma*, p. 375.
76 See Peter G. Edwards, *Crises and Commitments: The Politics and Diplomacy of Australia's Involvement in South East Asia, 1948–1965* (Sydney: Allen & Unwin, 1992).
77 Pfennigwerth, *Missing Pieces*, p. 196.
78 Pfennigwerth, *Missing Pieces*, p. 198.
79 Blaxland, *Strategic Cousins*, pp. 129–34.

80 Ferris, *Behind the Enigma*, p. 378.
81 Ferris, *Behind the Enigma*, p. 380.
82 Pfennigwerth, *Missing Pieces*, p. 200.
83 Pfennigwerth, *Missing Pieces*, p. 199.
84 Pfennigwerth, *Missing Pieces*, p. 200.
85 Blaxland, *Swift and Sure*, p. 94.
86 Blair Tidey, *Forewarned Forearmed: Australian Specialist Intelligence Support in South Vietnam, 1966–1971* (Canberra: Strategic and Defence Studies Centre, ANU, 2007), p. 78.
87 James and Shiel-Small, *The Undeclared War*, p. 110, cited in Peter Dennis and Jeffrey Grey, *Emergency and Confrontation Australian Military Operations in Malaya and Borneo 1950–1966,* The Official History of Australia's Involvement in Southeast Asian Conflicts, 1948–1975 (Sydney: Allen & Unwin, 1996), p. 248.
88 Dennis and Grey, *Emergency and Confrontation*, pp. 248–49.
89 Dennis and Grey, *Emergency and Confrontation*, p. 249.
90 Pfennigwerth, *Missing Pieces*, p. 207.
91 Pfennigwerth, *Missing Pieces*, p. 214.
92 John Blaxland, 'The Role of Signals Intelligence in Australian Military Operations, 1939–1972,' *Australian Army Journal*, II:2 (2008), pp. 203–16.
93 General Sir Walter Walker, letter to the author, 17 September 1992 in Dennis and Grey, *Emergency and Confrontation,* p. 249.
94 Ron Ratchford, Interview with John Blaxland, August 1993, cited in Blaxland, 'The Role of Signals Intelligence in Australian Military Operations, 1939–1972', p. 208.
95 Dennis and Grey, *Emergency and Confrontation*, p. 249.
96 Dennis and Grey, *Emergency and Confrontation*, p. 78.
97 See Ian McNeill, *The Team: Australian Army Advisors in Vietnam, 1962–1972* (Canberra: Australian War Memorial, 1984).
98 See Bob Breen, *First to Fight: Australian Diggers, N.Z. Kiwis, and U.S. Paratroopers in Vietnam, 1965–66* (Sydney: Allen & Unwin, 1988).
99 Tidey, *Forewarned Forearmed*, p. 2.
100 Pfennigwerth, *Missing Pieces*, p. 225.
101 Pfennigwerth, *Missing Pieces*, p. 362.
102 Tidey, *Forewarned Forearmed*, p. 78.
103 Tidey, *Forewarned Forearmed*, p. 23.
104 Tidey, *Forewarned Forearmed*, p. 27; and Defence Honours and Awards Tribunal, Inquiry into Recognition for Service with 547 Signal Troops in Vietnam from 1966 to 1971, at <https://defence-honours-tribunal.gov.au/wp-content/uploads/2019/11/547-Signal-Troop-Inquiry-Report.pdf>, accessed 4 January 2023.
105 J. Fenton, H. O'Flynn, S. Hart, P. Murray and M.J. Davies (eds), *The Unclassified History of 547 Signal Troop in South Vietnam*, at <www.angelfire.com/empire/547sigs/History%20of%20547%20Signal%20Troop.pdf>, accessed 15 August 2021.
106 See Ian McNeill, *To Long Tan: The Australian Army and the Vietnam War, 1950–1966* (Sydney: Allen & Unwin, 1993).
107 Tidey, *Forewarned Forearmed*, p. 35.
108 Trevor Richards Interview Transcript, 3. Transcript of tape received October 1966. ASD Declassified. 547 Signal Troop in Vietnam, 1966 to 1971.
109 Tidey, *Forewarned Forearmed*, p. 37.
110 ASD Narrative on the Deployment of 547 Signal Troop to Vietnam, 1966 to 1971. Declassified by ASD 15/12/2020, at <www.asd.gov.au/sites/default/files/2022-03/

asd_narrative_on_the_deployment_of_547_signal_troop_to_vietnam_1966_-_1971.pdf>, accessed 15 August 2021.
111 Fenton et al., *The Unclassified History of 547 Signal Troop in South Vietnam*, p. 13.
112 Steve Hart with Ernie Chamberlain, 'Story 3 – A Tactical SIGINT Success Story', Pronto in South Vietnam, at <www.pronto.au104.org/547Sigs/547story3.html>, accessed 31 October 2022.
113 Murray, Interview, August 1989, cited in Blaxland, 'The Role of Signals Intelligence in Australian Military Operations', p. 208.
114 Blaxland, 'The Role of Signals Intelligence in Australian Military Operations,' p. 211.
115 Blaxland, *Swift and Sure*, p. 250.
116 See Jeffrey Grey, *Up Top: The Royal Australian Navy and Southeast Asian Conflicts, 1955–1972* (Sydney: Allen & Unwin, 1998).
117 Pfennigwerth, *Missing Pieces*, p. 245.
118 Pfennigwerth, *Missing Pieces*, p. 226.
119 Ferris, *Behind the Enigma*, p. 376.
120 Ferris, *Behind the Enigma*, p. 376.
121 Ferris, *Behind the Enigma*, p. 376.

9   Reform, computers & military Sigint since the 1970s

1 Jennifer Bussell, 'Cyberspace', *Britannica*, at <www.britannica.com/topic/cyberspace>, accessed 1 November 2022.
2 Richard J. Aldrich, *GCHQ: The Uncensored Story of Britain's Most Secret Intelligence Agency* (London: Harper Press, 2010), pp. 349–50.
3 Aldrich, *GCHQ*, p. 350.
4 HPE, 'From the Cray-1 to HPE Today', at <www.hpe.com/us/en/compute/hpc/cray.html>, accessed 1 November 2022.
5 Australian Signals Directorate, 'Decoded: 75 Years of the Australian Signals Directorate,' <www.asd.gov.au/75th-anniversary/events/2022-04-01-decoded-exhibition-national-museum-australia>, accessed 1 November 2022.
6 ASD REDSPICE Blueprint, p. 8.
7 HPE, 'A Supercomputing Journey Inspired by Curiosity: History of Cray Supercomputers,' at <www.hpe.com/us/en/compute/hpc/cray.html>, accessed 1 November 2022.
8 HPE, 'The 2010s: Harnessing Big Data', at <www.hpe.com/us/en/compute/hpc/cray.html>, accessed 1 November 2022.
9 Government of New Zealand, Government Communications Security Bureau, 'History of the GCSB', at <www.gcsb.govt.nz/about-us/history-of-the-gcsb/>, accessed 1 November 2022.
10 Government of New Zealand, Government Communications Security Bureau, 'UKUSA partners', at <www.gcsb.govt.nz/about-us/ukusa-allies/>, accessed 1 November 2022.
11 '*USS Buchanan* refused entry to New Zealand, 4 February 1985', at <www.nzhistory.govt.nz/page/uss-buchanan-refused-entry-new-zealand>, accessed 1 November 2022.
12 Richard Tanter, 'WikiLeaks, Australia and Empire,' in Felicity Ruby and Peter Cronau (eds), *A Secret Australia* (Melbourne: Monash University Publishing, 2020), p. 25.
13 Tanter, 'WikiLeaks, Australia and Empire,' p. 25.
14 Government Communications Security Bureau, 'About Us', at <www.gcsb.govt.nz/about-us/>, accessed 1 November 2022.
15 'National Cyber Security Centre', at <www.ncsc.govt.nz>, accessed 1 November 2022.
16 See Kevin Burnett, 'The Significance of New Zealand's Contribution to INTERFET', in

John Blaxland (ed.), *East Timor Intervention: A Retrospective on INTERFET* (Melbourne: Melbourne University Press, 2015), pp. 209–26.
17 See John Blaxland, *The Protest Years: The Official History of ASIO*, vol. II, 1963–1975 (Sydney: Allen & Unwin, 2015), chapters 14 & 18.
18 See David Horner, *The Spy Catchers: The Official History of ASIO, 1949–1963*, vol. I (Sydney: Allen & Unwin, 2014), Part 3.
19 Robert Hope, *Royal Commission on Intelligence and Security (RCIS)*, Sixth Report, item 6A, para 926, NAA: A8908.
20 Hope, *RCIS*, Third Report (abridged version), item 3B, para 45, NAA: A8908.
21 Hope, *RCIS*, Sixth Report, item 6A, para 948, NAA: A8908.
22 These incidents are considered in John Blaxland and Rhys Crawley, *The Secret Cold War: The Official History of ASIO*, vol. III, 1975–1989 (Sydney: Allen & Unwin, 2016).
23 Hope, *RCIS*, Sixth Report, item 6A, para 734, NAA: A8908.
24 Hope, *RCIS*, Sixth Report, item 6A, para 798, NAA: A8908
25 Hope, *RCIS*, Sixth Report, item 6A, para 768, NAA: A8908.
26 Richard Tranter, 'Hiding from the Light: The Establishment of the Joint Australia–United States Relay Ground Station at Pine Gap,' NAPSNet Policy Forum, 2 November 2019, p. 16, at <www.nautilus.org/napsnet/napsnet-policy-forum/hiding-from-the-light-the-establishment-of-the-joint-australia-united-states-relay-ground-station-at-pine-gap/?view=pdf>, accessed 1 November 2022.
27 Historic Hansard, House of Representatives, 25 October 1977, 30th parliament, 2nd session, at <www.historichansard.net/hofreps/1977/19771025_reps_30_hor107/#subdebate-38-16>, accessed 1 November 2022.
28 Philip Flood, *Report of the Inquiry into Australian Intelligence Agencies, July 2004* (Canberra: Government of Australia, 2004), p. 135, at <www.fas.org/irp/world/australia/flood.pdf>.
29 Hope, *RCIS*, Sixth Report, item 6A, para 775, NAA: A8908.
30 Hope, *RCIS*, Sixth Report, item 6A, para 917, NAA: A8908.
31 Department of Defence, *Annual Report 2000–01* (Canberra: Australian Government Publishing Service, 2001), p. 43.
32 Robert Hope, *Royal Commission into Australia's Security and Intelligence Agencies* (RCASIA) (Canberra, Australian Government Publishing Service, 1984), General Report, chapters 2, 2.30.
33 Hope, *RCASIA*, General Report, chapters 2, 2.31.
34 Hope, *RCIS*, Sixth Report, item 6A, para 752, NAA: A8908
35 Hope, *RCIS*, Sixth Report, item 6A, para 738, NAA: A8908.
36 Hope, *RCIS*, Sixth Report, item 6A, para 761, NAA: A8908.
37 Flood, *Report of the Inquiry into Australian Intelligence Agencies*, p. 135.
38 Allan Gyngell and Michael Wesley, *Making Australian Foreign Policy* (Melbourne: Cambridge University Press, 2007, 2nd edn), p. 121
39 Peter Few, 'JTUM Unit History', at <www.3teluunitassn.com/files/ex-P-Few-JTUM-history-article.pdf>, accessed 1 November 2022.
40 Few, 'JTUM Unit History'.
41 Dennis Moore, 'CB: Central Bureau Intelligence Corps Association Inc.', June 2000, at <www.ozatwar.com/sigint/2000_jun_cbic.pdf>, accessed 1 November 2022.
42 Stephen Merchant, LinkedIn, <www.linkedin.com/in/stephen-merchant-5141bb120/>, accessed 1 November 2022.
43 Department of Prime Minister and Cabinet, '2017 Independent Intelligence Review', at <www.pmc.gov.au/national-security/2017-independent-intelligence-review >, accessed 1 November 2022.

44 DFAT, 'Agreement with the Government of the United States of America relating to the Establishment of the Joint Defence Space Research Facility', Australian Treaty Series Number (1966) ATS 17, at <www.info.dfat.gov.au/info/Treaties/Treaties.nsf/AllDocIDs/675B157B757190FDCA256B59007D0CF3>, accessed 1 November 2022.
45 Stephen Smith, Ministerial statements, Australian Parliament House, 26 June 2013, at <www.parlinfo.aph.gov.au/parlInfo/search/display/display.w3p;query=Id:%22chamber/hansardr/4d60a662-a538-4e48-b2d8-9a97b8276c77/0016%22;src1=sm1>, accessed 1 November 2022.
46 Australian Government, 'Naval Communications Station Harold E Holt (AREA A), Exmouth, WA, Australia', Department of Agriculture, Water and the Environment, at <www.environment.gov.au/cgi-bin/ahdb/search.pl?mode=place_detail;place_id=103552>, accessed 1 November 2022.
47 Desmond Ball, 'The Strategic Essence', *Australian Journal of International Affairs*, 55:2 (2001), pp. 235–48.
48 Desmond Ball, *Australia and the Global Strategic Balance,* Canberra Papers on Strategy and Defence, No. 49 (Canberra: Strategic and Defence Studies Centre, 1989), abstract.
49 Ball, 'The Strategic Essence,' pp. 240, 242.
50 Kim Beazley, 'The Joint Facilities in the 1980s,' in Patrick Walters (ed.), *ANZUS at 70: The Past, Present and Future of the Alliance* (Canberra: Australian Strategic Policy Institute, 2021), p. 33.
51 Ruby and Cronau, *A Secret Australia*, p. xvii.
52 Beazley, 'The Joint Facilities in the 1980s', p. 33.
53 See Guy Rundle, 'Darkening Ecliptic', in Ruby and Cronau, *A Secret Australia*, p. 187.
54 John Blaxland, *The Protest Years: The Official History of ASIO,* vol. II (Sydney: Allen & Unwin, 2015), chapter 18.
55 Stephen Smith, Ministerial statements, Australian Parliament House, 26 June 2013, at <www.parlinfo.aph.gov.au/parlInfo/genpdf/chamber/hansardr/4d60a662-a538-4e48-b2d8-9a97b8276c77/0016/hansard_frag.pdf;fileType=application%2Fpdf>, accessed 1 November 2022.
56 Desmond Ball, *A Suitable Piece of Real Estate: American Installations in Australia* (Sydney: Hale & Iremonger, 1980).
57 Beazley, 'The Joint Facilities in the 1980s', p. 33.
58 Beazley, 'The Joint Facilities in the 1980s', p. 33.
59 Jeffrey Richelson and Desmond Ball, *The Ties That Bind: Intelligence Cooperation Between the UKUSA Countries* (Sydney: George Allen & Unwin, 1985); Desmond Ball, *A Base for Debate: The US Satellite Station at Nurrungar* (Sydney: Allen & Unwin, 1987); and Desmond Ball, *Pine Gap: Australia and the US Geostationary Signals Intelligence Satellite Program* (Sydney: Allen & Unwin, 1988).
60 David Rosenberg, *Inside Pine Gap: The Spy who Came in from the Desert* (Melbourne: Hardie Grant Books, 2011).
61 Tom Gilling, *Project Rainfall: The Secret History of Pine Gap* (Sydney: Allen & Unwin, 2019).
62 Gilling, *Project Rainfall.*
63 Smith, Ministerial statements, 26 June 2013.
64 Desmond Ball, *Code 777: Australia and the US Defense Satellite Communications System (DSCS),* Canberra Papers on Strategy and Defence, No. 56 (Canberra: Strategic and Defence Studies Centre, 1989), abstract.
65 Beazley, 'The Joint Facilities in the 1980s', p. 35.
66 Beazley, 'The Joint Facilities in the 1980s', p. 35.

67 Desmond Ball, *The Intelligence War in the Gulf*, Canberra Papers on Strategy and Defence, No. 75 (Canberra: Strategic and Defence Studies Centre, 1991).
68 RAAF, *Aircraft of the Royal Australian Air Force* (Canberra, Air Force History Branch, 2021), p. 358; and Stewart Wilson, *Catalina, Neptune and Orion in Australian Service* (Canberra: Aerospace Publications, 1991), p. 107.
69 Wilson, *Catalina, Neptune and Orion in Australian Service*, pp. 166–68.
70 See Andrew McLaughlin, 'RAAF's 10SQN AP3C (EW) Orions transferred to 42WG', *Australian Defence Business Review*, 26 June 2019, at <www.adbr.com.au/raafs-10sqn-ap-3cew-orions-transferred-to-42wg/>, accessed 1 November 2022.
71 Jeff Malone, '"A Tale of the Kangaroo and the Crow": Electromagnetic Spectrum Operations Challenges for the Australian Defence Force', 26–28 May 2015, Stockholm, Sweden, at <www.repository.jeffmalone.org/files/personal/A%20Tale%20of%20the%20Kangaroo%20and%20the%20Crow.pdf>, accessed 1 August 2022.
72 Anon., '$20million JEWOSU redevelopment opened,' *Australian Defence Magazine*, 10 January 2008, at <www.australiandefence.com.au/D719AFE0-F806-11DD-8DFE0050568C22C9>, accessed 1 November 2022; and Department of Defence, 'Electronic Warfare Operations', Defence Science and Technology Group, at <www.dst.defence.gov.au/capability/electronic-warfare-operations>, accessed 1 August 2022.
73 See Department of Defence 'AIR 3503', *Australian Tenders*, at <www.australiantenders.com.au/tenders/472718/air3503/>, accessed 1 August 2022; Rodney Barton, 'Australia (DGS-AUS) Operations in 2030', at <www.defense.info/williams-foundation/2019/07/australia-dgs-aus-operations-in-2030>, accessed 1 August 2022; and APDR staff, 'Veteran Owned-and-operated Allectum wins RAAF Support Contract', at <www.asiapacificdefencereporter.com/veteran-owned-and-operated-allectum-wins-raaf-support-contract>, accessed 1 November 2022.
74 Geoffrey Barker, 'The Mystery Boats', *Australian Financial Review*, 28 November 2003, at <www.afr.com/companies/manufacturing/the-mystery-boats-20031128-j77yd>, accessed 1 August 2022.
75 Henry George Barnard, 'James Gerald Bridges Armstrong', Geni, 26 May 2009, at <www.geni.com/people/James-Gerald-Bridges-Armstrong/6000000003982737032>, accessed 1 August 2022.
76 Paul Kirk, 'Navy Information Warfare: Developing an Information Warfare Strategy for the Royal Australian Navy to Support Contemporary Maritime Operations', MPhil thesis, UNSW, Canberra, 2015, p. 79.
77 Barker, 'The Mystery Boats'.
78 Barker, 'The Mystery Boats'.
79 Barker, 'The Mystery Boats'.
80 Barker, 'The Mystery Boats'.
81 Siegphyl, Guest Author, 'Mission of Secret Australian Submarines during the Cold War Revealed,' *War History Online*, 14 November 2013, at <www.warhistoryonline.com/war-articles/missions-secret-australian-submarines-cold-war-revealed.html>, accessed 1 November 2022.
82 'OSIS Oceanic Surveillance Information System', GlobalSecurity.org, at <www.globalsecurity.org/intell/systems/obu.htm>, accessed 1 August 2022.
83 'OSIS Oceanic Surveillance Information System', GlobalSecurity.org, at <www.globalsecurity.org/intell/systems/obu.htm>, accessed 1 August 2022.
84 Walter S. Mossber, 'U.S. Intelligence Agencies Triumphed in Gulf War Despite Some Weak Spots,' *The Wall Street Journal*, 18 March 1991, sec. A, 10, cited in Lieutenant Colonel John J. Bird, 'Analysis of Intelligence Support to the 1991 Persian Gulf War:

Enduring Lessons,' U.S. Army War College, Carlisle, Pennsylvania, March 2004 at <www.apps.dtic.mil/sti/pdfs/ADA423282.pdf>, accessed 1 August 2022.
85 John J. Bird, 'Analysis of Intelligence Support to the 1991 Persian Gulf War: Enduring Lessons,' U.S. Army War College, Carlisle, Pennsylvania, March 2004, pp. 7–8, at <www.apps.dtic.mil/sti/pdfs/ADA423282.pdf>, accessed 1 August 2022.
86 Ball, *The Strategic Essence*, pp. 238–40.
87 See John Blaxland, *The Australian Army from Whitlam to Howard* (Melbourne: Cambridge University Press, 2014), pp. 88–89; and David Horner, *Australia and the 'New World Order': From Peacekeeping to Peace Enforcement: 1988–1991, The Official History of Australian Peacekeeping, Humanitarian and Post-Cold War Operations* (Melbourne: Cambridge University Press/Australian War Memorial, 2011), pp. 424–26.
88 David Horner, *The Gulf Commitment: The Australian Defence Force's First War* (Melbourne: Melbourne University Press, 1992), pp. 2, 210.
89 Ian Dudgeon, Discussions with author, 1 August 2022.
90 Ian Dudgeon, 'The National Information Infrastructure' briefing, October 1997 (unclassified copy supplied by Dudgeon to authors). This concept is elaborated upon in Ian Dudgeon, 'Targeting Information Infrastructures' in Gary Waters, Desmond Ball and Ian Dudgeon, *Australia and Cyber-Warfare* (Canberra: ANU E-Press, 2008), pp. 59–84.
91 Ian Dudgeon, Discussions with author, 1 August 2022; and 'The Dudgeon Report,' February 1997 (unclassified copy supplied by Dudgeon to authors) and Dudgeon, 'Targeting Information Infrastructures', pp. 59–84.
92 See, for instance, Alan D. Campen and Douglas H. Dearth, *Cyberwar 2.0: Myths, Mysteries & Reality* (Fairfax, VA: AFCEA International Press, 1998).
93 Ian Dudgeon, 'A National Approach to Information Operations', SCD Information Operations Symposium, 11 November 1997.
94 Dudgeon, 'A National Approach to Information Operations'.
95 Gough Whitlam, ABC Interview with Graeme Dobell, 'Whitlam reveals his East Timor Policy', *The World Today* Archive – Monday, 6 December 1999.
96 See Desmond Ball and Hamish McDonald, *Death in Balibo, Lies in Canberra* (Sydney: Allen & Unwin, 2000). See also Bernard Collaery, *Oil Under Troubled Water: Australia's Timor Sea Intrigue* (Melbourne: Melbourne University Press, 2020), pp. 133–73; and Craig Stockings, *Born of Fire and Ash: Australia's operations in response to the East Timor crisis 1999–2000* (Sydney: UNSW Press, 2022), pp. 38–40.
97 Hope, *RCASIA*, General Report Ch.2, 2.19.
98 Hope, *RCASIA*, General Report Ch.2, 2.18, 2.19.
99 Desmond Ball, 'Silent Witness: Australian Intelligence and East Timor', in R. Tanter, D. Ball and G. van Klinken (eds), *Masters of Terror: Indonesia's Military and Violence in East Timor* (Lanham, MD: Rowman & Littlefield, 2006), pp. 177–202.
100 Collaery, *Oil Under Troubled Water*, pp. 152–61; and Stockings, *Born of Fire and Ash*, pp. 38–40
101 Collaery, *Oil Under Troubled Water*, p. 155.
102 Desmond Ball, 'Silent Witness: Australian intelligence and East Timor', *Pacific Review*, 14:1 (2001), p. 43.
103 Paul Barratt, 'The Right to Know and the Role of the Whistleblower', in Ruby and Cronau, *A Secret Australia*, p. 139.
104 Cited in Desmond Ball, 'Silent Witness: Australian intelligence and East Timor', *Pacific Review*, 14:1 (2001), pp. 45–46.
105 Ball, 'Silent Witness', p. 51.
106 Ball, 'Silent Witness', p. 54.

107 Ball, 'Silent Witness', p. 60.
108 For a fresh and authoritative account of the INTERFET mission, see Stockings, *Born of Fire and Ash*.
109 John Blaxland, *Information Era Manoeuvre: the Australian Led Mission in East Timor* (Canberra: Land Warfare Studies Centre, 2002). See also John Blaxland (ed.), *East Timor Intervention: A Retrospective on INTERFET* (Melbourne: Melbourne University Press, 2015).
110 Flood, *Report of the Inquiry into Australian Intelligence Agencies*, p. 137.
111 Desmond Ball and Hamish McDonald, *Death in Balibo, Lies in Canberra* (Sydney: Allen & Unwin, 2000); and Philip Knightley, 'More About Death in Balibo, Lies in Canberra by Desmond Ball and Hamish McDonald', *BAM – Books-a-Million*, January 2010, at <www.booksamillion.com/p/Death-Balibo-Lies-Canberra/Desmond-Ball/Q7386056#overview>, accessed 1 August 2022.
112 Flood, *Report of the Inquiry into Australian Intelligence Agencies*, p. 48.
113 See John Blaxland, 'On Operations in East Timor: The Experiences of the Intelligence Officer, 3rd Brigade', *Australian Army Journal* (2000), pp. 1–2; and John Blaxland, *Information Era Manoeuvre* (Canberra: Land Warfare Studies Centre, Working Paper, 2002), pp. 46–47.
114 Blaxland, *Information Era Manoeuvre*, p. 47.
115 Mick Lehmann, 'Intelligence in Afghanistan', in John Blaxland, Marcus Fielding and Thea Gellerfy (eds), *Niche Wars: Australia in Afghanistan and Iraq, 2001–2014* (Canberra: ANU Press, 2020), pp. 173–86.
116 Mark Willacy, *Rogue Forces: An Explosive Insiders' Account of Australian SAS War Crimes in Afghanistan* (Sydney: Simon & Schuster, 2021), pp. 40–41, 43.
117 Willacy, *Rogue Forces*, p. 86.
118 Willacy, *Rogue Forces*, p. 47.
119 'Heron Tactical Unmanned Aerial System (UAS)', at <www.iai.co.il/p/tactical-heron>, accessed 1 November 2022.
120 See 'Australia extends Heron aircraft deployment in Afghanistan', *Airforce Technology*, 11 December 2013, at <www.airforce-technology.com/news/newsaustralia-extends-heron-aircraft-deployment-in-afghanistan-4144870/>, accessed 1 November 2022.
121 Australian Government, 'Australian Heron mission ends in Afghanistan', *Defence News*, 3 December 2014, at <www.news.defence.gov.au/media/media-releases/australias-heron-mission-ends-afghanistan>, accessed 1 August 2022.
122 See Senator the Hon. Linda Reynolds, 'E7-A Wedgetail joins US-led Global Coalition against Daesh', Minister for Defence Media Release, 14 September 2019, at <www.minister.defence.gov.au/minister/lreynolds/media-releases/e-7a-wedgetail-joins-us-led-global-coalition-against-daesh>, accessed 1 August 2022 .
123 Reynolds, 'E7-A Wedgetail joins US-led Global Coalition against Daesh'.
124 This is addressed in several works, including Blaxland et al., *Niche Wars*.

10 Legislative reform & the coming of cyber

1 Gordon J. Samuels and Michael H. Codd, *Commission of Inquiry into the Australian Secret Intelligence Service, Report on the Australian Secret Intelligence Service*, Public Edition, March 1995 (Canberra: Australian Government, 1995).
2 Intelligence Services Bill 2001, Parliament of the Commonwealth of Australia, House of Representatives Explanatory Memorandum, at <www.aph.gov.au/Parliamentary_Business/Bills_Legislation/Bills_Search_Results/Result?bId=r1350>, accessed 2 November 2022.

3   Intelligence Services Bill 2001, Bills Digest No. 11, 2001–02, Background.
4   Guy Rundle, 'Darkening Ecliptic: Radical Melbourne and the Origins of WikiLeaks', in Felicity Ruby and Peter Cronau (eds), *A Secret Australia* (Melbourne: Monash University Publishing, 2020), p. 185.
5   Intelligence Services Bill 2001, Bills Digest No.11, 2001–02, Background.
6   CBR Staff Writer, 'Australia Admits to Monitoring Electronic Signals', TechMonitor, 27 May 1999, at <www.techmonitor.ai/technology/australia_admits_to_monitoring_electronic_signals>, accessed 2 November 2022.
7   Jane Perrone, 'The Echelon Spy Network', *The Guardian*, 30 May 2001, at <www.theguardian.com/world/2001/may/29/qanda.janeperrone>, accessed 2 November 2022.
8   Franco Piodi and Iolanda Mombelli, *The ECHELON Affair: The European Parliament and the Global Interception System* (Luxembourg: European Parliamentary Research Service, 2014).
9   *Intelligence Services Act 2001*, Section 7, 1 (a).
10 *Intelligence Services Act 2001*, Section 8, 1 (a).
11 *Intelligence Services Act 2001*, Note to Section 8.
12 *Intelligence Services Act 2001*, Section 15.
13 *Intelligence Services Act 2001*, Section 15, 3 (c).
14 *Intelligence Services Act 2001*, Sections 28 (1) and 29 (1)a.
15 Desmond Ball, *Australia's Secret Space Programs, Canberra Papers on Strategy and Defence* (Canberra: Strategic and Defence Studies Centre, 1988), abstract.
16 Desmond Ball, 'Silent Witness: Australian Intelligence and East Timor', *Pacific Review*, 14:1 (2001), p. 40.
17 Gary Adshead, 'Spy Base in Our Backyard', *The West Australian*, 14 October 2013, at <www.thewest.com.au/news/mid-west/spy-base-in-our-backyard-ng-ya-267224>, accessed 2 November 2022.
18 Adshead, 'Spy Base in Our Backyard'.
19 Defence Connect, 'Spy Base Looking for Contractors, Upgrades', 30 January 2017, <www.defenceconnect.com.au/intel-cyber/233-spy-base-looking-for-contractors-upgrades>, accessed 2 November 2022.
20 Minister for Defence Senator Linda Reynolds CSC, 'New Satellite Ground Station in WA Fully Operational', media release, Department of Defence, 12 October 2020, at <www.minister.defence.gov.au/minister/lreynolds/media-releases/new-satellite-ground-station-wa-fully-operational>, accessed 2 November 2022.
21 Richard Tanter, 'WikiLeaks, Australia and Empire', in Ruby and Cronau, *A Secret Australia*, p. 27.
22 Mark Corcoran, 'The Chinese Embassy Bugging Controversy', *ABC News*, 8 November 2013, at <www.abc.net.au/news/2013-11-08/the-chinese-embassy-bugging-controversy/5079148>, accessed 2 November 2022.
23 AAP, 'Spy Sex Scandal Dossier to Be Examined: Hill', *The Sydney Morning Herald*, 26 September 2002, at <www.smh.com.au/national/spy-sex-scandal-dossier-to-be-examined-hill-20020926-gdfo4w.html>, accessed 2 November 2022.
24 John Miller, 'NSA Speaks Out on Snowden, Spying', *CBS News*, 15 December 2013, at <www.cbsnews.com/news/nsa-speaks-out-on-snowden-spying/>, accessed 2 November 2022.
25 Kyle Balluck, 'Hayden: NSA "Infinitely Weaker" after Snowden', *The Hill*, 29 December 2013, at <www.thehill.com/policy/technology/194098-hayden-nsa-infinitely-weaker>, accessed 2 November 2022.
26 Ewan MacAskill and Lenore Taylor, 'Australia's spy agencies targeted indonesian president's mobile phone', *The Guardian*, 18 November 2013, at <www.theguardian.

com/world/2013/nov/18/australia-tried-to-monitor-indonesian-presidents-phone>; and Michael Brissenden, 'Australia Spied on Indonesian President Susilo Bambang Yudhoyono, Leaked Edward Snowden Documents Reveal', *ABC News*, 5 December 2014, at <www.abc.net.au/news/2013-11-18/australia-spied-on-indonesian-president,-leaked-documents-reveal/5098860>, accessed 2 November 2022.

27 Philip Dorling, 'Exposed: Australia's Asia Spy Network', *The Sydney Morning Herald*, 31 October 2013, at <www.smh.com.au/politics/federal/exposed-australias-asia-spy-network-20131030-2whia.html>, accessed 2 November 2022.

28 Dorling, 'Exposed: Australia's Asia Spy Network'.

29 Lenore Taylor, 'Tony Abbott: No Explanation, No Apology to Indonesia for Spying', *The Guardian*, 19 November 2013, at <www.theguardian.com/world/2013/nov/19/tony-abbott-no-apology-explanation-indonesia-spying>, accessed 2 November 2022.

30 Michael Bachelard and Mark Kenny, 'Indonesia Suspends Police Co-Operation in Phone-Tapping Fallout,' *The Sydney Morning Herald*, 3 November 2013, at <www.smh.com.au/politics/federal/indonesia-suspends-police-co-operation-in-phone-tapping-fallout-20131122-2y1p6.html>, accessed 2 November 2022.

31 Sabrina Siddiqui, 'Congress Passes NSA Surveillance Reform in Vindication for Snowden', *The Guardian*, 3 June 2015, at <www.theguardian.com/us-news/2015/jun/02/congress-surveillance-reform-edward-snowden>, accessed 2 November 2022.

32 Malcolm Fraser, 'Spying Row: Australians Deserve Accountable Intelligence Services', *The Guardian*, 22 November 2013, at <www.theguardian.com/commentisfree/2013/nov/22/spying-row-australians-deserve-accountable-intelligence-services>, accessed 2 November 2022; see also Christian Leuprecht and Hayley McNorton, *Intelligence as Democratic Statecraft Accountability and Governance of Civil-Intelligence Relations Across the Five Eyes Security Community – the United States, United Kingdom, Canada, Australia, and New Zealand* (Oxford: Oxford University Press, 2021).

33 Stephen Smith, Ministerial statements, Australian Parliament House, 26 June 2013, at <www.parlinfo.aph.gov.au/parlInfo/search/display/display.w3p;query=Id:%22chamber/hansardr/4d60a662-a538-4e48-b2d8-9a97b8276c77/0016%22;src1=sm1>, accessed 2 November 2022.

34 Smith, Ministerial statements, 26 June 2013.

35 Smith, Ministerial statements, 26 June 2013.

36 Smith, Ministerial statements, 26 June 2013.

37 Dylan Welch, 'Top Intelligence Analyst Slams Pine Gap's Role in American Drone Strikes', *730 Report*, ABC TV, 13 August 2014, at <www.abc.net.au/7.30/top-intelligence-analyst-slams-pine-gaps-role-in/5669322>, accessed 2 November 2022.

38 Peter Cronau, 'Pine Gap Plays Crucial Role in America's Wars, Leaked Documents Reveal,' Background Briefing, *ABC News*, 20 August 2017.

39 Richard Tanter, 'Fifty Years On, Pine Gap Should Reform to Better Serve Australia', *The Conversation*, 9 December 2016, at <www.theconversation.com/fifty-years-on-pine-gap-should-reform-to-better-serve-australia-65650>, accessed 2 November 2022.

40 Kim Beazley, 'Pine Gap at 50: The Paradox of a Joint Facility', The Strategist, Australian Strategic Policy Institute, 30 August 2017, at <www.aspistrategist.org.au/pine-gap-50-paradox-joint-facility/>, accessed 2 November 2022.

41 Kim Beazley, 'The Joint Facilities in the 1980s', in Patrick Walters (ed.), *ANZUS at 70: The Past, Present and Future of the Alliance* (Canberra: Australian Strategic Policy Institute, 2021), p. 36.

42 Richard Tanter, 'Pine Gap – An Introduction,' Nautilus Institute, 21 February 2016, at <www.nautilus.org/publications/books/australian-forces-abroad/defence-facilities/pine-gap/pine-gap-intro>, accessed 2 November 2022.

43 Desmond Ball, 'The Strategic Essence,' *Australian Journal of International Affairs*, 55:2 (2001), p. 240.
44 Desmond Ball, Bill Robinson and Richard Tanter, 'Australia's Participation in the Pine Gap Enterprise,' Nautilus Institute, NAPSNet Special Report, 9 June 2016, at <www.nautilus.org/napsnet/napsnet-special-reports/australias-participation-in-the-pine-gap-enterprise>, accessed 2 November 2022.
45 Emma Shortis, *Our Exceptional Friend: Australia's Fatal Alliance with the United States* (Melbourne: Hardie Grant Books, 2021), p. 40.
46 'Accountability,' Australian Signals Directorate, <www.asd.gov.au/about/accountability>, accessed 2 November 2022.
47 Justin Hendry, 'Mike Burgess Returns to ASD', itnews, 1 December 2017, at <www.itnews.com.au/news/mike-burgess-returns-to-asd-478976>, accessed 2 November 2022.
48 Australian Information Security Association, 'Security Perspectives of a Former Spy Chief', AISA Melbourne Branch meeting, 13 February 2019, at <www.aisa.org.au/Public/Events/Event_Display.aspx?EventKey=d6acbf83-ed03-4590-abe0-a21b722b4cb6>, accessed 2 November 2022.
49 Andrew Tillet, 'Why Australia's cybersecurity chief isn't a fan of social media', *Australian Fianncial Review*, 2 December 2022, <afr.com/oilitics/federal/why-australia-s-cybersecurity-chief-isn-t-a-fan-of-social-media-20221110-p5bx85>.
50 Joseph S. Nye, Jr, 'The End of Cyber-Anarchy? How to Build a New Digital Order,' *Foreign Affairs*, January/February 2022, p. 4.
51 Simone Fox Coob and Colin Kruger, 'Medibank faces $1 billion bill as hackers release 1500 more sensitive records', *The Sydney Morning Herald*, 20 November 2022, at <www.smh.com.au/business/companies/medibank-hackers-release-1500-more-sensitive-medical-records-20221120-p5bzpk.html>, accessed 29 November 2022.
52 Nye, Jr, 'The End of Cyber-Anarchy?'.
53 Nye, 'The End of Cyber-Anarchy?,' p. 2.
54 Malcolm Turnbull, 'Speech: Launch of Australia's Cyber Security Strategy Sydney', 21 April 2016, at <www.parlinfo.aph.gov.au/parlInfo/search/display/display.w3p;query=Id:%22media/pressrel/4513168%22>, accessed 2 November 2022.
55 Malcolm Turnbull, *Australia's Cyber Security Strategy* (Canberra: Department of Home Affairs, 2016), pp. 2–3.
56 Turnbull, *Australia's Cyber Security Strategy*, p. 3.
57 Turnbull, *Australia's Cyber Security Strategy*, p. 3.
58 Michael L'Estrange and Stephen Merchant, '2017 Independent Intelligence Review, June 2017,' Commonwealth of Australia 2017, Recommendation 6,16, <www.pmc.gov.au/national-security/2017-independent-intelligence-review>, accessed 2 November 2022.
59 L'Estrange and Merchant, '2017 Independent Intelligence Review,' Recommendation 4.55, 66.
60 Department of Defence, 'Information Warfare Division', 3 March 2021, at <www.defence.gov.au/jcg/iwd.asp#:~:text=The%20Information%20Warfare%20Division%20%28IWD%29%20was%20formed%20in,to%20Australia%E2%80%99s%20national%20interests%20in%20the%20information%20environment>, accessed 2 November 2022.
61 Malcolm Turnbull, 'Speech at the Opening of the Australian Cyber Security Centre Canberra – 16 August 2018', at <www.malcolmturnbull.com.au/media/speech-at-the-opening-of-the-australian-cyber-security-centre-canberra-16-a>, accessed 2 November 2022.
62 Turnbull, 'Speech at the Opening of the Australian Cyber Security Centre.'
63 Turnbull, 'Speech at the Opening of the Australian Cyber Security Centre.'

64 The Intelligence Services Amendment (Establishment of the Australian Signals Directorate) Act 2018, at <www.legislation.gov.au/Details/C2018A00025>, accessed 2 November 2022.
65 Dennis Richardson, Anna Harmer and Tara Inverarity, *Comprehensive Review of the Legal Framework of the National Intelligence Community*, December 2019 (Canberra: Australian Government, 2020), vol. 3, p. 153.
66 'New Top Secret Sydney Cyber Security Centre Operational', Cyber Security Connect, 30 August 2022, at <www.cybersecurityconnect.com.au/critical-infrastructure/8202-new-top-secret-sydney-cyber-security-centre-operational>, accessed 2 November 2022.
67 David Jones, 'Banks Outpace Other Industries in Cyber Investments, Defense Strategies: Report', Cybersecurity Dive, 15 November 2021, at <www.cybersecuritydive.com/news/banks-cyber-security-investments/610045/>, accessed 2 November 2022.
68 'Ransomware Attacks Have Increased 24% in the Last Quarter', Cyber Security Connect, 29 August 2022, at <www.cybersecurityconnect.com.au/commercial/8194-ransomware-attacks-have-increased-24-in-the-last-quarter>, accessed 2 November 2022.
69 See Anthea Roberts, Henrique Choer Moraes and Victor Ferguson, 'The Geoeconomic World Order', Lawfare blog, 19 November 2018, at <www.lawfareblog.com/geoeconomic-world-order>, accessed 2 November 2022.
70 That is, the leading Chinese telecommunications company Huawei.
71 Anthea Roberts and Nicolas Lamp, *Six Faces of Globalisation* (Cambridge, MA: Harvard University Press, 2019), pp. 133–34.
72 Roberts and Lamp, *Six Faces of Globalisation*, p. 142.
73 See Kai-Fu Lee, *AI Superpower: China, Silicon Valley, and the New World Order* (Boston, MA: Mariner Books, 2018)
74 Danielle Cave, 'Australia and the Great Huawei Debate: Risks, Transparency and Trust', The Strategist, Canberra, ASPI, 11 September 2019; and Roberts and Lamp, *Six Faces of Globalisation*, p. 134.
75 See Anthea Roberts, Henrique Choer Moraes and Victor Ferguson, 'The U.S.-China Trade War is a Competition for Technological Leadership', Lawfare blog, 21 May 2019, at <www.lawfareblog.com/us-china-trade-war-competition-technological-leadership>, accessed 2 November 2022.
76 Annika Smethurst, 'Spying Shock: Shades of Big Brother as Cyber-security Vision Comes to Light,' *Daily Telegraph*, 29 April 2019, <www.dailytelegraph.com.au/news/nsw/spying-shock-shades-of-big-brother-as-cyber-security-vision-comes-to-light/news-story/bc02f35f23fa104b139160906f2ae709>, accessed 2 November 2022.
77 Felicity Ruby and Peter Cronau (eds), *A Secret Australia* (Melbourne: Monash University Publishing, 2020), p. xix.
78 Richardson et al., *Comprehensive Review of the Legal Framework*, vol. 1, p. 89.
79 Richardson, et al., *Comprehensive Review of the Legal Framework*, vol. 3, p. 151.
80 Richardson, et al., *Comprehensive Review of the Legal Framework*, vol. 3, p. 152.
81 Richardson, et al., *Comprehensive Review of the Legal Framework*, vol. 3, p. 156.
82 Richardson, et al., *Comprehensive Review of the Legal Framework*, vol. 1, p. 47.
83 Australian Cyber Security Centre, 'About the ACSC', Australian Signals Directorate, Department of Defence, October 2020, at <www.cyber.gov.au/acsc>, accessed 2 November 2022.
84 Australian Cyber Security Centre, 'About the ACSC'.
85 Richardson, et al., *Comprehensive Review of the Legal Framework*, vol. 3, p. 216.
86 Richardson, et al., *Comprehensive Review of the Legal Framework*, vol. 3, p. 220.
87 Richardson, et al., *Comprehensive Review of the Legal Framework*, vol. 3, p. 220.

88 Australian Government, *Reform of Australia's Electronic Surveillance Framework: Discussion Paper* (Canberra: Department of Home Affairs, 2021).
89 Australian Government, *Reform of Australia's Electronic Surveillance Framework*, p. 2.
90 See Karen Middleton, 'AFP's New Power to Spy on Australians,' *The Saturday Paper*, 12–18 December 2020, at <www.thesaturdaypaper.com.au/news/politics/2020/12/12/afps-new-power-spy-australians/160769160010864>, accessed 2 November 2022.
91 Sarah Basford Canales, 'Australian Signals Directorate and ANU on the Hunt for "Spy Kids" Through Co-Lab Program', *The Canberra Times*, 18 November 2021, at <www.canberratimes.com.au/story/7516431/intelligence-agency-looks-to-anu-for-its-next-batch-of-spy-kids/>, accessed 2 November 2022.
92 Mike Burgess, 'Speech to the Lowy Institute, 27 March 2019,' at <www.asd.gov.au/publications/director-general-asd-speech-lowy-institute>, accessed 2 November 2022.
93 Burgess, 'Speech to the Lowy Institute'.
94 Matthew Knott, 'Cyber Defence Bolstered by $10 Billion via Project REDSPICE', *The Sydney Morning Herald*, 29 March 2022, <www.smh.com.au/politics/federal/cyber-defence-bolstered-by-10-billion-via-project-redspice-20220323-p5a79m.html>, accessed 2 November 2022.
95 Robert Hope, *Royal Commission on Intelligence and Security (RCIS)* (Canberra: Australian Government Publishing Service, 1977), Sixth Report, item 6A, para 768, NAA: A8908.
96 Hope, *RCIS*, Sixth Report, item 6A, para 769, NAA: A8908.
97 Hope, *RCIS*, Sixth Report, item 6A, para 770, NAA: A8908.
98 Hope, *RCIS*, Sixth Report, item 6A, para 770, NAA: A8908.
99 Australian Government, *Australia's Cyber Security Strategy 2020* (Canberra: Commonwealth of Australia, 2020), p. 26, para 50.
100 Department of Foreign Affairs and Trade, *Australia's International Cyber and Critical Tech Engagement Strategy 2021* (Canberra: Commonwealth of Australia, 2021), p. 41.
101 Edward Zheng, Kevin Stewart and Michael Caplan, 'A Spice Up to Australia's Cyber Capabilities?' Gilbert+Tobin, Digital Hub, 13 April 2022, at <www.gtlaw.com.au/knowledge/spice-australias-cyber-capabilities-what-we-know-so-far-about-project-redspice>, accessed 2 November 2022.
102 Casey Tonkin, '$1.3 Drone Scrapped for REDSPICE,' Information Age, 5 April 2022, at <www.ia.acs.org.au/article/2022/1-3b-drone-scrapped-for-redspice.html>, accessed 2 November 2022.
103 *Security Legislation Amendment (Critical Infrastructure Protection) Act 2022*, Cyber Infrastructure and Security Centre Factsheet, Department of Home Affairs, March 2022, at <www.cisc.gov.au/critical-infrastructure-centre-subsite/Files/cisc-factsheet-security-legislation-amendment-critical-infrastructure-protection-act-2022.pdf>, accessed 2 November 2022.
104 Rohit Ranjan, 'Quad Senior Cyber Group Meets in Sydney to Strengthen Cybersecurity Cooperation says WH', RepublicWorld.com, 26 March 2022, at <www.republicworld.com/world-news/rest-of-the-world-news/quad-senior-cyber-group-meets-in-sydney-to-strengthen-cybersecurity-cooperation-says-wh-articleshow.html>, accessed 2 November 2022.
105 Richard A. Clarke and Robert K. Knake, *The Fifth Domain: Defending Our Country, Our Companies and Ourselves in the Age of Cyber Threats* (New York: Penguin, 2020), p. 15.
106 The Soufan Center, 'IntelBrief: Could Confrontation in Cyberspace Escalate the War in Ukraine?', The Soufan Center, 10 June 2022, at <www.thesoufancenter.org/intelbrief-2022-june-10/>, accessed 2 November 2022.

107 Brad Williams, 'Nakasone: Cold War-style Deterrence "Does Not Comport to Cyberspace"', *Breaking Defense*, 4 November 2021, at <www.breakingdefense.com/2021/11/nakasone-cold-war-style-deterrence-does-not-comport-to-cyberspace/>, accessed 2 November 2022.
108 Jason Healey, 'Research Paper: The Implications of Persistent (and Permanent) Engagement in Cyberspace,' *Journal of Cybersecurity*, 5:1 (2019), pp. 6–7.

Conclusion & looking ahead

1 This theme is explored in Daniel Baldino and Rhys Crawley (eds), *Intelligence and the Function of Government* (Melbourne: Melbourne University Press, 2018).
2 Mike Burgess, 'Director-General ASD Speech to the Lowy Institute, 27 March 2019,' at <www.asd.gov.au/publications/director-general-asd-speech-lowy-institute>, accessed 2 November 2022.
3 Rachel Noble, 'Director-General ASD Speech at National Press Club, 18 November 2021,' at <www.asd.gov.au/publications/director-general-asd-speech-national-press-club>, accessed 2 November 2022.
4 Dennis Richardson, Anna Harmer and Tara Inverarity, *Comprehensive Review of the Legal Framework of the National Intelligence Community*, December 2019 (Canberra: Australian Government, 2020), vol. 1, p. 162.
5 John le Carré, *Tinker, Tailor, Soldier, Spy* (London/Sydney: Pan Books, 1975), p. 306.
6 Alex Younger, Intelligence Services Speech at St Andrews, Scotland, 3 December 2018, at <www.gov.uk/government/speeches/mi6-c-speech-on-fourth-generation-espionage>, accessed 2 November 2022.
7 Nigel de Grey, 'Allied Sigint-Policy and Organisation', Chapter VIII, 'Australia', p. 19, TNA (UK) HW 43/78.
8 De Grey, 'Allied Sigint-Policy and Organisation', p. 49.
9 De Grey, 'Allied Sigint-Policy and Organisation', p. 50.
10 Memorandum for Commander E.W. Travis RN signed by C.F. Holden, USN, Director of Naval Communications, 2 October 1942. C-in-C signified his concurrence on 20 October 1942, cited in Frank Birch, *GC&CS*, vol. V (a) 'The Organisation and Evolution of British Naval Sigint', p. 181, TNA (UK) HW 43/13.
11 'The Holden Agreement 2 October 1942,' Appendix B, 556, TNA (UK) HW 43/14.
12 Minute, F.G. Cresswell, Director Signal Section to Head, N Branch of 8 June 1921. NAA: MP 1049/9. Control Symbol 1997/5/196, Item no 505956, p. 311.
13 John Fahey, *Australia's First Spies: The Remarkable Story of Australia's Intelligence Operations, 1901–1945* (Sydney: Allen & Unwin, 2018), pp. 111–13.
14 Robert Hope, *Royal Commission on Intelligence and Security* (Canberra: Australian Government Publishing Service, 1977), Sixth Report, item 6A, para 798, NAA: A8908.
15 Major General Steve Meekin, correspondence with authors.
16 Rishi Iyengar, 'Why Ukraine is Stuck with Elon for Now', *Foreign Policy*, 22 November 2022, at <www.foreignpolicy.com/2022/11/22/ukraine-internet-starlink-elon-musk-russia-war>, accessed 29 November 2022.
17 Richardson, Harmer and Inverarity, *Comprehensive Review of the Legal Framework of the National Intelligence Community*, vol. 1, 10. Recommendation 1, p. 166.

INDEX

1st Australian Imperial Force (AIF)
    organisational hierarchies 43, 52
    receipt of Sigint 62
    Sigint experience 48–49, 52–53, 66–67, 69–72
1st Commonwealth Division 236–43, 245–46
2nd Australian Imperial Force (AIF)
    8th Division 105, 139
    9th Division 115–16, 204
    combat experiences 102–5, 115–16
    organisational hierarchies 102–4
    publications on 103
    raising of 102, 104
    Special Wireless Groups 112–13
3 Telecommunications Unit (3TU) 224–25, 273
3rd Brigade 105, 296
5G networks 319–21
7th Signal Regiment (formerly 101 Wireless Regiment)
    72 Electronic Warfare Squadron 295–96
    121 Signal Squadron (formerly 201 Signal Squadron) 249–51
    547 Signal Troop 252–56
    origins of 224, 281
9/11 297
200 and 201 Flight 184–85

ABC-1 talks 134
ABDA (American–British–Dutch–Australian) Command 140
Adelklassen 156
Admiralty (UK) *see also* Director of Naval Intelligence (DNI) (UK); Jellicoe, John; Room 40; Royal Navy
    control of RAN 43
    GC&CS links 75, 90–92
    response to Nave 153
    Singapore Naval Conference 95
    wireless radio usage 33
    WWI preparations 43–46
    WWII preparations 96, 100
Aeneas the tactician 9
Afghanistan 297–98
Afrika Korps 114–16

AFSA *see* National Security Agency
Aid, Matthew 212–13
air forces *see* German Air Force; Royal Air Force; Royal Australian Air Force; United States Army Air Force
Air Ministry (UK) 77, 96
airborne radio direction finder (ARDF) 255–56
air-to-surface vessel radar 117
Akagi 199
Akin, Spencer B. 84, 162–63, 182
Albanese, Anthony 340
Albatross, HMAS 95, 284
Alberti, Leon Battista 11–12
alcohol 82–83
al-Kindi, Yaqub ibn Ishaq 10
Allied Intelligence Bureau 164, 184–85, 214, 234 *see also* Services Reconnaissance Department
Allied Translator and Interpreter Service (ATIS) 164, 170
Allies *see* World War, First; World War, Second
ALP *see* Labor Party, Australian
America *see* United States of America
The American Black Chamber 82
American Expeditionary Force (AEF) 63–65 *see also* G2-A6
American Revolutionary War 16–17
American–British–Dutch–Australian (ABDA) Command 140
Amiens, attack on 61
Anderson, Brigadier-General 53
Anglo–Boer War 30–32
Anglo–Japanese Alliance 88
Anti-Ship Missile Defence (ASMD) 284
ANZAC Wireless Signal Squadron 4–5, 67–69
ANZUS (Australia New Zealand United States) Treaty 242, 267
AP-3C aircrafts 282
ARDF (airborne radio direction finder) 255–56
Arlington Hall *see* Signal Intelligence Service
Armed Forces Security Agency (AFSA) *see* National Security Agency

armies 43 *see also* Australian Army; British Army; German Army ...; Imperial Japanese Army; New Zealand Army; United States Army
Armstrong, James 284
The Art of Decyphering 15
The Art of War 10
Asama, HIJMS 94
ASD *see* Australian Signals Directorate
Asia 259–60, 331 *see also* Confrontation; Pacific War
ASIO *see* Australian Security Intelligence Organisation
ASV (air-to-surface vessel radar) 117
Atlantic Charter 126
Atlas 1 and 2 264
Attlee, Clement 229–30
Australia, HMAS 47, 94
Australian Army *see also* 1st Australian Imperial Force; 2nd Australian Imperial Force; 3rd Brigade; 7th Signal Regiment; Australian Women's Army Service; Central Bureau; Special Wireless Groups
  1ATF 252–53
  1RAR 240, 251
  1st Divisional Intelligence Unit 252–53
  3 RAR 236
  6RAR 255
  COIC 111, 234
  Intelligence Corps 36–38
  during the Korean War 236, 240, 245–46
  in the Middle East 298
  Special Air Service 256
  support for Sigint 106
  during the Vietnam War 251–58
Australian Commonwealth Naval Board (ACNB) 39, 87, 89, 98
Australian Criminal Intelligence Commission (ACIC) 324, 340
Australian Cyber Security Centre (ACSC)
  Australia's Cyber Security Strategy 316–18
  establishment of 291, 313, 318–19
  expansion 343
  Richardson Review on 321–23
  role and partnerships 315, 319–20
Australian Defence Force *see also* 1st Commonwealth Division; Australian Army; Combined Operations Intelligence Centre; Defence Department (Australia); Royal Australian Air Force; Royal Australian Navy; South West Pacific Area General Headquarters
  ASD and 312–13
  Boer War involvement 30–32
  Commonwealth Military Forces 35
  JEWOSU 283
  in the Middle East 280, 297–99
  Pacific involvement 294–97, 330–31
  peacekeeping missions 296
  WWI Sigint 56
Australian Defence Satellite Communications Station (ADSCS) 304–5
Australian Federal Police (AFP) 323–24, 340
Australian Federation 29–30, 35
Australian Flying Corps 59, 116
Australian Geospatial-Intelligence Organisation (formerly Defence Imagery and Geospatial Organisation) 234, 236, 302–3
Australian Imperial Force (AIF) *see* 1st Australian Imperial Force; 2nd Australian Imperial Force
Australian Labor Party *see* Labor Party, Australian
Australian National University (ANU) 324
Australian Naval and Military Expeditionary Force (ANMEF) 47
Australian Security Intelligence Organisation (ASIO)
  formal reviews 268–69
  independence of 245
  move to Canberra 273
  origins of 229–30
  Petrovs and 245
  PJCIS 302–3
  role of 340
  staffing 312
Australian Security Intelligence Service (ASIS) 234, 270, 300–303
Australian Sigint Liaison Officer Washington (AUSLOW) 244
Australian Siginters, traits of 333–36
Australian Signals Directorate (ASD) *see also* Australian Cyber Security Centre; Burgess, Mike; Noble, Rachel
  computers 178, 265
  control of communications units 224–25, 274
  digital strategies 289–91, 313, 315–16, 319–20, 324–30
  East Timor and 292–97
  facilities 225, 274, 303–5
  formal reviews 268–73, 275, 321–22, 326
  formation and evolution 2, 75, 103, 223–24, 227, 234–35, 247, 311–13
  international relationships 243–44,

434 REVEALING SECRETS

246–47, 259, 301–3
move to Canberra 273–74, 290
organisational hierarchies 244, 314–15
peacekeeping missions 296
PJCIS 302–3
staffing 244, 247, 274–75, 286, 305–6, 311–13, 324
statutory foundations 301–3, 328
Vietnam War role 252, 254
work culture 335
Australian Theatre Joint Intelligence Centre (ASTJIC) 234
Australian Transactions Reporting Centre (AUSTRAC) 340
Australian Women's Army Service (AWAS) 123, 178–80
Australian Women's Flying Corps 121
Australia's Cyber Security Strategy 316–18, 326
Australia's First Spies 94
Australia's International Cyber and Critical Technology Engagement Strategy 326–27
Australia's National Intelligence Community 233–36, 314–15, 338, 343–44 *see also* Australian Federal Police; Australian Security Intelligence Organisation; Australian Security Intelligence Service; Australian Signals Directorate; Australian Transactions Reporting Centre; Defence Intelligence Organisation; Home Affairs Department
Austria's black chamber 14, 23
AWAS (Australian Women's Army Service) 123, 178–80
Axis powers 110, 128 *see also* Germany during WWII; Japan during WWII

Babbage, Charles 23
Babington Plot 13–14
Baldwin, Stanley 78–79
Balibo Five 6, 292–93
Ball, Desmond
　on East Timor 293–94
　on Iraq 280
　on SBRS 304
　on scope of Sigint 306–7
　on self-sufficiency in Sigint 220
　on UKUSA agreement 243
　on US-Australian facilities 276–79, 309
Ballard, Geoffrey 208
Barker, Geoffrey 284–85
Barnes, Harold 97–98
Barratt, Paul 293
Bartomello Colleoni 111

Barton, Edmund 35
Battle of the Bismarck Sea 202
Baudot code 132
B-Dienst 110
Bean, C.E.W. 7
Beatty, David 51
Beazley, Kim 277–80, 285
Bell Telephone Laboratories 174
Bellamy, Craig 118
Betrayal at Pearl Harbor 138
Birch, Frank 90, 106
black chambers 12–13, 328
　Allied 109. *see also* Allied...; Central Bureau; Fleet...; Special Intelligence Bureau
　Australian 283, 302–3. *see also* Australian Criminal Intelligence Commission; Australian Cyber Security Centre; Australian Geospatial-Intelligence Organisation; Australia's National Intelligence Community; Combined Operations Intelligence Centre; Defence Intelligence Organisation; Special Intelligence Bureau
　Austrian 14, 23
　Canadian 120, 231–32
　Dutch 136, 206
　French 12–13, 17, 23–24, 69
　German 110
　NZ 266–67
　Russian 19, 23–24, 34, 70–71, 212–13, 229–30
　UK 13–14, 21, 29, 31, 60, 96, 158–60. *see also* GCHQ; London Signal Intelligence Board; MI...; Wireless Experimental Centre
　US 241, 278. *see also* Cipher Bureau; MI-8; National Security Agency; Radio Intelligence Section; Signal Intelligence Service

Blarney, Thomas 104, 193, 210
Blencowe, William 14
Bletchley Park, UK 113, 132
Blick, Bill 295, 303
Boer War (Anglo–Boer War) 30–32
Bolshevik Revolution 70–71
Bombe machines 130–31, 150
Bonaparte, Napoleon 17–20
Bonegilla, VIC 155–57, 180
Bonighton, Ron 274–75
Booth, Henry Roy 163
Borneo *see* Confrontation
Boye-Jones, Ruby 5, 99–100
Brady, Martin 289–90, 301

Brest Litovsk, Treaty of 69
Bridges, William Throsby 36–37
Brisbane, QLD 164, 180
Britain *see also* British Armed Forces;
  BRUSA Agreement; Foreign Office (UK);
  South East Asia Command
  1844 Deciphering Branch closure 21
  American Revolutionary War 16–17
  Anglo–Boer War 30–32
  Anglo–Japanese Alliance 88
  Australian defence relationship 30, 32,
    71–73, 80, 103–6, 108–10, 144–45,
    160–61, 211, 219–20, 225–30, 238–
    39, 242–43, 245–49, 259, 336–37
  Babington Plot 13–14
  black chambers 13–14, 21, 29, 31, 60, 96,
    158–60 *see also* GCHQ; MI1b; MI5;
    MI6; Wireless Experimental Centre
  Confrontation 248–51
  Crimean War 22–24
  cryptanalysis 60, 66
  empire cable networks 25, 40
  empire decline and Atlantic Charter
    126–27
  empire defence arrangements 29–30, 40,
    42–43, 101, 116–17
  empire Sigint and Ultra security
    regulations 160–62
  Far East Strategic Reserve 246–47
  intelligence accountability 3
  during the Korean War 238, 245–46
  Malayan Emergency 247
  Napoleonic Wars 17–20
  relationship with Russia 34–36, 70–71,
    77–79, 229–30 *see also* Venona
    program
  relationship with US 126–28, 144–45,
    154–55, 160–61, 229–30, 238–39,
    243, 247–48 *see also* Venona program
  relationships with Allies 275
  routine intelligence collection and
    sharing 27
  Suez Crisis 247–51, 259
  WWI, Russian collaboration during 70
  WWI Sigint evaluated 70–72
  WWII, Central Bureau role 189–92
  WWII, support for D Special section
    109–10
  WWII ABC-1 talks 134
  WWII Cocos Island disinformation
    194–95
  WWII collaboration with US 126–28,
    134–36, 144–45, 154–55, 160–61
  WWII declaration 103–4
  WWII intelligence security regulation
    158–61, 211
  WWII Sigint evaluated 114–15
  WWII Sigint sharing 144–45
British Armed Forces *see also* 1st Australian
  Imperial Force; 2nd Australian Imperial
  Force; Anglo–Boer War; British Army;
  Interpreter Operators; Royal Air Force;
  Royal Navy; War Office
  ABDA Command 140
  cipher machines 80
  Corps of Guides 18
  Far East Strategic Reserve 246–47
  pre-WWI Sigint development 40–41
  radio interception 75
British Army 32, 58, 60, 68–69 *see also*
  British Expeditionary Force; Squires,
  Ernest; War Office
British Commonwealth Far East Strategic
  Reserve 246–47
British Expeditionary Force (BEF) 43,
  53–56, 60–61, 66 *see also* 1st Australian
  Imperial Force; 2nd Australian Imperial
  Force
British Phosphate Commission 97
British Tabulating Machines (BTM) 130–31,
  175
Brooks, Thomas A. 127
Brown, Jack 201, 220
BRUSA Agreement 154–55, 226–27, 243–
  44 *see also* Five Eyes/UKUSA Agreements
Buchanan, USS 267
Burgess, Mike 312, 325, 327, 333
Burley, Sidney F. 161–62
Burma 144, 207–8
Burns Philp Pty Ltd 35

C130 aircrafts 282
cabinet noir *see* black chambers
Caesar, Julius 9
Cairo 115
Canada *see also* Canadian Army; Five Eyes/
  UKUSA Agreements
  CANUSA agreement 231–33
  interwar lack of Sigint 74
  WWII black chambers 120, 231–33
  WWII talks and collaboration 134, 151
Canadian Army
  Canadian Expeditionary Force (CEF) 43
  Special Wireless Group 207
Canberra, ACT
  leaks from 209–11, 229–30, 334–35
  Petrov affair 245
  Sigint agencies' relocation to 273

Canberra, HMAS 94
CANUSA agreement 231–32
Cartier, François 65–66
CAST (USN detachment) 84, 136, 148
Catalina flying boats 182–83
cell phones 288, 297–98 *see also* digital communications
Central Bureau *see also* Special Wireless Groups
  closure and legacy 211
  consolidation 218
  D Special Section and 160, 223
  formation of 163, 164
  intelligence security role 169
  international relationships and liaison 161–62, 170–72, 175–80, 189–93, 208, 334
  isolation of 149, 189–93
  Japanese communication cryptanalysis role 169–71, 179, 188–91, 203–5
  organisational layout 166–67, 170–72, 179–81, 189, 345–49
  relocation 178–79
  staffing 145, 153, 157, 164–69, 192–93, 208
  technology 172–78
  Training Group 180
  women and 178–79, 181
Central Photographic Establishment (CPE) 234
Central War Room 111
chaff strategy 118
Charles I, King 14
Chauvel, Harry 66
Chieffriermachinen AG company 79
Chifley, Ben 219, 227, 229–30, 335
Chile 47–48
China
  Allied information leaked via Chinese Nationalist channels 209
  ancient cryptology 10
  Chinese Embassy bugging 305
  Chinese telegraph code 25–26
  control of 5G networks 319–20
  East Timor and 291
  Japanese aggression during WWII 104
  during the Korean War 236, 240–41
  relationship with Russia 241
  during the Vietnam War 252
Churchill, Winston 126, 134, 208
Cipher Bureau (US) 81–82
cipher disks 11
cipher machines 79–80, 128–30, 151, 173–74 *see also* Bombe machines; electronic computers; Hagelin machines; Heath Robinson machines; Lorenz machines; 'Purple' material and machines
ciphers *see* codes and codebreaking
Civil War, English 14–15
Civil War, US 24–25
Claret 250
Clark, A.B. 174
Clark, S.R.I. 'Pappy' 207
Clarke, Richard 328
Clausewitz, Carl von 20
Clauson, Gerard 68
Coast Guard (US) 82–83
Coastwatchers 5, 87–88, 99–100
Cobra, Operation 215–17
Cocos (Keeling) Islands 48–49, 194
Codd, Michael H. 300
codes and codebreaking *see also* black chambers; cipher machines; Holden Agreement; Morse Code; signals intelligence; Venona program; Zimmerman telegram
  1300s–1600s Europe 11–14
  1700s Europe 14–16
  1800s Europe 17–20, 22–24
  American Revolutionary War 16–17
  ancient Eastern 10–11
  ancient Western 8–10
  British 13–16, 18–20, 60–61
  Chinese telegraph code 25–26
  female codebreakers 124–26
  German codes 65–66
  Hypo station 194
  Japanese codes 158, 188–89, 203–5 *see also* Four Winds message decryption; JN-25; 'Purple' material and machines
  Napoleonic Wars 17–20
  place in Sigint 85
  substitution ciphers 9, 11–12, 23
  transposition ciphers 9, 188
  Vienna Congress 20
  Vietnamese codes 253–55, 258
  wireless radio integrates Sigint aspects 34
COIC (Combined Operations Intelligence Centre) 111, 234
Cold War *see also* Gulf War; Korean War; Venona program
  Australian agencies during 233
  East Timor and 291
  Malayan Emergency 247
  monitoring and leaking of Allied information 209–13, 229–32, 334–35
Colegrave, E. 152

Collins, John 110–11
Collins submarines 285–86
Colossus computers 132–33
Colvin, Ragnar 105
Combes, Bertrand 226–27
Combined Operations Intelligence Centre (COIC) 111, 234
Commander-in-Chief Far East (CINCFE) 140
Committee for State Security (KGB) 229–30, 245
Commonwealth *see* Britain
Commonwealth Division 236–43, 245–46
Commonwealth Military Forces 35
communications intelligence (Comint) *see* digital communications; OP-20-G; signals intelligence
Communications Security Establishment of Canada (CSE) 232
communism 76–79 *see also* Cold War; Russia
Comprehensive Review of the Legal Framework of the National Intelligence Community 7, 321–24, 343
Computer Emergency Response Teams (CERT) 291
computers *see* technology
Concerning the Solution of Ciphers 11
Confrontation 248–51
Conradus, David Arnold 15
continuous wave radio 33, 81
Cook, Ralph E. 148
Coonawarra, HMAS 98, 225
Coral Sea, Battle of the 194–97
Coronel, Battle of 48–49
Corps of Guides and Intelligence Corps 36–38
Corregidor Island 147
Coxeter, Harold 120
Cray supercomputers 264–66
Creed Relay (wireless recorders) 92
Creswell, William R. 94
Crete 112–14
crime 322–25, 327–28 *see also* Australian Federal Police
Crimean War 22–24
cryptanalysis *see* codes and codebreaking
Cryptography Unmasked 15
CSE (Communications Security Establishment of Canada) 232
Curtin, John 144, 209
Curzon, Lord 77–79
Cutlack, F.M. 59
Cyber Security Operations Centre (CSOC) *see* Australian Cyber Security Centre
cyberspace *see* digital communications
cyphers *see* codes and codebreaking

D Special Section *see also* Holden Agreement
  formation of 107–10
  intelligence security regulations 159
  Japanese Sigint decryption 108–10, 136–39, 155–58
  oversight 156–57, 159–60, 211, 223
  Russian traffic interception 211
  staffing 107–9, 125
Daesh 298–99, 325
Daily Herald 77–78
Daily Telegraph 321
Darwin, NT 98, 225, 303–4
De Componendis Cifris 11
de Grey, Nigel
  on Australian initiative 171
  on intelligence security regulation 172
  on Nave 108
  on traffic analysis 4–5, 86, 188
  Zimmerman telegram decryption 62–63
deciphering *see* codes and codebreaking
Defence Department (Australia) 118, 224, 272–73, 313 *see also* Information Warfare Division
Defence Imagery and Geospatial Organisation (DIGO) *see* Australian Geospatial-Intelligence Organisation
Defence Intelligence Organisation (DIO) (formerly Joint Intelligence Bureau) 227, 234, 236, 247, 302–3
Defence Signals Bureau (DSB) *see* Australian Signals Directorate
Defense Satellite Communications System (DSCS) (US) 278
Denniston, Alastair 78–79, 151
Department of Defence (Australia) 118, 224, 272–73, 313 *see also* Information Warfare Division
Department of External Affairs (Australia) 35, 211, 229–30
Department of Home Affairs (Australia) 323–24, 340–41
Department of Naval Intelligence (RAN) 38–40, 87
Department of the Prime Minister, Australia's Cyber Security Strategy 316–17
Department of War (US) 65, 190
Derby, Charles 183
Desch, Joseph 131
Deuxieme Bureau 69
Dictaphone 92–93

digital communications *see also* North West Cape facility, WA; Nurrungar facility, SA; Pine Gap facility, NT
   5G regulation 319–21
   ASD and ACSC adaptations 289–91, 301–5, 313–30, 337–38, 343
   Australia's International Cyber and Critical Technology Engagement Strategy 326–27
   commercial and outsourced intelligence 341–42
   complicating Sigint procedures 262–63
   development of 288
   Echelon network 301–2, 304–5
   future of 336–44
   international partnerships in 342–43
   laws relevant to 300–303, 306–7, 312, 318, 328
   Middle Eastern Sigint and 297–98
   Quad Senior Cyber Group 328
   transparency and accountability in 306–7, 313–16, 322–24, 338–41
   US strategy 301–2, 304–5, 328–29
Dili, East Timor 156, 215
Diplomatic Special Section *see* D Special Section
direction finding and radar
   Australian interwar 85
   RANTEWSS 284–86
   Special Operations Task Group 297–98
   Vietnam War technology 253, 255–56
   wireless radio integrates Sigint aspects 34
   WWI RAN initiatives 89
   WWII 95, 98, 110, 117–18, 175, 181–85, 223–25
Director of Borneo Operations (DOBOPS) (UK) 250
Director of Military Intelligence (DMI) (Australia) 156–59, 163
Director of Naval Intelligence (DNI) (Australia) 104
Director of Naval Intelligence (DNI) (UK) 75, 96–99, 136, 241
Director of Naval Operations (DNO) (UK) 51
Directorate of Military Intelligence (DMI) (UK) 29, 53, 158–59
Distributed Ground Station – Australia 283
Dollis Hill 132
Donovan, William 'Wild Bill' 127–28
Drea, Edward 149, 170
Dudgeon, Ian 289–91
Duke of Wellington 18–19
Dumas, Charles William Frederic 16–17

Dutch East Indies 136, 140, 146, 178, 206

East Timor 291–97, 330–31 *see also* Balibo Five; Operation Cobra
EATS (Empire Air Training Scheme) 117–18
Echelon network 301–2, 304–5
ECM and ECCM (electronic countermeasures and counter-countermeasures) 2, 117–18, 181–83, 284, 297–98
economics 319–21
Eden, Anthony 247–48
Egypt *see* Suez Crisis
Eisenhower, Dwight D. 248
El Alamein, Battle of 115–16
electricity, discovery of 21
electronic computers 173, 264–66, 291 *see also* digital communications
electronic countermeasures and counter-countermeasures 2, 117–18, 181–83, 284, 297–98
electronic intelligence (Elint) 118, 284
   see also direction finding and radar; electronic countermeasures and counter-countermeasures; signals intelligence
electronic warfare (EW) *see also* electronic countermeasures and counter-countermeasures
   72 Electronic Warfare Squadron 295–96
   RAAF systems 282–83
   RANTEWSS 284
   role of Sigint 2
   Special Operations Task Group 297–98
Elizabeth I, Queen 13
Ellwood, A. J. 214–15
Elwell, Cyril 33
Emden, SMS 48–49
Empire Air Training Scheme (EATS) 117–18
English Civil War 14
Enigma machines 79–80, 84, 128–30, 151
   see also Bombe machines; Lorenz machines; Ultra reports
equipment *see* technology
ESM (electronic support measures) *see* electronic intelligence
espionage *see* intelligence
Europe 150, 155–56 *see also* World War ...
Evatt, H.V. 'Doc' 211, 229–30
Examination Unit of the National Research Council 231–32
External Affairs, Department of (Australia) 35, 211, 229–30

Fabian, Rudolph 148–49, 152–53

Fabyan, George 65
Fahey, John 94
Far East Combined Bureau (FECB)
  collaboration with US 135–36, 150
  decline of 145–47
  interception and DF stations 97–98
  Japanese codes and 97, 99, 135–36, 150, 206
  Kamer 14, 206
  Nave in 108, 110
  origins of 77, 95–96
  relocation to Hong Kong 104
  SEAC assignment 144
Far East Strategic Reserve 246–47
Federation, Australian 29–30, 35
Fellers, Bonner 115–16
'Ferrets' 182, 184–85
Ferris, John 43, 54–56, 246, 248, 259
Fetterlein, E.C. 70, 77–78
The Fifth Domain 328
Findlay, A. 237–38
First World War *see* World War, First
'Fish' devices 129, 131–33 *see also* Enigma machines
Five Eyes/UKUSA Agreements *see also* BRUSA Agreement; *Echelon* network
  ASD and ACSC under 247, 259, 269, 342
  Australian deployments under 247, 297
  benefits and disadvantages 226, 331, 336–37
  Canada's membership 231–33
  cyber security and 342
  GCHQ-NSA relationship 248
  Huawei and 320
  Tripartite Conference 243–44
flag vocabulary 16, 19
Fleet, A. James Van 239–40
Fleet Radio Unit Melbourne (FRUMEL)
  Battle of Midway 194, 197–200
  Battle of the Bismarck Sea 202
  Battle of the Coral Sea 194–97
  consolidation 218
  D Special Section and 156–57
  death of Yamamoto and 203
  equipment 147–49
  formation of 147–48
  intercept stations 149
  moves focus to naval communication 164
  personal tensions affecting 148–49, 152–53
  staffing 145, 147–48, 152–54, 237
Fleet Radio Unit Pacific (FRUPAC) 148
Flood, Phillip 303

Flowers, Tommy 132–33
Foreign Office (UK) 14, 75 *see also* GCHQ
Four Winds message decryption 136–39
France
  1593 flag vocabulary 16
  adopts wireless radio 34
  black chambers 12–13, 17, 23–24, 69
  Napoleonic Wars 17–20
  Suez Crisis 247–48
  in Vanuatu 35–36
  WWI cable control 45
  WWI code quality 34
  WWI monitoring of Germany 54, 58–60, 65–66, 69–70
Franklin, Benjamin 21
Franklin, SY 93
Fraser, J. Malcolm 271–72, 301
Freakes, H.J. 95
Fremantle, WA 98
frequency principle 10–11, 18
Frewen, John 312
Freyberg, Bernard 113–14
Friedman, Elizebeth Smith 65, 83
Friedman, William F.
  cipher machine development 128, 173–74
  Sigint teaching 65, 83–84
  Sinkov mission 134–35
Fry, Ken 292
Frydenberg, Josh 325
Fuji cypher 158

G2-A6 64–65, 171
GCHQ (formerly GC&CS) *see also* BRUSA Agreement; Far East Combined Bureau; Hut 6
  areas of mastery 76
  armed services sections 75
  Cryptography and Interception Committee 76
  decryption of Russian communication 78
  intelligence security regulation 159–61, 172
  interception stations 76–77, 97
  interwar decryption 78–79
  monitors Japan 91, 97, 158, 190–92
  organisational collaboration 74–76, 90, 190–91, 243–44, 246
  origins of 74–75
  relationship with Australia 74–75, 105–6, 159–61, 166, 172, 207, 243–44, 246
  relationship with US 127–28, 134–35, 243–44, 248
  staffing 91, 130, 132, 153–54

technology 83, 130–33, 264
Tripartite Conference 243–44
WWII IJN communications decryption 90, 97
WWII successes and failures 76
Geheime Kabinet-Kanzlei 14
Geheimschreiber devices 129, 131–33 *see also* Enigma machines
gender rights *see* women
General Staff, Military Intelligence Division 64–65, 171
George I, King 14
geospatial intelligence (Geoint) 24, 234, 236, 295, 299, 302–3
Geraldton, WA 303–5
German Air Force 58–60, 117
German Army, Battle of El Alamein 115–16
German Navy during WWI
   Allied direction finding and 47–51
   Battle of Coronel 48
   Battle of Jutland 50–52
   codes 45–47
   Emden's defeat 48–49
German Navy during WWII
   Allied direction finding and 110
   B-Dienst's cryptanalysis 110
   sinks HMAS *Sydney* 111
   technology 117
Germany *see also* Enigma machines; Germany during WWI; Germany during WWII; Lorenz machines
   early 1900s Sigint 33–34
Germany during WWI *see also* German Navy during WWI
   Battle of the Somme 55–56
   capture of German New Guinea 47
   codes 45–47, 79
   German Air Force Sigint 58–60
   interception of and by 54–58, 61
   merchant ships 45–47
   WWI codes 53–54, 63–66, 69
   WWI deceptive Sigint 61
Germany during WWII *see also* Enigma machines; German Navy during WWII; Lorenz machines; Zimmerman telegram
   2nd AIF confrontations 104–5
   Adelklassen attempt to assassinate Hitler 156
   Battle of El Alamein 115–16
   Battle of Kursk 133
   Britain declares war on 103
   cable destruction and 45
   codes 128–30, 135, 150–51
   in Crete 113–14

   detects Allied shipping 80
   German Air Force Sigint 117
   Hut 6 monitors 113–14
   invasion of Russia 211–12
G-H systems 117
Ghormley, Robert 128
Gilding, Simeon 319–20
Gilling, Tom 278
Goble, Stanley 105
Gouzenko, Igor 231–32
Government Communications Headquarters (formerly Government Code & Cypher School) *see* GCHQ
Government Communications Security Bureau (GCSB) 266–67
Great Britain *see* Britain
Great Paris Cipher 18
Greece 9, 112
Gregory, Ancell C. 88
GRU (Main Intelligence Directorate) 212–13
Guadalcanal, Solomon Islands 5
Guam Doctrine 259
Gulf War 280, 287

H2S devices 117
Habibie, B.J. 293
Hagelin machines 79–80, 84
Hallett, H.A. 183
Halsey, Admiral 5
Handelsschiffsverkehrsbuch (HVB) 46–47
Hanyok, Robert J. 138–39
Harbin, China 209–10
Harman, HMAS 98, 122–23, 125
Hastie, Andrew 324
Hawaii 148
Hawke, Bob 278–79, 285
Headquarters Joint Operations Command 286
Heath Robinson machines 132–33
Hebern, Edward 173
Hercules aircrafts 282
Heron UAVs 298
high frequency direction finding (HFDF) 95, 98, 175
Hinsley, F.H. 'Harry' 5, 76
Hiroshi, Oshima 135
Hitler, Adolf *see* Germany during WWII
Hobart, SS 46
Holden Agreement 150–54, 156–57, 334
Hollandia 178
Hollerith, Herman 28–29 *see also* IBM
Hollis, Roger 229
Home Affairs Department (Australia) 323–24, 340–41

Hong Kong 246, 259
Hooper, Dr 191–92
Hoover, Herbert 81
Hope, Robert
  on ASD 268–73, 292, 326, 335, 340
  on ASIO 269
  on ASIS 270
  demarcation of Sigint organisations 340–41
  on women in Sigint 125
Horner, David 228
Howard, John 293
Howe, Richard 16
Huawei 320
Hughes, Billy 123
Huguenots 13
human intelligence (Humint) 82–83, 295, 299 *see also* Coastwatchers
Hunt, Atlee 35
Hussein, Saddam *see* Gulf War
Hut 6 (Bletchley Park) 113, 132
Hutton, Edward 35
Hypo stations 194

Iberian Peninsula 18–19
IBM (International Business Machines)
  adoption in cryptology 83–84
  ASD receipt of 178
  capabilities 176
  Central Bureau and 175–78
  FRUMEL and 147–49
I(e) (cryptanalysis organisation) 60, 66
IJA *see* Imperial Japanese Army
IJN *see* Imperial Japanese Navy
imagery intelligence (Imint) *see* geospatial intelligence
Imperial German Navy *see* German Navy during WWI
Imperial Japanese Army (IJA) 84–85, 163, 191, 203–5
Imperial Japanese Navy (IJN) *see also* Pearl Harbor attack
  Battle of Midway 194, 197–200
  Battle of the Bismarck Sea 202
  Battle of the Coral Sea 194–97
  Cocos Island bombing 194
  codes 84–85, 91, 94, 99 *see also* JN-25
  Combined Fleet 195–96, 198–200
  mapping military elements in Japanese Mandated Territories 93–95
Independent Intelligence Review 275, 307, 317
India 190–91 *see also* South East Asia Command

Indonesia *see also* Dutch East Indies
  Australian monitoring of presidential communications 306–7
  Confrontation 248–51
  East Timor and 291–95, 330–31
  Information Operations concept 291
  Information Warfare Division 317
  Information Warfare Wing (RAAF) 283
Ingersoll, Royal 127
Inspector-General of Intelligence and Security (IGIS) 302–3
intelligence *see* communications intelligence; electronic intelligence; geospatial intelligence; human intelligence; interception; signals intelligence
Intelligence Corps 36–38
Intelligence Services Act 2001 300–303, 307, 318
interception *see also* black chambers; IToc stations; Moritz stations; Special Wireless Groups
  1800s British interception practices 21
  Australian interception during the Korean War 238, 240
  Australian interception during the Vietnam War 253–55, 258
  Australian interception in the Middle East 297–98
  Britain advises empire on 100–101
  of British traffic 61
  Canadian stations 231
  by Central Bureau 166–67, 169
  by D Special Section 211
  by FECB 97–99
  by FRUMEL 149, 195–200, 203
  by GC&CS 76, 97–98
  by Germany 57–58
  of Japanese traffic 90–95, 99, 166–67, 169, 194–200, 203, 209–11
  nature of interception in WWI 43
  by RAN 94–95, 100–101
  by RN 90–92
  role in Sigint 2
  of Russian traffic 57–58, 77–78, 209–11
  stations 44, 64–65, 75, 97–99, 194, 223–25
  technology assisting 22–24, 92–93, 297–98
  by USN 194
  Venona program 212–13, 229–30
  wireless radio integrates Sigint aspects 34
INTERFET (International Force East Timor) 294–97

internet *see* digital communications
Interpreter Operators 56, 66
Iraq
　Gulf War 280, 287
　Operation Okra 297–99
　in WWI 67–69
ISIS/ISIL (Islamic State in the Levant/Syria) 298–99, 325
Israel 247
Italy 110, 128
IToc stations 54–57

JAA *see* 'Purple' material and machines
Jackson, Thomas 51
James, William Milbourne 95–97
Jamieson, A.B. 'Jim' 206
Jandakot, WA 98
Japan *see also* Japan during WWII
　Anglo–Japanese Alliance 88
　codes 81–82, 89, 91, 99–100
　interception by 97
　interception of messages from 90–97, 108, 110
　Japanese language courses 119–20
　RAN and 89–90
　Russo–Japanese War 33
　Washington Disarmament Conference 81–82, 88, 91
Japan during WWII *see also* Pacific War; Pearl Harbor attack
　aggression in China 104
　attack on *Panay* 126
　codes 83–85, 108–10, 118–20, 155–56
　　see also 'Purple' material and machines
　partition of Korea 236
　surrender 222
Japanese Mandated Territories 93, 95
Jayapura, Indonesia 178
Jellicoe, John 51, 87–89
Jiangyong, China 11
Jiaozi currency 11
JN-25 (main Japanese fleet cipher) 151, 194, 203
John XXII, Pope 11
Joint Defence Facility Pine Gap *see* Pine Gap facility, NT
Joint Discrimination Unit 232
Joint Electronic Warfare Operational Support Unit (JEWOSU) 283
Joint Intelligence Bureau *see* Defence Intelligence Organisation
Joint Intelligence Committee 213, 227
Joint Planning Committee 213
Joint Telecommunications Unit Melbourne (JTUM) 273–74
Jose, Arthur W. 49–50
Jutland, battle of 50–52

Kamer 14 136, 206
kana alphabet 89, 175, 202, 204
Kasiski, Friedrich 23
Katakana Man 201–3
Keegan, John 49
Keeling (Cocos) Islands 48–49
Kellogg-Briand pact 81
KGB (Komitet Gosudarstvennoy Bezopasnosti) 229–30, 245
Kilindini, Kenya 147
Kim Il Sung *see* Korean War
Kitchener, Lord 31
Knake, Robert 328
Kojarena, WA 304
Konfrontasi 248–51
Korean War 236–43, 245–46
Kormoran, HSK 111
Kruger, President 32
Kuibyshev (Samara), Russia 156
Kuwait, invasion of 280, 287

Labor Party, Australian 144, 209, 245, 278–80, 340 *see also* Beazley, Kim; Chifley, Ben; Evatt, H.V. 'Doc'; Hawke, Bob; Smith, Stephen; Whitlam, Gough
Lagarto, Operation 214–17
Lake, Margaret Ethel 119–20
Lamsdorff, V.N. 34
Lange, David 267
Le Carre, John 333
Le Couteur, Wilson 35
Lehmann, Mick 297
Liberal Party 293, 316–18, 320, 323 *see also* Fraser, J. Malcolm; Menzies, Robert
Liberator (B-24) aircraft 182–83
Little, Robert 153, 156–58, 164, 223
Lockheed P2V Neptune aircrafts 281–82
London Signal Intelligence Board (LSIB) 158–60
Long, Gavin 8
Long, Rupert 'Cocky' 104–5
Long Tan, Battle of 254–55
loop and pin technique 55
Lorenz machines 129, 131–33
Louis XIV 15
Ludendorff, Eric Von 61
Lyons, Richard J. 107

M-209 79
MacArthur, Douglas *see* South West Pacific

Area General Headquarters
MacLean, Donald 284
'Magic' reports 144, 161 *see also* 'Purple' material and machines
Main Intelligence Directorate (GRU) 212–13
Malaya
    Confrontation 248–51
    Far East Strategic Reserve 246–47
    Malayan Emergency 247
Maleme airfield 113
Manila, Philippines 193, 200
Manoora, HMAS 110
Marconi, Guglielmo 27, 81
Mary, Queen of Scots 13
Mauborgne, Joseph O. 16
Maude, Stanley 67–68
McCay, James Whiteside 36–38
McEwen, John 123
McKenzie, Florence Violet 121–23, 125
McKenzie, Ian 313
McLaughlin, W.E. 94
Mediterranean Sea 44, 93, 110, 241
Meekin, Steve 168, 339
Melbourne, Victoria 224, 243–44, 303 *see also* Fleet Radio Unit Melbourne; Joint Telecommunications Unit Melbourne
Menzies, Robert
    1949 election 230
    espionage commission 245
    Suez Crisis and 247
    during WWII 103–6, 109
Menzies, Stewart 127–28, 134
Merchant, Stephen 275
merchant ships 45–47
Mesopotamia 67–69 *see also* Iraq
Metox search-receivers 117
Metternich, Prince 20
Mexico 62–64
MI1b (formerly MO5B) 44, 53, 58, 61–62
MI1e 58–59
MI5 (formerly MO5) 41, 76
MI6 76, 96
MI-8 (US Codes and Cipher Section) 64–65, 171
Michael Offensive 61
Middle East 67–69, 104–5, 112–13 *see also* Afghanistan; Daesh; Iraq; Suez Crisis
Midway, Battle of 194, 197–200
Military Intelligence Division 64–65, 171
MO5 (now MI5) 41, 76
MO5b (now MI1b) 44, 46–47, 53, 61–62
mobile phones 288, 297–98 *see also* digital communications

Monash, John 37, 66
mono-alphabetic ciphers 9
Monterey Apartments *see* Fleet Radio Unit Melbourne
Montgomery, General Bernard 116
Moreton base 125
Moritz stations 54–56
Morrison, Scott 323
Morse, Samuel 21–22
Morse Code *see also* telegraphy
    Chinese telegraph code and 25–26
    Coastwatchers and 99
    invention of 21–22
    mechanics of 21–22
    training in 89, 92, 121–23
    WWII monitoring of Japan 169–70
Moscow *see* Russia
Mount Olympus 112
movable type printing 10–11
Mowry, David P. 138–39
Murray, Peter 255–56
Myanmar *see* South East Asia Command

Nagara 199
Nairnville Park, NZ 155
Nakasone, Paul 328–29
Napoleonic Wars 17–20
NASA (National Aeronautics and Space Administration) 265
Nasser, Gamal Abdel *see* Suez Crisis
National Cash Register Company (NCR) 131
National Intelligence Community *see* Australia's National Intelligence Community
National Research Council 232
National Security Agency (NSA) (formerly Armed Forces Security Agency) *see also USA Freedom Act 2015*
    international collaboration 243, 246, 248, 267
    origins of 240
    technology and 264, 328
    Vietnam War and 255
Nauru 97–98
Naval Communication Station Harold E. Holt 275–77
Naval Intelligence Department (NID) (RAN) 38–40, 87
Naval Intelligence Directorate (NID) (RN) 97 *see also* Director of Naval Intelligence (DNI) (UK); Room 40
Nave, Eric
    Central Bureau role 157, 164–66, 168

death 169
Four Winds message decryption 136–39
GC&CS work on IJN Sigint 91
Holden Agreement and 150–54, 168
Newman and 146–47
RN and RAN work on IJN Sigint 90–92, 94, 107–8
Sigint conference 206
Special Intelligence Bureau and D Special Section 107–10, 118, 136–39, 334
Washington Disarmament Conference 91
navies *see* German Navy ...; Imperial Japanese Navy; Royal Australian Navy; Royal Navy; United States Navy
Nazism *see* Germany during WWII
Nelson, Horatio 16, 19
Neptune aircrafts 281–82
Netherlands East Indies 136, 140, 146, 178, 206
New Guinea 47, 93, 99–100, 181
New Hebrides, Australian espionage in 35
New South Wales Police Network 319
New Zealand *see also* Five Eyes/UKUSA Agreements; New Zealand Army
GCSB 266–68
links to Australia and US 242, 246–47, 267–68, 330
RNZN 98
New Zealand Army 113–14, 155 *see also* ANZAC Wireless Signal Squadron
Newman, Jack 122, 146–47, 152, 197, 206
Newman, Max 132
Niche Wars 297
NID *see* Naval Intelligence ...
Nimitz, Chester 148–49
Nixon, Richard 259
NKVD (People's Commissariat for Internal Affairs) 212–13
Noble, Rachel 312, 327, 333
North Korea *see* Korean War
North Vietnamese Army *see* Vietnam War
North West Cape facility, WA 275–77
Northcott, John 158–59
Nosov, Feodor 211
NSA *see* National Security Agency
Nurrungar facility, SA 275–76, 279–80
Nushu script 11
Nye, Joseph 315

Oberon submarines 284–86
Ocean Surveillance Information System (OSIS) 286

Office of National Intelligence (ONI) 302–3
Office of Naval Intelligence (US) 65, 241
Official History of British Intelligence in the Second World War 5
Official Secrets Act 1920 75
Okra, Operation 298–99
Olympus, Mount 112
one-time pads 78–80
OP-20-G (US Navy Sigint unit) 83
Operation Cobra 215–17
Operation Lagarto 214–17
Operation Okra 298–99
Operational Intelligence Centre 97
Orange Free State, war in 30–32
Orion aircrafts 282, 294
Orione 45
Ottawa, Canada 207, 231
Ottoman empire 4–5, 67–68

P2V Neptune aircrafts 281–82
P3-B and P3-C Orion aircrafts 282, 294
Pacific, Australia's relationship with 331
Pacific War *see also* BRUSA Agreement; Central Bureau; Far East Combined Bureau; Holden Agreement; Operation Cobra; Operation Lagarto; Pearl Harbor attack; South East Asia Command; South West Pacific Area General Headquarters
ABDA Command 140
Allied role evaluated 185–86
ATIS, AGS and AIB 164, 170
attack on Malaya 136
Australian Army role 105, 139, 155, 169–70
Australia's role evaluated 102, 185–86
Battle of Midway 194, 197–200
Battle of the Coral Sea 194–97
beginning of 102, 126, 135–36
Coastwatchers 5, 87–88, 99–100
Cocos Island bombing 194
FRUMEL role 147–49, 164, 179, 197–200, 202
FRUPAC role 148–49
intelligence security regulation 145, 158–61, 172
Japanese advances towards Australia 162
Japanese codes 84–85, 136, 155–58 *see also* JN-25; 'Purple' material and machines
Japanese surrender 222
Kamer 14 136
NZ role 155–56
Pacific Ocean Area 144
partition of Korea 236

RAAF role 163, 179–85
RAN, D Special Section and SIB 136–40, 146–47, 155–60, 181
RCM and Radar 181–85
Russian leaks 209–11
SLUs and SLOs 161–62
Special Wireless Groups 155–57, 162–63
UK role 135–36, 139–41, 144–47, 158–61, 172
US role 135–36, 139–41, 144–45, 147–49, 170, 172, 176–79, 181–84
women's roles during 178–81
Palestine 67
Palmer, Bruce 254
Panay 126
paper currency 11
Parliamentary Joint Committee on Intelligence and Security 302–3
Parramatta, HMAS 44–45
PBY Catalina flying boats 182–83
Peach, Stuart 238
Pearl Harbor attack 85, 138–39, 150, 202 *see also* Four Winds message decryption
Peloponnesian War 9
People's Commissariat for Internal Affairs (NKVD) 212–13
Persian Gulf War 280, 287
petits chiffres 18
Petrov, Vladimir and Evdokia 245
Pfennigwerth, Ian 39, 257
Phelippes, Thomas 13
Philippines 181, 192–93, 200
phones 174–75, 288, 297–98 *see also* digital communications
Phuoc Tuy, Vietnam 253–54
Piesse, Edmund 52
pin and loop technique 55
Pine Gap facility, NT
 ALP stances on 276–80
 development of 275–76
 DSCS 279
 East Timor involvement 294
 Iraq involvement 280
 public opinion on 277–78, 301, 330
 value of 280, 330
Pires, M. de J. 214–15
Polish codebreaking 128–30
poly-alphabetic ciphers 11–12, 23
Polygraphiae 12
Port Moresby, PNG 195–96
Portugal 214–17, 291–92 *see also* Iberian Peninsula
Poulden, J.E. 'Teddy' 228

Prime Minister's Department, Australia's Cyber Security Strategy 316–17
printing, development of 10–11
prisoners of war 200
Prohibition 82–83
Project AIR 3503 283
Project Peacemate 282
Project Rainfall 278
Prussia 20
Public Governance, Performance and Accountability Act 2013 312
'Purple' material and machines
 analogue Allied machines 84, 158
 development of 80, 83–84
 GC&CS cryptanalysis work 158
 Hiroshi's traffic 135
 intelligence security regulation 144, 161, 208
 limited use of 133, 138–39
 SIS cryptanalysis work 83–84, 128
 trade of information on 128, 134–35

Quad Senior Cyber Group 328

RAAF *see* Royal Australian Air Force
radio 262–63 *see also* direction finding and radar; wireless telegraphy
Radio Intelligence Section 64–65, 171
RAF *see* Royal Air Force
RAN *see* Royal Australian Navy
Rankin, Ronald 238
recorders 92–93
Red code 84
REDSPICE 325–30
Reform of Australia's electronic surveillance network 324
Rejewski, Marian 129
Report on Joint Intelligence Organisation 226–27
Rhee, Syngman *see* Korean War
Richards, Trevor 254
Richardson Review 7, 321–24, 343
Richelieu, Cardinal 12–13
RN *see* Royal Navy
RNZN (Royal New Zealand Navy) 98
Roberts, Caleb 163
Robinson, Gilbert 120
Rockex machines 80, 174
Rogers, Bruce 'Buck' 240
Rogers, John 158–59
Rommel, Erwin 114–16
Romolo, MV 110
Room, Thomas Gerald 107, 120, 157, 164, 206

Room 14 136, 206
Room 40 (NID-25) *see also* Zimmerman telegram
  analysis of German Navy Sigint 29, 46–47, 50–51, 61–62
  establishment of 44
  MI1b and 53, 61–62
  Sigint areas of focus 61–62, 96
  staff movement to GC&CS 75
  traffic analysis 96
Roosevelt, Franklin 126–27, 134
Rossignol, Antoine 13
Rowlett, Frank 173
Royal Air Force (RAF) 75, 80, 105–6, 246 *see also* Air Ministry; Empire Air Training Scheme; Goble, Stanley; Winterbotham, Frederick W.
Royal Australian Air Force (RAAF) *see also* Central Photographic Establishment; Combined Operations Intelligence Centre; Distributed Ground Station – Australia; Empire Air Training Scheme; 'Ferrets'; Women's Australian Auxiliary Air Force
  200 and 201 Flight 184–85
  367 Signals Unit 246
  aerial photography 234
  aircrafts of 182–83, 234, 281–83
  JEWOSU 283
  during the Korean War 237
  in the Middle East 297–99
  Section 22 and 183, 185
  Special Wireless Units 180–81, 224–25, 273, 350–52
  Vietnam War role 251, 253
  WWII Central Bureau role 163, 166, 201–2
  WWII female recruitment 123
  WWII intelligence collaboration 184–85, 201–2, 246
  WWII interception 163, 180–81
  WWII training 182
Royal Australian Navy (RAN) *see also* Fleet Radio Unit Melbourne; Special Intelligence Bureau
  ACNB 39, 87, 89, 98
  ANMEF 47
  coastguard 87–88, 99–100
  COIC role 111
  DNI 104
  Eric Nave and 108
  interwar W/T developments 92–95
  during the Korean War 237
  Malayan Emergency 247
  Morse code training 89–91, 334

Nauru station 97
NID 38–40, 87
OSIS 286
RANTEWSS 284–86
reports on Eric Nave 91–92
SBRS 225, 303–4
submarines 284–86
Vietnam War role 251, 253, 257
WRANS 100, 123–25, 237
WWI confrontations 44–45, 51
WWI position in hierarchy 43–44
WWI role legacy 71, 86
WWI Sigint 40, 89–91
WWII confrontations 110–11, 116
WWII Sigint 100–101
WWII telegraphist jobs 121–23
Royal Commission into Intelligence and Security (RCIS) 125, 268–72
Royal Commission on Australia's Security and Intelligence Agencies (RCASIA) 272, 292
Royal Navy (RN) *see also* Admiralty (UK); Colvin, Ragnar; Naval Intelligence Directorate (NID) (RN)
  British Pacific Fleet 218
  China Fleet 90–92, 127
  collaborative Pacific stations 98
  Napoleonic Wars 19
  Section 22 and 183
  SIB role 109–10, 118
  WRENS 122, 124–25
  WWI Sigint 48–50
  WWII Sigint 127, 151, 218
Royal New Zealand Navy (RNZN) 98
Rozycki, Jerzy 129
RTZ messages 156
Rusbridger, James 138
Rushcutter, HMAS 181–83
Russia *see also* Cold War
  adopts one-time pads 78–79
  black chambers 19, 23–24, 34, 70–71, 212–13, 229–30
  Bolshevik Revolution 70–71
  codebreaking 34–35, 79
  interwar relationship with UK 77–79
  Napoleonic War Sigint 19–20
  Russo-Japanese War 33
  Venona program 212–13, 229–30
  WWI, unencrypted messages of 35, 57
  WWI alliances 70
  WWI Treaty of Brest Litovsk 69
  WWII, monitoring and leaking of Allied information 209–13, 229–32, 334–35
Ryan, Jack 49, 112, 162

Sabah *see* Confrontation
Salamanca, Battle of 19
Samara, Russia 156
Samsonov, General 57
Samuels, Gordon J. 300
San Miguel, Philippines 181
Sandford, Alistair 'Mic'
   with ASWS in Crete 112–13, 206
   Central Bureau and 153–54, 160, 162–63, 166, 168, 192–93
   D Special Section role 211, 223
   links to GC&CS 154, 159, 166, 206
   lobbying for women in the ADF 179
   Nave and 153, 168
   personality 112–13
Sandford, Wallace 112
Sandwith, Captain 124–25
Sarawak *see* Confrontation
satellite communications facilities 275–80
   see also *Echelon* network; Pine Gap facility, NT
Savinsky, Aleksandr 34
SBIRS (Space Based Infrared System) 276
Scharnhorst, SMS 47
Scovell, George 18–19
Second World War *see* World War, Second
Secret Intelligence Service 96
Section 22 182–85
Section H 31
Security Legislation Amendment (Critical Infrastructure) Act 2022 328
Security Service *see* MI5
semaphore 16, 19
September 11 297
Services Reconnaissance Department (SRD) 214–17
sexism and women's rights *see* women
Shedden, Frederick 179, 224, 226–27, 230
Sheraton Hotel 270
Shoal Bay Receiving Station (SBRS) 225, 303–4
Shoho 195–96
Shokaku 195–96
SIGABA machines 173–74
Signal Book for Ships of War 16
Signal Intelligence Board *see* London Signal Intelligence Board
Signal Intelligence Service (SIS) (US) 83–84, 128, 173, 212
Signalbuch der Kaiserlinchen Marime (SKM) 46–47
Signals Intelligence Centre *see* Australian Signals Directorate

signals intelligence (Sigint) *see also* black chambers; codes and codebreaking; digital communications; direction finding and radar; interception; traffic analysis; transparency and accountability
   definition 2
   Elint 118, 284
   present and future 336–44
   traits of Siginters 333–36
Sigsaly telephones 174–75
Sillitoe, Percy 229
Silvestri, Jacopo 12
Singapore 146–47, 200, 259 *see also* Confrontation
Singapore Naval Conference 95, 97
single station locator 256
Sinkov, Abraham 134–35, 167–68, 205
skytales 9
Smart, E.K. 153, 158, 164
Smith, Stephen 279–80, 307–8
Snowden, Edward 306–7, 317
Solomon Islands 99–100
Somme, Battle of the 55–56
South East Asia Command (SEAC) 144, 172, 208
South Korea *see* Korean War
South Vietnamese Army *see* Vietnam War
South West Pacific Area General Headquarters (SWPA GHQ) *see also* Central Bureau
   COIC 111, 234
   collaboration with other organisations 84–85, 148–49, 164, 201–2, 208
   Curtin and 144
   establishment of role 141, 162–65
   opening of Sigint results to 172
   Section 22 182–85
Southeast Asia 259–60 *see also* Confrontation
Soviet Union *see* Russia
Space Based Infrared System (SBIRS) 276
Spain, Iberian Peninsula conflict 18–19
Spartan cryptology 9
Special Duties Section *see* MI5
Special Intelligence Bureau (SIB) (RAN) *see also* D Special Section
   dissolution 223
   formation of 108–10, 118
   FRUMEL and 147
   staff 108–9, 118–20, 153–54
Special Liaison Unit (SLU) 161–62
Special Operations Task Group 297
Special US Liaison Officer Melbourne (SUSLOM) 244

Special Wireless Groups *see also* Joint Telecommunications Unit Melbourne
ANZAC Wireless Signal Squadron 4–5, 67–69
    Australian Army 56, 69, 112–13, 155–57, 162–63, 179–80, 206, 224, 350–52 *see also* 7th Signal Regiment
    Canadian Army 207
    RAAF 180–81, 350–52 *see also* 3 Telecommunications Unit
    UK Wireless Observation Groups 57
    US 64, 352
Squires, Ernest 106
SRD (Services Reconnaissance Department) 214–17
Standard Telegraphic Code (STC) 25–26
statecraft, Sigint as part of 7–8, 13, 15
Stevens, Major 189–90
Stimson, Henry L. 81–82
Strachey, Oliver 120
Strong, George V. 128, 154
'Sturgeon' machines *see* Enigma machines
submarines 62, 80, 117, 284–86
substitution ciphers 9, 11–12, 23
Suez Crisis 247–51, 259
A Suitable Piece of Real Estate 277–78
Sukarno 248–49
Sun Tzu 10
Sunday program 301
Sunzi bingfa 10
support measures, definition 2
Surveillance Legislation Amendment (Identify and Disrupt) Bill 323
Sutherland, Richard K. 170–71
Sydney, HMAS 47–49, 110–11
Sydney University Group 106–8
Syngman Rhee *see* Korean War
Syria 112–13, 298–99, 325

tabulators 28–29
Tamar, HMS 96
Tannenberg, Battle of 57
Tanter, Richard 309
technology *see also* cipher machines; digital communications; electronic computers; electronic countermeasures and counter-countermeasures; radio; satellite communications facilities; telegraphy
    air force 281–83, 294, 298
    electricity discovered 21
    phones 174–75, 288, 297–98
    recorders 92–93
    submarines 62, 80, 117, 284–86
    tabulators 28–29

WWI interception stations 54–57
Telconia 45
Telefunken 33
telegraphy *see also* Morse Code; wireless telegraphy
    arrival in Australia 26–27
    Boer War 31–32
    British empire and 22–23, 25–27, 31–32, 75–76
    Crimean War 22–23
    development of 21–22
    enables espionage 22–24
    expansion of 22–27
    invention of 21–22
    mechanics of 21–22
    Nauru station 97–98
    Sepoy Mutiny 23
    US Civil War 24
    Zimmerman telegram 62–63
telephones 174–75, 288, 297–98
terrorism 297–99, 325
Thompson, Ralph 244, 246, 270
Tiltman, Colonel 26, 189–90
The Times 22
Timor 214–17 *see also* East Timor; Indonesia
Tin Tei Wui 11
Tinker Tailor Soldier Spy 333
Tokyo, Japan 156
Toronto, University of 120
traffic analysis (TA)
    ANZAC Wireless Signal Squadron 67–68
    Australian 85–87, 188, 198
    Central Bureau 166–67, 169–70
    Room 40 96
Traicté des Chiffres 12
transparency and accountability *see also* *Comprehensive Review of the Legal Framework of the National Intelligence Community*; Independent Intelligence Review; *Intelligence Services Act 2001*; Richardson Review; Royal Commission into Intelligence and Security
    of ASD 271–73, 292, 305–6, 312–15, 333, 337, 340
    Balibo Five controversy 6, 292–93
    commercial and outsourced intelligence 341–42
    development of concepts 337–39
    embassy bugging controversy 305
    NIC 314–15, 338, 340–43
    RCASIA 272, 292
    regarding international collaboration 336–39, 342–43

Snowden 306–7, 317
transposition ciphers 9, 188
Transvaal, war in 30–32
Travis, Edward 154, 159, 207–8, 228
Treatise of Ciphers 12
Treaty of Brest Litovsk 69
Treaty of Vereeniging 32
Trendall, Arthur Dale 107, 119, 157
Treweek, Athanasius Pryor 107, 119–20, 157
Tripartite Conference 243–44
Trithemus, Johannes 12
Truman, Harry 229–30
Trump, Donald 328
'Tunny' machines *see* Lorenz machines
Turing, Alan 130, 174
Turkey 4–5, 67–68
Turnbull, Malcolm 316–18, 320
Tutte, Bill 132
Typex machines 80

U-boats 62, 80, 117
Ukraine 329
UKUSA Agreement *see* Five Eyes/UKUSA Agreements
Ultra reports *see also* Enigma machines
　access to 144–45
　Central Bureau analysis 171
　regulating protection of 158–62, 208–9
United Kingdom (UK) *see* Britain
United Nations (UN)
　East Timor and 293–96
　Iraq and 280
　during the Korean War 236–39, 241–42
United States Armed Forces *see also* Cipher Bureau; MI-8; United States Army; United States Navy
　Coast Guard 82–83
　interwar Sigint development 80–81
　Pacific War 102, 145
　Radio Intelligence Service 64
United States Army *see also* American Expeditionary Force; MacArthur, Douglas; Signal Intelligence Service; UKUSA Agreement; Venona program; War Department
　Central Bureau role 163, 166, 190
　intelligence security regulation 159
　Military Intelligence Division 64–65, 171
　monitors Japan 190
　relationship with USN 148–49, 163
　Signals Security Agency 231
　Special Wireless Group 352
　Station 6 site 136
　in the Vietnam War 252–54, 256
　WAC 181
United States Army Air Force (USAAF) 163, 182–84, 201–3
United States Coast Guard 82–83
United States Navy (USN) 83, 124, 127–28, 163, 194, 218, 267 *see also* Bombe machines; CAST; Fleet Radio Unit Melbourne (FRUMEL); Holden Agreement; Office of Naval Intelligence
United States of America *see also* BRUSA Agreement; Cipher Bureau; Cold War; National Security Agency; Washington Disarmament Conference and Naval Treaty
　in Afghanistan 297–99
　American Revolutionary War 16–17
　ANZUS Treaty 242, 267
　Australian-US facilities 275–77, 304–5
　　see also Nurrungar facility, SA; Pine Gap facility, NT
　Battle of the Coral Sea 195
　CANUSA 231–33
　Central Bureau and 172, 189–93
　Chinese Embassy bugging 305
　Cyber Command 328–29
　DSCS 279
　interwar Sigint 73–74, 80–83
　in Iraq 280
　relationship with NZ 267
　Section 22 182–85
　Snowden revelations 306–7, 317
　Soviet interception 79
　Suez Crisis role 247–48
　US–Australian defence facilities 308
　Vietnam War role 251–55, 258
　WWII involvement 62–65, 84–85, 115–16, 134–36, 139–41, 144–45, 160–62, 172, 195, 207, 211, 217–19, 229–30, 244, 294
United States Signal Corps 79
University of Toronto 120
USA Freedom Act 2015 306–7
USAAF (United States Army Air Force) *see* United States Army Air Force

Van Deman, Ralph 64
Van Fleet, A. James 239–40
Vanikoro, Solomon Islands 99
Vanuatu, Australian espionage in 35
Varcoe, John 44–45
Venona program 212–13, 229–30
Vereeniging, Treaty of 32
Verkerhrsbuch (VB) 47
Vernam, Gilbert 16

Victoria Barracks, Melbourne 111, 273–74
Victoria Barracks, Sydney 106–8
Victorian Sigint 27
Vienna Congress 20
Vietnam War
   Sigint contributions 251–57
   Sigint limitations 257–58
Vigenère ciphers 12, 23
Von Müller, Karl 48–49
Von Spee, Graf 49

Waller, Captain 97
Wallis, John 14
Walsingham, Francis 13–14
War Cabinet (Australia) 123–24, 179
War Department (US) 65, 190
War Office (UK) 31, 40–41, 44, 53, 76, 96
   see also Directorate of Military Intelligence; MI5
Wark, Wesley 74, 232–33
WARSPITE 196
Washington, George 16–17
Washington, US see Australian Sigint Liaison Officer Washington
Washington Disarmament Conference and Naval Treaty 81–82, 88, 91
Water Transport Code 204
Waterloo, Battle of 20
Wavell, Archibald 114, 140
web see digital communications
Weber, Ralph 17
Weisband, William 212–13
Welchman, Gordon 130
Wellesley, Arthur (Duke of Wellington) 18–19
whistleblowing 306–7
Whitlam, Gough 259, 268, 276–77, 292
Wiesband, William 79
William III of Orange 14
Willoughby, Charles A. 170–71
Window strategy 118
Winds message decryption 136–39
Winterbotham, Frederick W. 208, 210
Wireless Experimental Centre (WEC) 144, 191–92, 206
Wireless Observation Groups 57
wireless telegraphy (W/T) see also direction finding and radar; Special Wireless Groups
   Australian wireless radio telegraphy 38–40
     early 1900s Sigint 32–34
     intercept stations 98–99
     interwar preparations 88–89, 100–101
     Prohibition and 82

radio countermeasures (RCM) 117–18, 181–82
technological development 27–28, 33, 81, 92–93
   during WWI 48–49, 57–60, 66
   during WWII 89–95, 99–101, 121–25, 156, 169–70
women 11, 100, 121–26, 181, 221, 237 see also AWAS; Women's Australian Auxiliary Air Force
Women's Army Corps (WAC) (US) 181
Women's Australian Auxiliary Air Force (WAAAF) 126, 178–79, 181
Women's Emergency Signalling Corps 121–23, 125
Women's Royal Australian Naval Service (WRANS) 100, 123–25, 237
Women's Royal Naval Service (WRENS) (UK) 122, 124–25
World War, First see also 1st Australian Imperial Force; ANZAC Wireless Signal Squadron; British Expeditionary Force; Jutland, Battle of; Michael Offensive; Somme, Battle of the
   armistice 69
   Australian Flying Corps role 59–60
   Australian role, disregard for 7
   British Army and MI1b role 43–44, 53–55, 58–62, 66
   British oversight 42–43
   British role, evaluated 70–71
   codebook decryption 46–47
   French Sigint 45, 54, 58–60, 65–66, 69–70
   German Air Force role 58–60
   German Army role 61
   lax communications security 57–58, 66
   RAN role 43–47, 71–72, 86
   RN and Room 40 role 43–48, 58, 61–62
   Russian role 69–71
   trench Sigint 53–55
   US entry 62–64
World War, Second see also 2nd Australian Imperial Force; BRUSA Agreement; Germany during WWII; Japan during WWII; Pacific War
   ASD and 103
   Australian Army role 86, 106, 123
   Australian women's corps 121–26
   beginning of 102
   Canadian role 120, 231
   Central Bureau role 84–85
   Central War Room 111
   COIC 111, 234

Crete operations 113–14
end of 222
FECB role 104
GC&CS role 127–28, 130–35
Hut 6 113
Italy/Australia conflicts 104–5, 110–11
Middle East operations 112–13
NZ role 113–14
Polish codebreaking 128–30
publications on 8, 103
RAAF role 116–18, 123
RAF role 80, 106, 117–18
RAN, D Special Section and SIB 108–11, 116, 118–23, 125
RN role 127
Sydney University Group 106–9
technology 117–18
UK role 102–6, 109–10, 114–15, 120, 126–28, 134
US role 83–85, 102–3, 115, 120, 124, 126–28, 131, 134–35
World Wide Web *see* digital communications

Wormald, Berenice 126
WRANS (Women's Royal Australian Naval Service) 100, 123–25, 237
WRENS (Women's Royal Naval Service) 122, 124–25
Wullenweber system 256
WWI and WWII *see* World War, First; World War, Second

Y Board *see* London Signal Intelligence Board
Yamamoto, Isoroku 197, 199–200, 202–3
Yardley, Herbert O. 64, 81–82
Yarra 47
Yoshino Maru 205
Younger, Alex 333
Yudhoyono, Susilo Bambang 306–7

Zimmerman telegram 62–63
Zuikaku 195, 198
Zygalski, Henryk 129